Front Cover Caption:
Heliophysics image highlights from 2009. For details of these images, see the key on Page v.

The NASA STI Program Office ... in Profile

Since its founding, NASA has been dedicated to the advancement of aeronautics and space science. The NASA Scientific and Technical Information (STI) Program Office plays a key part in helping NASA maintain this important role.

The NASA STI Program Office is operated by Langley Research Center, the lead center for NASA's scientific and technical information. The NASA STI Program Office provides access to the NASA STI Database, the largest collection of aeronautical and space science STI in the world. The Program Office is also NASA's institutional mechanism for disseminating the results of its research and development activities. These results are published by NASA in the NASA STI Report Series, which includes the following report types:

- TECHNICAL PUBLICATION. Reports of completed research or a major significant phase of research that present the results of NASA programs and include extensive data or theoretical analysis. Includes compilations of significant scientific and technical data and information deemed to be of continuing reference value. NASA's counterpart of peer-reviewed formal professional papers but has less stringent limitations on manuscript length and extent of graphic presentations.

- TECHNICAL MEMORANDUM. Scientific and technical findings that are preliminary or of specialized interest, e.g., quick release reports, working papers, and bibliographies that contain minimal annotation. Does not contain extensive analysis.

- CONTRACTOR REPORT. Scientific and technical findings by NASA-sponsored contractors and grantees.

- CONFERENCE PUBLICATION. Collected papers from scientific and technical conferences, symposia, seminars, or other meetings sponsored or cosponsored by NASA.

- SPECIAL PUBLICATION. Scientific, technical, or historical information from NASA programs, projects, and mission, often concerned with subjects having substantial public interest.

- TECHNICAL TRANSLATION. English-language translations of foreign scientific and technical material pertinent to NASA's mission.

Specialized services that complement the STI Program Office's diverse offerings include creating custom thesauri, building customized databases, organizing and publishing research results ... even providing videos.

For more information about the NASA STI Program Office, see the following:

- Access the NASA STI Program Home Page at http://www.sti.nasa.gov/STI-homepage.html

- E-mail your question via the Internet to help@sti.nasa.gov

- Fax your question to the NASA Access Help Desk at (443) 757-5803

- Telephone the NASA Access Help Desk at (443) 757-5802

- Write to:
  NASA Access Help Desk
  NASA Center for AeroSpace Information
  7115 Standard Drive
  Hanover, MD 21076

NASA/TM–2010–215854

GSFC Heliophysics Science Division 2009 Science Highlights

Holly R. Gilbert
NASA Goddard Space Flight Center, Greenbelt, Maryland

Keith T. Strong and Julia L.R. Saba
SP Systems, Inc., Greenbelt, Maryland

Yvonne M. Strong
American Society for Microbiology

National Aeronautics and
Space Administration

Goddard Space Flight Center
Greenbelt, Maryland 20771

December 2009

Available from:

NASA Center for AeroSpace Information
7115 Standard Drive
Hanover, MD 21076-1320

National Technical Information Service
5285 Port Royal Road
Springfield, VA 22161

GSFC Heliophysics Science Division 2009 Science Highlights

Table of Contents

FRONT COVER KEY v
FOREWORD vi
PREFACE ix
INTRODUCTION 1
    The Sun 1
    The Inner Heliosphere 3
    Geospace 3
    The Outer Heliosphere 5
THE HSD ORGANIZATION 6
FACILITIES 8
2009 SCIENTIFIC HIGHLIGHTS 9
    IBEX Discovers Bright Ribbon of ENA Emission 9
    Stereo Views of the Ring Current from TWINS 11
    Cluster Spacecraft Discover How Solar Wind Turbulence Dissipates 12
    STEREO and SOHO Monitor Extended Solar Minimum 13
HSD HONORS AND AWARDS 14
HSD PROJECT LEADERSHIP 17
DEVELOPING FUTURE HELIOPHYSICS MISSION CONCEPTS 19
HELIOPHYSICS EDUCATION AND PUBLIC OUTREACH 21
    Advanced Student Research Programs 22
    K-12 E/PO Programs 22
    Elementary / Secondary Education 23
    Informal Education 23
    Public Outreach 24
    Public Affairs 24
    The E/PO Team 26
    International Heliophysical Year (IHY) 27
SCIENCE INFORMATION SYSTEMS 28
COMMUNITY COORDINATED MODELING CENTER 31
    Science Support 32
    Space Weather Modeling 33
TECHNOLOGY DEVELOPMENT 35
HSD LINE OF BUSINESS 37
    FY09 LOB Internal Research and Development Priorities 37
    FY09 LOB Internal Research and Development Achievements 37
APPENDIX 1: INDIVIDUAL SCIENTIFIC RESEARCH 38
APPENDIX 2: HSD PUBLICATIONS AND PRESENTATIONS 142
    Journal Articles 142
    Submitted / In Press 172
    Presentations 181
APPENDIX 3: OPERATIONAL HSD MISSIONS 204
    Interstellar Boundary Explorer (IBEX) 204
    Communications/Navigation Outage Forecasting System (C/NOFS) 205
    Aeronomy of Ice in the Mesosphere (AIM) 207

Time History of Events and Macroscale Interactions During Substorms 208
Solar Terrestrial Relations Observatory (STEREO) 209
Hinode 210
Solar Radiation and Climate Experiment (SORCE) 211
Ramaty High Energy Solar Spectroscopic Imager (RHESSI) 212
Thermosphere, Ionosphere, Mesosphere, Energetics, and Dynamics (TIMED) 213
Cluster 214
Two Wide-Angle Imaging Neutral-Atom Spectrometers (TWINS) 215
Transition Region And Coronal Explorer (TRACE) 216
Advanced Composition Explorer (ACE) 217
Solar and Heliospheric Observatory (SOHO) 218
Wind 219
Geotail 220
Voyager 221
APPENDIX 4: FUTURE MISSIONS 222
Solar Dynamics Observatory (SDO) 222
Radiation Belt Storm Probes (RBSP) 223
Magnetospheric MultiScale (MMS) 224
Solar Orbiter 225
Solar Probe Plus (SP+) 226
APPENDIX 5: ACRONYM LIST 227
BACK COVER KEY 239

GSFC Heliophysics Science Division 2009 Science Highlights

FRONT COVER KEY

	Superposed solar active region images from the Hinode X-Ray Telescope show coronal loop temperature structure. The blue and green regions are hot coronal plasma near 10 MK.
	The launch of a sounding rocket carrying an HSD payload to look at particle and field interactions. HSD uses suborbital experiments to develop and test instruments as well as to conduct scientific research.
	Technology development is a key aspect of the work done at HSD. The Plasma Impedance Spectrum Analyzer (PISA) being integrated onto the FASTSat spacecraft at GSFC.
	Solar Probe Plus will be the first mission to skim the outermost layers of the Sun's atmosphere to discover how plasma processes in the corona accelerate the solar wind to speeds from 200 to 2000 km/s.
	Comparing model results directly with observations is key to validating the models and understanding of what the observations can reveal. Here a 3D loop simulation has been used to generate the signal that would be seen in Fe IX and X at 171A (~1 MK).
	A spectacular result from the IBEX spacecraft. This is an all-sky map of the interaction of the Sun's magnetic fields with those of interstellar space.
	Sophisticated 3D visualizations have become indispensible tools for modelers in HSD. A magnified view of a model of two vortices in the Earth's magnetosphere shows their rotation direction and speed.
	Coronal mass ejections are some of the most violent and spectacular phenomena in our solar system. All the complex physical processes involved can only be truly understood by using 3D MHD simulations in conjunction with observations.
	New materials can lead to advances in instrumentation. A very robust telescope mirror made from simulated lunar regolith materials opens the possibility of producing large optics on the Moon.
	Education and public outreach is a central theme to the work that we do in HSD. We organize events that illustrate to the public the importance of the work that is done by the heliophysics community.
	The thermosphere is a layer of the Earth's atmosphere that absorbs the highly variable solar UV radiation, which can significantly change its temperature structure. The effect of a solar eclipse on the thermosphere was captured by TIMED.

GSFC Heliophysics Science Division 2009 Science Highlights

FOREWORD

This report presents the scientific, technological, and flight project achievements of the Goddard Space Flight Center (GSFC) Heliophysics Science Division (HSD) for FY09. HSD consists of 299 scientists, technologists, engineers, education and public outreach specialists, and administrative personnel dedicated to advancing our knowledge and understanding of the Sun and the wide variety of domains influenced by its variability.

> Heliophysics is one of the four primary disciplines that form the framework for NASA's Strategic Plan for Earth and Space Science.
>
> GSFC has critical enabling roles in nearly all Heliophysics science missions including mission formulation and development, instrumentation, data systems, and science theory, analysis, and modeling.
>
> HSD made important contributions to the C/NOFS, STEREO, TWINS, IBEX, Cassini, and MESSENGER missions in 2009.

HSD's Mission is to explore the Sun's interior and atmosphere, discover the origins of its temporal variability, understand its influence over the Earth and the other planets, and determine the nature of the interaction between the heliosphere and the local interstellar medium.

Our major activities include:
- Leading science investigations involving flight hardware, theory, modeling, and data analysis that will answer the strategic questions posed in the Heliophysics Roadmap;
- Leading the development of new solar and space physics mission concepts and supporting their implementation as Project Scientists;
- Providing access to measurements from the Heliophysics Great Observatory (HGO) through our Science Information Systems;
- Communicating science results to the public and inspiring the next generation of scientists and explorers.

Our strategic goals are to:
- Open the frontier to space environment prediction;
- Understand the nature of our home in space;
- Safeguard the journey of exploration.

This year HSD scientists completed and delivered three new flight instruments – the Thermospheric Temperature Imager (TTI), the Plasma Impedance Spectrum Analyzer (PISA), and the Miniature Imager for Neutral Ionospheric atoms and Magnetospheric Electrons (MINI-ME). The Instrument Scientists, J. Sigwarth (670), D. Rowland (674), and M. Collier (673), respectively, are to be congratulated on this important milestone. These instruments were developed jointly with the US Naval Academy under the aegis of the Department of Defense's Space Test Program. They will fly on the MSFC FASTSat which will be launched in May 2010.

The Magnetospheric MultiScale (MMS) mission, Project Scientist M. Goldstein (673), is now in Phase C. Critical to its success will be the Fast Plasma Instrument led by T.

GSFC Heliophysics Science Division 2009 Science Highlights

Moore (670). The Critical Design Review for MMS and the Fast Plasma Instrument will be held later this year.

The Solar Dynamics Observatory (SDO), Project Scientist D. Pesnell (671), has completed all environmental tests and is at Cape Canaveral awaiting launch.

FY09 also saw the selection of HSD scientists and their strategic partners at other institutions for many new flight instrument, sounding rocket, science data system, and theory, modeling, and analysis tasks. Chief among these excellent results was the selection of the Spectral Imaging of the Solar Environment (SPICE) instrument for the Solar Orbiter mission. SPICE is the result of strategic partnership between J. Davila and D. Rabin (both 671) and D. Hassler of the Southwest Research Institute in Boulder. When Solar Orbiter is on-station later in the next decade, SPICE will provide the first close-up views of the effects of the explosive release of magnetic energy in the Sun's atmosphere during flares and CMEs.

HSD scientists published 298 papers in refereed scientific journals in FY09 of which 39% had one of our scientists as first author. We also gave 230 presentations at 74 different conferences (see Appendix 2 for details).

Noteworthy among the publications are the IBEX and MESSENGER articles in *Science* which announced discoveries that are pushing back the frontiers of heliophysics. The IBEX team has produced the first all-sky map of the outer limits of our solar system using energetic neutral-atom (ENA) imaging techniques. It shows a bright band of ENA emission believed to be caused by the impact of the galactic magnetic field on the heliosphere. These results give the large-scale context to the point measurements recently made by Voyager. The MESSENGER mission to Mercury has provided the first observations of magnetic reconnection for the extreme solar wind conditions typical of the inner solar system. Analysis of these new measurements has led to the conclusion that magnetic reconnection at Mercury proceeds at a rate that is nearly an order of magnitude higher than at Earth. These results are of particular importance because magnetic reconnection occurs throughout the heliosphere and it is the primary energy source for all of the forms of space weather that are most important for our nation's space infrastructure.

Other important science results included the initial published images of the global distribution of precipitating energetic ions in the Earth's upper atmosphere, using the new ENA cameras flying on the TWINS spacecraft. Cluster scientists also discovered that solar wind turbulence dissipates via kinetic Alfvénic waves, a result that may have profound implications for our understanding of how the solar corona is heated. SOHO and STEREO measurements have been used to probe the exciting and perplexing extended solar minimum. The EUV telescopes on these missions provide independent views of the inner solar corona from three different vantage points covering (currently) over 300-degrees of the solar surface. Similarly their coronagraphs work together to define the structure of the outer corona from just above the solar surface out to beyond the Earth's orbit. C/NOFS observations of the terrestrial effects of the deepest solar minimum in 100

GSFC Heliophysics Science Division 2009 Science Highlights

years have revealed the weakest levels of ionization in the upper layers of the Earth's ionosphere ever measured.

Individual scientists and teams from HSD garnered many important professional honors in 2009. Keith Ogilvie was awarded the NASA Distinguished Service Medal, the highest honor that NASA confers. Tom Moore was elected a fellow of the American Geophysical Union, an honor conferred on not more than 0.1% of AGU members in any year. Roger Thomas and Doug Rabin received the NASA Exceptional Service Medal. HSD personnel were part of a number of teams that received NASA Group Achievement awards, including the C/NOFS VEFI, Cassini Plasma Spectrometer, MESSENGER, EUNIS, and Sun-Earth Connection Forum teams. Sten Odenwald won the American Astronomical Society Solar Physics Division Popular Science Writing Award. Alex Glocer received the Ralph Baldwin Award in Astrophysics and Space Science from the University of Michigan for his PhD thesis.

Last, but certainly not least, we are very pleased to have had seven new civil service scientists join HSD in FY09. They are Steven Christe (671), Adrian Daw (671), Errol Summerlin (672), John Dorelli (673), Deirdre Wendel (673), Peter Schuck (674), and Larry Kepko (674). Of the 299 people in HSD, 61 (20%) are civil servants, 23 (8%) are co-located civil servants from other divisions, 116 (39%) are contractors, 18 (6%) are emeritus, 12 (4%) are NASA Postdoctoral Fellows, and 66 (22%) are university scientists.

Further news and developments regarding our organization and facilities, instruments, missions, and science highlights can be found on our Website (http://hsd.gsfc.nasa.gov/). We thank you for your interest in our programs, and welcome feedback via the website.

—James A. Slavin, Director
NASA GSFC Heliophysics Science Division
December 2009

GSFC Heliophysics Science Division 2009 Science Highlights

PREFACE

This document summarizes the work performed in FY09 by members of HSD who conducted research, developed models, designed and built instruments, managed projects, and carried out numerous other activities that have made significant contributions to understanding the domain of the Sun. Unfortunately only a small fraction of these activities can be even briefly highlighted in this report.

In the main body of the report we describe the scientific studies that heliophysics encompasses and the components of the HSD program at GSFC that help the heliophysics community achieve its goals. As part of the HSD program, we do a great deal of scientific research. This is summarized in Appendix 1. Appendix 2 lists the publications and presentations made by members of HSD in FY09. In Appendixes 3 and 4, we give a brief description of both operational heliophysics missions and planned missions. We use many acronyms throughout the report, which are in common usage and so are not independently defined in each section; acronym definitions are given in Appendix 5.

The production of this report was guided by Holly Gilbert, HSD Associate Director for Science, who, together with Keith Strong (SP Systems), assembled the contributions and checked the report for accuracy, made suggestions regarding its content, and contributed to several sections. Julia Saba (SP Systems) and Yvonne Strong (American Society for Microbiology) helped with the scientific and technical editing, and Elizabeth Serago (ADNET Systems Inc) and Keith Feggans (Sigma Space Corp) contributed by compiling the large HSD bibliography. Graphics support was provided by Robert Kilgore (TRAX International and TIMS). Many others in HSD helped with useful and constructive suggestions concerning the organization of this report and its content.

Holly Gilbert and Keith Strong
December 2009

GSFC Heliophysics Science Division 2009 Science Highlights

> HSD's goal is to further our scientific understanding of the nature and origins of the ever-changing physical conditions throughout the vast volume of space controlled by the Sun, and in which humanity will function for the foreseeable future.

INTRODUCTION

Heliophysics is the study of the domain of the Sun—the heliosphere—from the nuclear core in the center of the Sun where hydrogen is transmuted into helium, producing the energy that drives changes throughout the entire solar system, to the edge of interplanetary space where the solar wind and magnetic fields cede control of the local physical conditions to the interstellar medium. That represents over 10^8 AU^3 suffused with outflowing plasma, magnetic fields, and solar radiation, with temperatures ranging from near absolute zero to over 20 MK.

The heliosphere is an interconnected network of physical processes driven by a relentless but varying outflow of energy from the Sun in the form of electromagnetic radiation from γ-rays to radio emissions, ionized and neutral particles, and magnetic fields. All these forms of solar emissions interact in different ways in the wide range of environments from the hot solar interior, through the Sun's thin surface layers, into the extended solar corona, and throughout interplanetary space to the very edge of the solar system. Along their tortuous path, these emissions interact with different planetary environments, comets, asteroids, and interstellar gas—each with its unique response to the changing solar stimuli. HSD's goal is to understand this system of systems.

Accomplishing this goal involves the study of the complex interactions between electromagnetic radiation, plasmas, and magnetic fields, with three principal objectives:
- To understand the changing flow of energy and matter throughout the Sun, solar atmosphere, heliosphere, and planetary environments
- To explore the fundamental physical processes that characterize space plasmas
- To define both the origins and the societal impacts of variability in the Sun–Earth system

There are four major physical domains that encompass HSD's mission: the Sun, the inner heliosphere, geospace, and the outer heliosphere.

The Sun

The Sun generates not only energy, but also magnetic fields. The solar dynamo, combined with both radial and latitudinal differential rotation, generates magnetic fields and stores vast amounts of energy in them. Convection in the outer layers of the Sun and the natural buoyancy of the flux ropes drag these strong fields (>10^4 G) to the surface, as evidenced by the presence of sunspots and faculae, which change the solar spectral irradiance that provides the energy to drive Earth's weather and climate system. Thus a substantial and persistent change in solar irradiance could significantly affect Earth's climate.

Energy is transported from the solar core region (the inner 20% of the solar radius) by photon radiation out to about 70% of the solar radius. Then convectional transport takes over, carrying most of the energy to the surface of the Sun—the photosphere—where the optical depth of the solar plasma drops, so that much of the energy can be radiated away into space. The falling temperature gradient as the energy flows outward seems well understood; however, just above the relatively cooler surface layers, the temperature of the plasma rises rapidly again to form a 1-MK corona. The physical processes involved in creating and maintaining the corona are not yet completely understood.

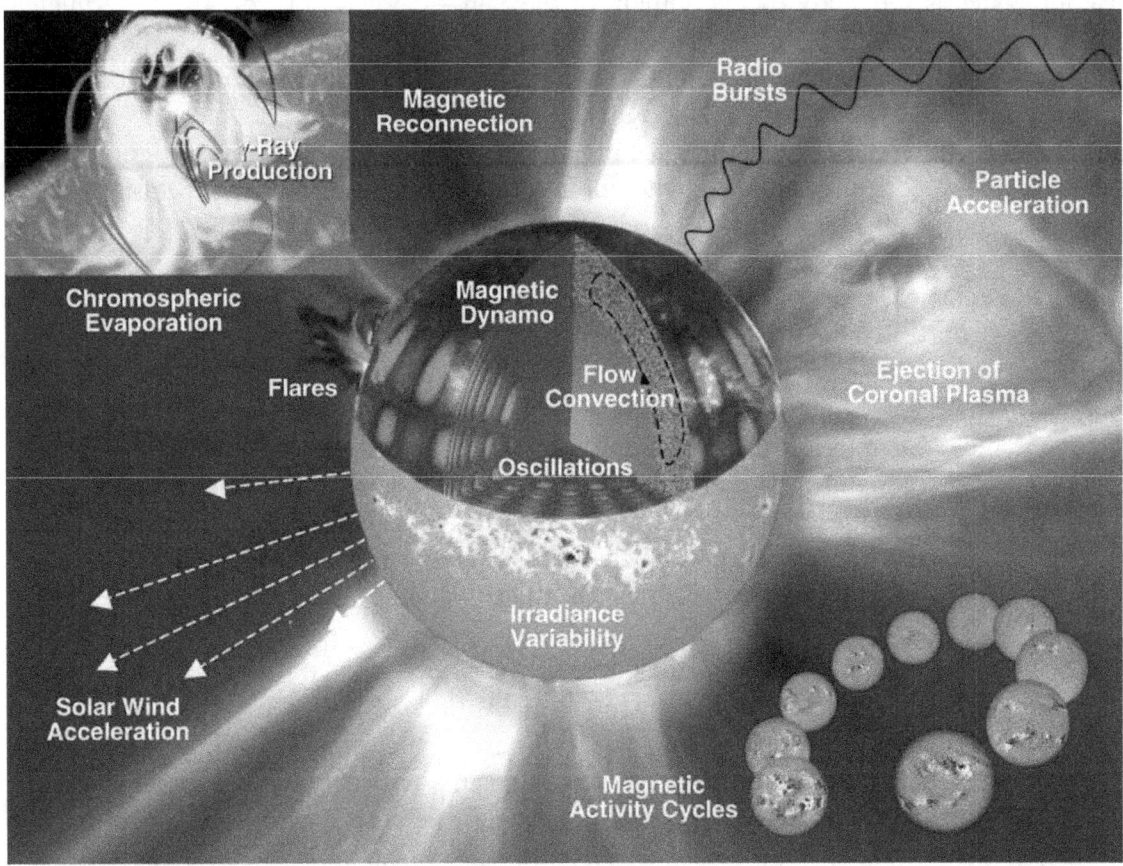

The anatomy of the Sun

The energy stored in the solar magnetic fields is often released suddenly by magnetic reconnection to produce flares and CMEs. Flares produce emissions from γ-rays to radio wavelengths, accelerate solar energetic particles, and transport material from the lower layers of the atmosphere up into the hot corona via chromospheric evaporation, accompanied by ejection of material away from the Sun. CMEs are vast ejection events that can grow to be many times the size of the Sun and at times move with velocities exceeding 2000 km/s.

Another type of mass outflow from the Sun is more continuous but also highly variable: the solar wind, which flows out along the spiraling solar magnetic field with velocities of between 300 and 700 km/s and having various temperatures, densities, and compositions. The manner of its acceleration is still not completely understood.

The Inner Heliosphere

This is the region between the Sun and Jupiter that is filled with outflowing, supersonic solar wind and frozen-in spiraling magnetic fields—the Parker Spiral. The streams of solar plasma evolve significantly as they pass through this region, where fast streams of solar wind plough into slower-moving streams, forming shocks. Transients, such as CMEs, reshape the ambient environment. Some CMEs move faster than the local solar wind, building up high-density fronts that form shocks where particles are accelerated to extremely high energies. CMEs can expand as they move outwards, leaving low-density regions behind the propagating front.

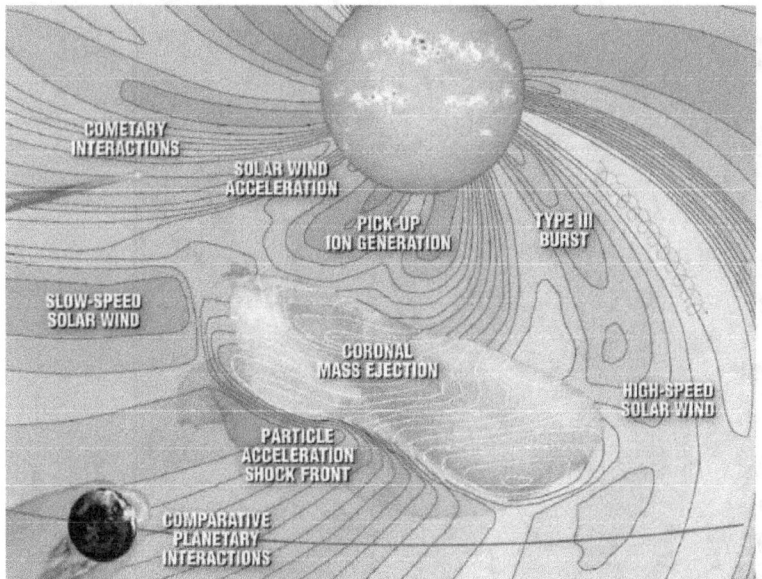

A CME propagates through the heliosphere towards Earth

Electrons flowing along the large-scale magnetic field lines, thus producing radio bursts of various types, show which field lines remain connected back to the Sun and which ones have reconnected.

Because of the spiral nature of the fields, Earth is better connected to the western hemisphere of the Sun; thus an event near the west limb of the Sun is more likely to be geoeffective than one in the eastern hemisphere. Photons take only eight minutes to arrive at Earth from the Sun, and high-energy protons can be detected a few minutes later, whereas material from a CME event seen on the Sun may take up to three days to arrive.

Geospace

Earth's magnetic field acts as a barrier to most of the harmful particle fluxes originating from the Sun. Much of the solar wind is deflected around the magnetosphere, which forms a teardrop-shaped shield around Earth. The shape and size of the magnetosphere change as solar wind conditions vary. Earth's magnetic field is compressed within about 10 Earth radii (R_E) on the sunward side of the planet and stretched out by many tens of R_E on the anti-sunward side.

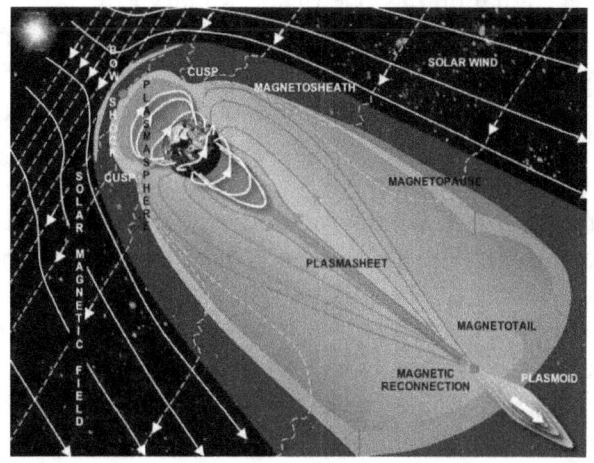

Structure of the magnetosphere

At the location where the solar wind and Earth's magnetic field collide there is a bow shock region. The faster and denser the solar wind is, the more the fields are compressed and stretched. The direction of the interplanetary field profoundly influences the effects seen; a CME with an oppositely directed magnetic field will more likely reconnect, allowing more energetic particles to enter geospace, and producing a more energetic geomagnetic storm. The fields become weak and disorganized as they interact in the magnetosheath. The surface where Earth's magnetic pressure is balanced by the solar wind is called the magnetopause; it often ripples and flaps in the solar wind, and parts constantly reconnect magnetically and break away. In the polar region, there is a path where the solar particles have easy access to Earth's magnetosphere—this is the cusp region.

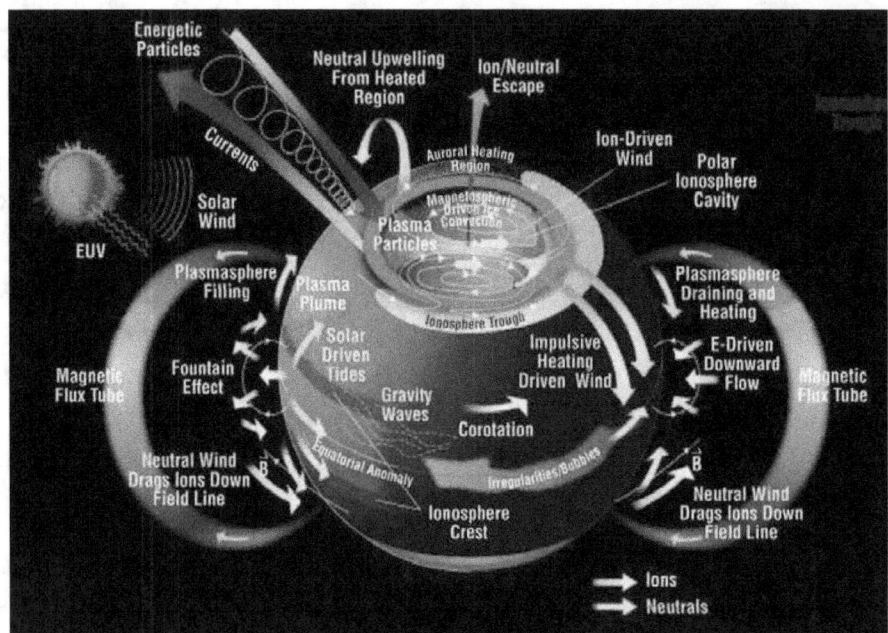

Some of the complex physical processes that occur when incoming radiation and charged particles interact with Earth's dynamic atmosphere

Closer to Earth, a completely different set of closely coupled processes occur. Here, solar UV radiation has one of its primary effects. UV light from the Sun, especially when enhanced by flare emissions, heats the neutral atmosphere thus increasing the temperature with height in the thermosphere. Gravity waves are also present in this layer and drive changes in the neutral atmosphere as they propagate up from below. The high-energy X-ray and UV photons partly ionize the neutral particles and create another intermittent layer of charged particles – the ionosphere. At times of high solar activity, these physical processes of heating and ionization are similarly intensified and cause increased scale height leading to increased satellite drag, interference with HF radio communications, and larger GPS errors.

High-energy electrons and protons become trapped in the Van Allen radiation belts. The outer belt gathers electrons from the aftermath of geomagnetic storms and substorms, accumulating enough plasma pressure to produce a ring-shaped current around Earth that substantially inflates the geomagnetic field. The inner belt originates from cosmic rays interacting with the upper atmosphere. Charged particles spiral along the field lines, mirroring as they descend into the stronger polar fields.

The Outer Heliosphere

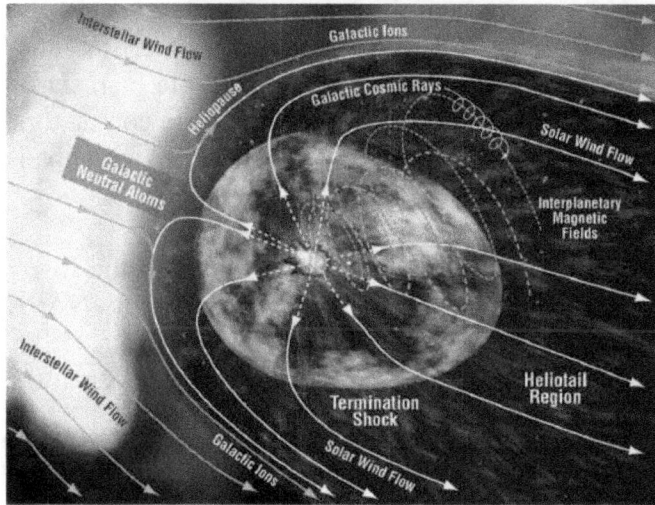

The boundary between the Sun's domain and interstellar space

The characteristics of the heliosphere change significantly past the orbit of Jupiter—this area is called the outer heliosphere. Here, the solar wind and transients interact with the gas-giant planets, which are very different from their rocky cousins in the inner heliosphere.

In this region, the nature of the outflowing plasma also changes; the interplanetary magnetic field becomes mostly azimuthal and hence is perpendicular to the solar wind flow; and solar transients and interplanetary shocks catch up with each other and form large, global merged interaction regions. In addition, a larger portion of the solar wind is composed of photoionized interstellar neutral particles, known as pickup ions.

With increasing distance from the Sun, the particle and field pressure of the solar wind decreases until it reaches pressure balance with the local interstellar medium. This boundary is called the heliopause. However, before reaching this boundary, the solar wind slows abruptly below its supersonic speed at the termination shock, and starts deflecting towards the heliotail, and continues slowing in the region known as the heliosheath. The interaction of the heliosphere with the interstellar medium is analogous to the deflection of the solar wind around Earth's magnetosphere. It is postulated that the interstellar plasma could also flow at supersonic speeds, necessitating the existence of an external bow shock and a pileup of interstellar particles upstream of the heliosphere, known as the hydrogen wall. In addition to low-energy neutral particles, the extremely high-energy galactic cosmic rays also enter the heliosphere, but not without first being modulated by the periodically varying heliospheric magnetic fields. The termination shock and heliosheath are thought to also be the source of an extra, anomalous component of the cosmic rays observed at Earth.

GSFC Heliophysics Science Division 2009 Science Highlights

HSD is an internationally recognized research organization dedicated to the furtherance of scientific understanding of all aspects of heliophysics.

THE HSD ORGANIZATION

At the end of FY09, HSD was composed of 299 scientists, engineers, and other staff supported by a small management and administrative team. Of the 299 people in HSD 61 (20%) are civil servants, 23 (8%) are co-located civil servants, 116 (39%) are contractors, 18 (6%) are emeritus, 12 (4%) are NASA Postdoctoral Fellows, and 66 (22%) are university scientists working under cooperative agreements.

HSD is divided into four Laboratories, each with its own Chief, as shown in the figure below. HSD is supported primarily by competitively awarded funding from the NASA Science Mission Directorate with the remainder made up of assigned NASA tasks, GSFC research and development investments, and funding provided by other Federal agencies.

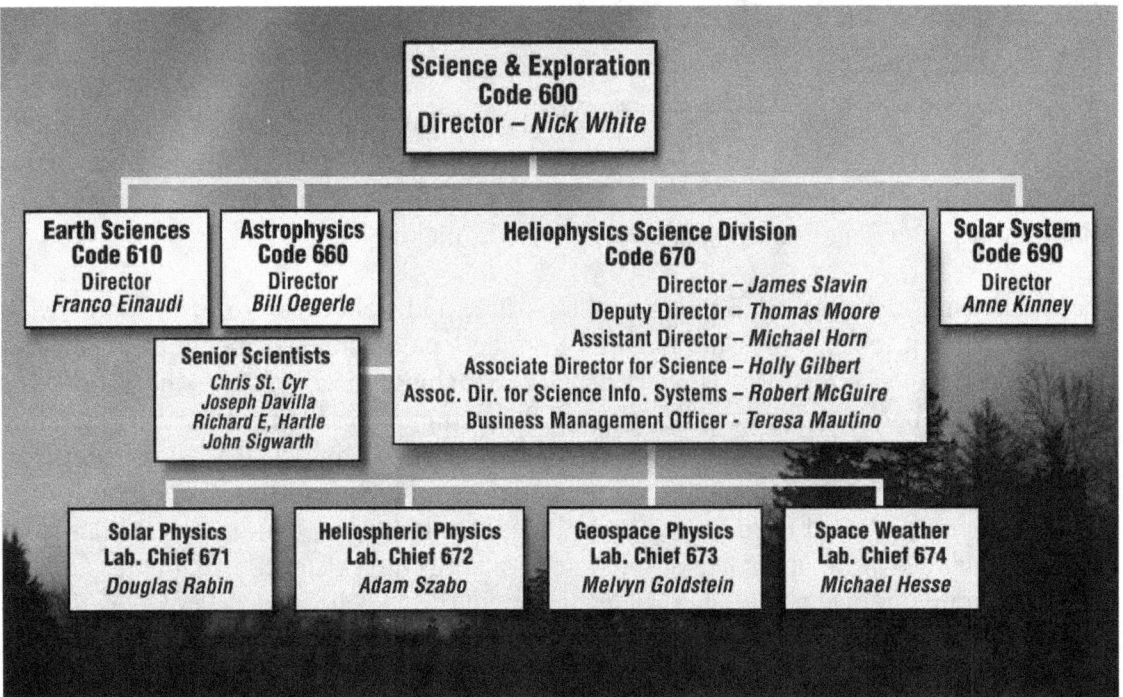

The responsibilities of HSD include:

- Scientific Research: HSD staff working as principal investigators (PIs), Co-investigators (Co-Is), instrument scientists, and flight team members have published 298 papers and given over 230 presentations at 74 different scientific meetings in 2009 (see Appendixes 1 and 2 for details).
- Project and Mission Scientist Assignments: HSD provides the project and mission scientists who manage operating heliophysics missions, as well as missions in development (see the Missions sections, Appendixes 3 and 4, for details).

- Future Mission and Instrument Concept Development: HSD provides scientific leadership and technical support for science mission concept development and formulation.
- Data and Modeling Centers: HSD scientists lead and operate four major centers that provide critical data services and simulation and modeling services to the heliophysics community. They are the Space Physics Data Facility, the Solar Data Analysis Center, the Heliophysics Data and Modeling Consortium, and the Community Coordinated Modeling Center. These centers are funded directly by NASA and reviewed periodically by NASA Headquarters-appointed external committees.
- Education and Public Outreach (E/PO): HSD scientists, led by the Associate Director for Science, carry out a variety of E/PO tasks supported by Project and competitively awarded funding.

Like many Government, academic, and industrial research laboratories that perform basic and applied research in specialized areas, HSD has experienced recruitment challenges in replacing retirees and recruiting new staff to attain its research goals while achieving greater ethnic and gender diversity in the HSD workforce. For this reason, HSD is actively recruiting within the university community to attract new postdoctoral, cooperative agreement, and civil service scientists. Seven new civil service scientists joined HSD in FY09. They are S. Christe (671), A. Daw (671), E. Summerlin (672), J. Dorelli (673), E.D. Wendel (673), P. Schuck (674), and L. Kepko (674). HSD also has other student development programs to attract more young researchers to space science and retain the most promising candidates.

GSFC Heliophysics Science Division 2009 Science Highlights

> HSD's goal is to provide a safe and efficient work environment, consistent with being a center of excellence for heliophysical research.

FACILITIES

HSD has people located in several buildings on the GSFC campus, primarily Buildings 21, 22, and 26. Following the opening of Building 34, plans are in place to consolidate the entire Division into Building 21 next year. Once this is completed, all HSD personnel, offices, laboratories, computer rooms, and other facilities will be housed in the same building for the first time, which will facilitate more efficient intra-divisional communications, critical for project and mission success.

Examples of the new clean rooms (left) and laboratory facilities (right) that are now installed in Bldg. 21

In FY09, the renovation of Building 21 continued. Initial design for all building renovations has been completed, and construction projects are scheduled for the next 18 months. The most extensive project is the expansion and rehabilitation of the Space Plasma Instrument Facility. The original laboratory configuration included a 753-square-foot (ft^2) clean room and a 935 ft^2 laboratory space, and housed an electron vacuum chamber and an ion vacuum chamber. Once the expansion is completed in late 2009, the Plasma Laboratory will have a 1200 ft^2 clean room, a 1000 ft^2 laboratory space, and an additional electron chamber. Completed renovation projects included updating two conference rooms and upgrading office space.

GSFC Heliophysics Science Division 2009 Science Highlights

With a strong scientific base, HSD can function effectively to support and promote a vital heliophysics program at NASA

2009 SCIENTIFIC HIGHLIGHTS

While it would not be practical to feature, in detail, all the scientific accomplishments of the HSD team in an annual report, a few outstanding examples of work that was successfully completed in FY09 are presented in this section. A short summary of individual scientific contributions is given in Appendix 1, and a more extensive list of publications and presentations can be found in Appendix 2.

IBEX Discovers Bright Ribbon of ENA Emission

The Interstellar Boundary Explorer (IBEX) spacecraft has released the first ever all-sky maps of the interactions occurring at the edge of the solar system, where the Sun's influence diminishes and meets with the interstellar medium. Because the region emits no light, IBEX collected ENAs traveling in from the interstellar boundary at velocities from 150,000 to more than 4 million kph.

The ribbon of ENAs observed by IBEX mapped to a model heliopause surface. The arrows show the postulated orientation of the local interstellar magnetic field lines.

ENAs are created in the heliosheath, where the solar wind interacts with neutral particles drifting in from the interstellar medium. ENAs can be created anywhere, but most come from this interaction region where the solar wind becomes slower, hotter, and more turbulent.

IBEX has two large particle detectors that collect ENAs from a range of velocities. As it spins, IBEX views an annulus of the sky that sweeps the entire sky in half a year, and then resweeps it during the next 6 months of its 12-month orbit around the Sun.

The surprising result is that there appears to be a strong impact of the galactic magnetic field on the heliosphere—in the form of a bright ribbon of ENA emission. The galactic magnetic field drapes around the heliosphere, and, as best as the scientists can tell, the bright ribbon is where the magnetic field lies parallel to the boundary.

The IBEX all-sky maps also put observations from the Voyager spacecraft in context. The twin Voyager spacecraft, launched in 1977, traveled to the outer solar system to explore Jupiter, Saturn, Uranus, and Neptune. In 2007, Voyager 2 followed Voyager 1 into the heliosheath. They are now in the midst of the region where the ENAs collected by IBEX originate.

However, whereas the Voyagers give two pinpoint "weather station" views of the interstellar boundary region, IBEX provides an all-encompassing "weather satellite" view. The IBEX results show that much of the activity is happening away from the two Voyagers and would not have been discovered without IBEX.

IBEX is the latest in NASA's series of low-cost, rapidly developed Small Explorer space missions. SWRI developed and leads the mission with a team of national and international partners. GSFC manages the Explorer Program for NASA's Science Mission Directorate.

For more details about IBEX, see Appendix 3.

Stereo Views of the Ring Current from TWINS

Early data from the TWINS spacecraft, the first ENA "stereo" mission, are showing the exciting scientific potential of stereoscopic observations using all-sky ENA imaging to continuously monitor the 1-60 keV ions in the Earth's ring current. The results of a recent analysis, independently confirmed by another mission, show the quality and utility of such observations.

When TWINS launched in June 2008, the Sun was in its extended solar minimum state but, on 2008 June 14/15, a small substorm was triggered by a recurring high-speed stream in the solar wind. SOHO EIT observed the large trans-equatorial coronal hole near the central meridian of the Sun that was its source region. A pair of simultaneous 40-minute exposures from TWINS 1 and 2 taken on 2008 June 15 starting at 0600 UT show the distribution of 10-keV hydrogen ENAs (the two images on the left side of the figure).

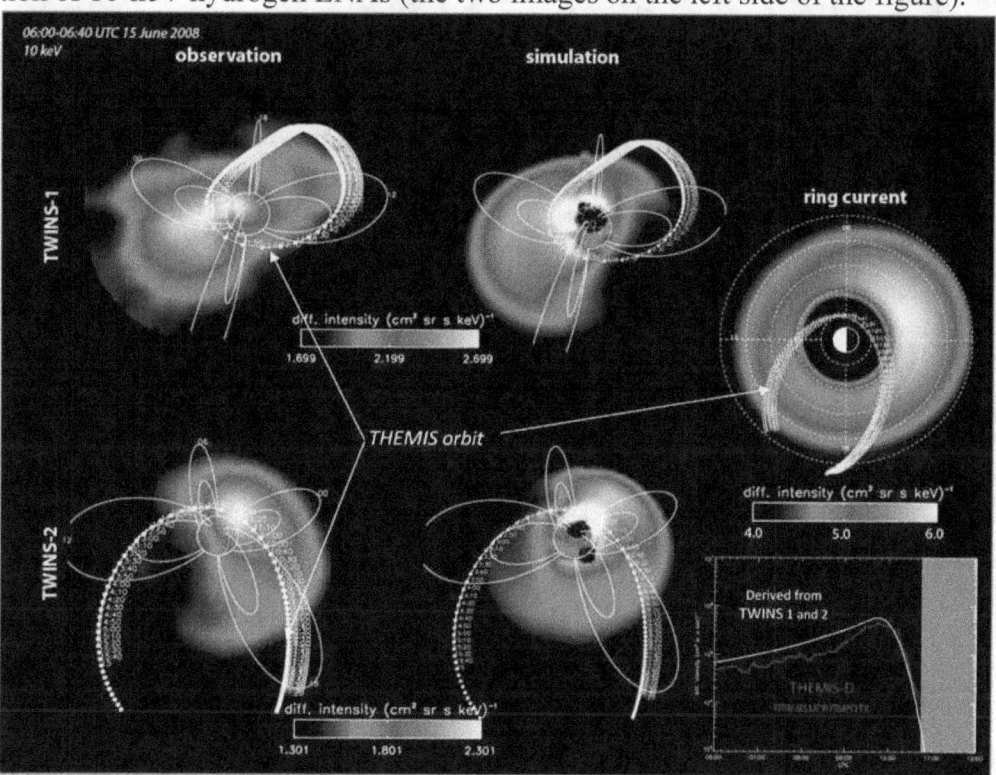

The parent ion (ring current) distribution is unfolded from the ENA images by forward modeling of the ENA emission. The equatorial ion distribution (upper-right panel) is specified using a 32-parameter model. The parameters are adjusted until the simulated ENA images (middle section of figure) match both of the TWINS 1 and 2 images.

In a further illustration of how various components of the HGO can work together, THEMIS-D passed through the ring current region at the time of TWINS observations. Its orbit (dotted line) is over-plotted on the ENA images and ion distributions. The graph (lower right) shows the excellent agreement between the ion flux profile from the THEMIS nightside passage (red line) with the equivalent profile from the TWINS forward-modeled data (white line).

Cluster Spacecraft Discover How Solar Wind Turbulence Dissipates

Measurements of the solar wind temperature, taken between 0.3 AU and more than 10 AU from the Sun, have shown that heat is being added continuously to the solar wind. It has been known for decades that the solar wind is turbulent. In turbulent fluids, large turbulent regions produce smaller regions in a cascade that proceeds to very small scales where the motions finally dissipate and the energy goes into heat.

The amount of heat that can be produced by dissipating turbulence in the solar wind is about what is needed to explain the observed decrease in temperature with distance from the Sun. Although there are many theories of how the waves are dissipated, earlier instrumentation was not able to observe the dissipation directly. Solving this problem is of fundamental importance to our understanding of how the solar corona is heated. New observations made by experiments on the ESA/NASA Cluster mission have indicated one significant channel by which the turbulence dissipates.

The analysis used several instruments on Cluster, including STAFF (Spatio-Temporal Analysis of Field Fluctuation), designed to measure high-frequency magnetic waves, the low-frequency (fluxgate) magnetometer (FGM), and the electric fields and wave experiment (EFW). Dr Fouad Sahraoui, a NASA Fellow (also LPP), Dr Goldstein (673), Dr Robert (LPP), and Dr Khotyaintsev (Institute of Space Physics, Sweden) found that solar wind turbulence extends down to scale sizes on the order of the electron gyroradius (from 10^5 km to ~2 km) and, that at least for the conditions observed, electrons efficiently damped the fluctuations.

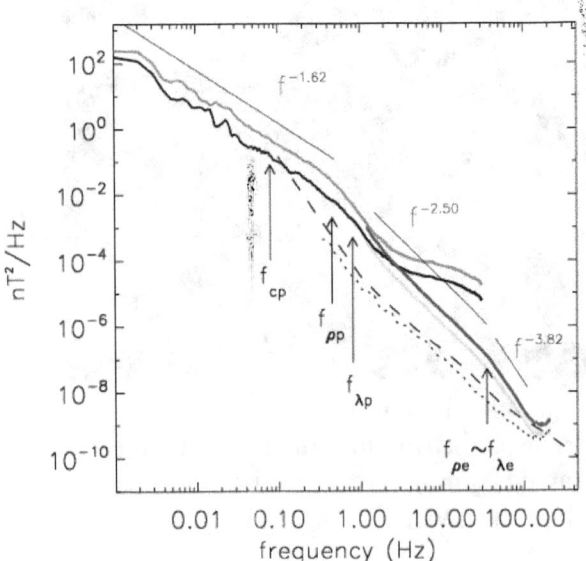

The fluctuations were found to be kinetic Alfvénic fluctuations that were in a state of strong turbulence. This is the first direct observation of the dissipation of solar wind turbulence at these small (electron) scales. The fact that the turbulence consists of kinetic Alfvén waves rather than, say, ion cyclotron waves that interact predominantly with the magnetic energy in the solar wind, may have profound implications for theories of how the solar corona itself is heated.

First measured spectrum of magnetic turbulence cascade in the solar wind, from large scale (10^5 km) down to the electron scales (~3 km). It reveals two new ranges of scales (blue and green curves) never before explored.

STEREO and SOHO Monitor Extended Solar Minimum

Since 2007, the Sun has been in the longest period of minimum-activity conditions in 100 years. The Sun was spotless for almost three quarters of 2008; magnetic flux and coronal activity levels dropped even further during the early part of 2009, although recently the Sun has shown signs of returning life with minor new-cycle activity.

Solar cycles during the first few decades of the Space Age had relatively active minimum phases. STEREO and SOHO together offer unprecedented coverage of the current longer, lower-level cycle minimum and the transition from solar cycle 23 to cycle 24. SOHO has provided a 13-year-plus baseline for solar coronal and magnetic activity from cycle minimum to minimum from L1. STEREO gives "behind the limb" views of the corona and CME activity that are invisible from a near-Earth or L1 perspective. Several GSFC scientists are using STEREO COR1 data to improve our understanding of the solar cycle by developing daily 3D tomographic models of the corona, which can be tested before the coronal structure becomes increasingly complex with rising activity. These models allow the unambiguous determination of the region of origin of CMEs in the streamer belt, more accurate computation of CME mass than previously possible, and clearer insight into streamer recovery. CMEs play a key role in determining the mission profiles of LEO spacecraft and in planning extended human spaceflight missions beyond the protective shield of the Earth's magnetosphere.

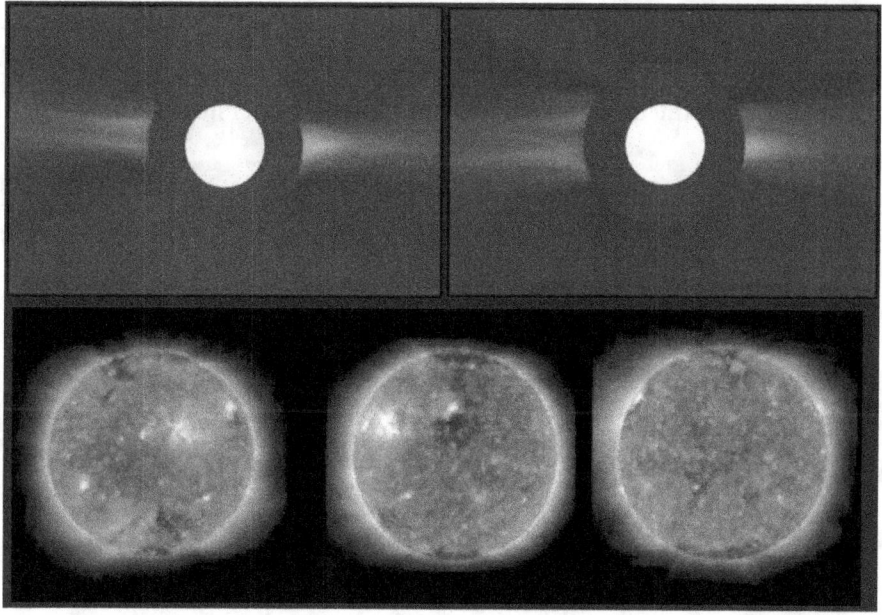

Top panel: Two views of coronal structure contrasting the last (1996) and the current solar minima (2009) as observed by SOHO. Note how during this extended minimum the corona has not simplified to a simple dipolar structure as it normally does. Lower Panel: Full-disk images of the solar corona taken at 195 A (STEREO B, SOHO, and STEREO A, respectively), on 2009 November 17.

GSFC Heliophysics Science Division 2009 Science Highlights

"O wad some Power the giftie gi' us; To see ourselves as others see us!" Robert Burns

HSD HONORS AND AWARDS

In FY09, members of HSD won a large number of honors both internal to NASA and from external organizations. HSD has a peer award system where members of the group are nominated and selected by their peers for an annual award. Here is a list of the awards won by members of HSD in FY09:

Tom Moore receiving his fellowship award at the AGU

- Keith Ogilvie was awarded the Distinguished Service Medal, the highest award that NASA confers
- Tom Moore was elected a fellow of the American Geophysical Union, awarded at the Spring Meeting of the AGU in Toronto
- Roger Thomas, Donald Fairfield, and Doug Rabin received the prestigious NASA Exceptional Service Medal
- Joe Davila received the Goddard Award for Leadership for his role in organizing the International Heliophysics Year
- Adolfo Viñas received the Goddard Award for Mentoring
- David Sibeck and Rob Pfaff received GSFC's Exceptional Achievement Award
- Sten Odenwald won the AAS Solar Physics Division Popular Science Writing Award for 2008 for the article "Bracing for a Solar Superstorm" (*Scientific American*, August 2008)
- Michael Horn received the Goddard Award for Management
- Aaron Roberts was awarded a NASA Special Service Award for his work on the Heliophysics Data Environment and the Data Policy
- Alex Glocer received the Ralph Baldwin Award in Astrophysics and Space Science from the University of Michigan for his PhD thesis.
- The EUNIS team, led by Doug Rabin, won a NASA Group Achievement Award
- John Cooper and Ed Sittler were part of the Cassini Plasma Spectrometer team, which won a NASA Group Achievement Award
- Eric Christian received a NASA Group Achievement Award as part of the Sun-Earth Connection Education Forum Team
- Doug Rowland was part of the C/NOFS VEFI team which was awarded a NASA Group Achievement Award
- Shing Fung accepted the NASA Group Achievement Award for the Radio Plasma Imager Team for the IMAGE mission
- Jim Slavin was part of the MESSENGER team which was awarded a NASA Group Achievement Award

GSFC Heliophysics Science Division 2009 Science Highlights

- Jack Ireland won an Outstanding Achievement award from ADNET Systems, Inc, and a NASA HGI award "Surveying active region oscillations"
- Ed Sittler received a NASA Special Act -- Individual Award from HSD for his lead authorship of the chapter "Energy Deposition Processes in Titan's Upper Atmosphere and Its Induced Magnetosphere" in the book "Titan from Cassini-Huygens"
- Tom Narock was awarded a HSD Peer Award, 2009, for critical support of multiple projects within HSD
- Jan Merka, Kathy Starling, and Zoe Rawlings won HSD Peer Awards
- Zoe Rawlins also received the Goddard Award for Secretarial/Clerical Staff
- Len Burlaga won a Superior-Performance Peer Award
- Dean Pesnell received the Editor's Citation for Excellence in Refereeing from *Geophysical Research Letters*
- Anand Bhatia, Richard Drachman, and Aaron Tempkin were named Outstanding Referees by the editors of *Physical Reviews* and *Physical Review Letters*.

HSD was awarded a number of new research grants:

GSFC-Led Proposals	
Proposal Title	**PI Name**
Multiscale Theory and Modeling of Solar Reconnection[1]	Spiro Antiochos
Establishing Links Between Solar-Wind and Topside-Ionospheric Parameters[1]	Robert Benson
Hybrid Kinetic Model for Europa Surface and Atmospheric Interaction with the Jovian Magnetosphere[2]	John Cooper
Lunar Surface Origins Exploration (LunaSOX)[1]	John Cooper
Fermi Solar Flare Observations[1]	Brian Dennis
Particle Transport in Quasistationary and Transient States of the Magnetosphere[1]	Mei-Ching Fok
The Community Coordinated Modeling Center: Supporting the Research Community and Space Weather Operations Through Ionosphere-Thermosphere-Mesosphere Modeling Services[3]	Michael Hesse
Neutral-Ion Coupling and Wind Shear Effects in the Daytime Lower Ionosphere[1]	Robert Pfaff
Analysis of C/NOFS Satellite Electric Field and Magnetic Field Observations in Earth's Ionosphere and their Space Weather Impact[4]	Robert Pfaff
VISIONS: Understanding the Sources of the Auroral Wind[1]	Doug Rowland
Understanding Interplanetary Shock Dynamics in the Inner Heliosphere with New Observations and Modeling Techniques[1]	Chris St Cyr
The Characteristics of Flux Transfer Events[1]	Dave Sibeck
Analysis of Titans Interaction with Saturn's Magnetosphere using Cassini Titan Flyby[1]	Ed Sittler

GSFC Heliophysics Science Division 2009 Science Highlights

These were won under a number of different opportunities:
1. ROSES 2008
2. ROSES 2007
3. Upper Atmospheric Research Section (UARS) Facility Program
4. Air Force Office of Scientific Research

HSD also participated with members of the heliophysics community as Co-Is on a number of successful proposals for the ROSES 2009 opportunity:

GSFC Co-I Proposals			
Proposal Title	**PI Name**	**PI Institution**	**GSFC Co-I**
Firefly on Demand	Joanne Hill	USRA	Doug Rowland
Optical Effects in NASA's Fleet of Solar Telescopes	Craig DeForest	SWRI	Jim Klimchuk
Modeling the Heating and the Acceleration of Multi-ion Solar Wind Plasma	Leon Ofman	CUA	Aldofo Figueroa-Vinas
Generalized 3D Magnetosheath Specification Model	Yongli Wang	UMBC	Dave Sibeck
Modeling the Effects of Energy Input from Solar Wind Dynamic Pressure Enhancements to the Magnetosphere-Ionosphere-Thermosphere System	Eftyhia Zesta	AFRL	Dave Sibeck

GSFC Heliophysics Science Division 2009 Science Highlights

Heliophysics missions consist of an intricate network of space observatories that monitor the Sun's variability and its effects on the entire solar system, life, and society.

HSD PROJECT LEADERSHIP

Heliophysics covers a vast volume of space from the center of the Sun to the edge of interstellar space. HSD looks at diverse physical processes from magnetic reconnection to particle acceleration, shock formation to convection, and thermal conduction to nonthermal heating. This requires the use of many different observing techniques, both remote and in situ, from a variety of vantage points. In recent years, it has become apparent that little further progress will be made unless these phenomena are studied as a "system of systems."

A fleet of spacecraft, referred to in NASA Headquarters parlance as the Heliophysics Great Observatory (HGO), is currently being flown. Each mission is addressing its own aspect of the problems that HSD is trying to unravel. HSD is the nexus of these projects, having a management role in all but three of these missions and scientific participation in all of them (see the table below).

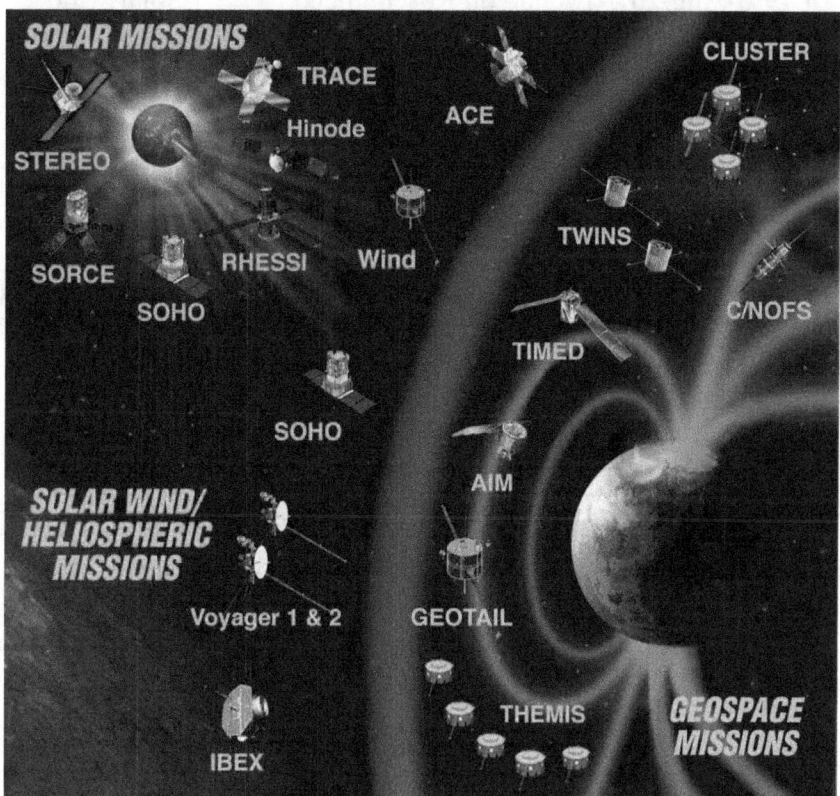

The HGO: Currently, heliophysics has three missions observing the ionosphere, mesosphere, and thermosphere regions; four missions observing the magnetosphere; six missions sampling the solar wind and heliosphere; and six missions observing the Sun. These make up a powerful combination of in situ and remote-sensing instruments.

GSFC Heliophysics Science Division 2009 Science Highlights

Mission	Launch Date	Project/Mission Scientist	Scientific studies of:
Heliophysics Flight Missions			
IBEX	2008 Oct 19	Bob MacDowall (GSFC)	Outer heliosphere
C/NOFS	2008-Apr-16	Rob Pfaff (GSFC)	Ionospheric scintillations
AIM	2007-Apr-02	Hans Mayr (GSFC)	Mesospheric clouds
THEMIS	2007-Feb-17	Dave Sibeck (GSFC)	Substorms
STEREO	2006-Oct-25	Joe Gurman (GSFC)	CMEs
Hinode	2006-Sep-22	John Davis (MSFC)	Solar magnetic fields
SORCE	2003-Jan-25	Robert Callahan (GSFC)	Solar irradiance
RHESSI	2002-Feb-05	Brian Dennis (GSFC)	High-energy flares
TIMED	2001-Dec-07	Dick Goldberg (GSFC)	Thermosphere, ionosphere, mesosphere
Cluster	2000-Jul-16 2000-Aug-09	Mel Goldstein (GSFC)	Particle / magnetic field interactions
TWINS	2000-Mar-25	Mei-Ching Fok (GSFC)	Magnetosphere
TRACE	1998-Apr-01	Joe Gurman (GSFC)	Solar atmosphere
ACE	1997-Aug-25	Tycho von Rosenvinge (GSFC)	Solar wind
SOHO	1995-Dec-02	Joe Gurman (GSFC)	Solar interior, magnetic activity cycle, corona, solar wind
Wind	1994-Nov-01	Adam Szabo (GSFC)	Solar wind
Geotail	1992-Jul-24	Guan Le (GSFC)	Dynamics of the magnetotail
Voyager 1 Voyager 2	1977-Sep-05 1977-Aug-20	Ed Stone (JPL)	Heliosphere
Heliophysics Programs			
Solar Terrestrial Probes		Jim Slavin (GSFC)	
Living With a Star		Chris St. Cyr (GSFC)	

The Ulysses, Polar, and FAST missions completed operations in FY09.

Appendix 3 briefly outlines (in reverse chronological order) the purpose and status of each of the current heliophysics missions and GSFC's role in them.

GSFC Heliophysics Science Division 2009 Science Highlights

"The best present is a future"—the next-generation instruments and spacecraft will gather data in new ways from new vantage points, creating a vibrant future for heliophysics.

DEVELOPING FUTURE HELIOPHYSICS MISSION CONCEPTS

At the moment, there are a number of future heliophysics missions in various stages of development. Key to the future of the HGO and the strength of this discipline are vitality of its strategic vision and the pipeline of new space missions that are required to realize the vision. These missions always build on past scientific achievements and technical capabilities, but it is essential that they produce more than incremental results and that some of the more challenging missions be undertaken not in spite of, but rather because of, the challenges they present to heliophysics and NASA. All of these missions are part of the Solar Terrestrial Probe (STP) or the Living With a Star (LWS) program. Heliophysics expects a steady stream of grassroots-developed, PI-led Explorer missions that will achieve more focused, nearer-term science objectives. There are mission-of-opportunity payloads on other government and commercial spacecraft, and collaborations involving international missions (as with Hinode and SOHO). GSFC HSD scientists serve as study scientists and as members of the Science and Technology Definition Teams of these missions.

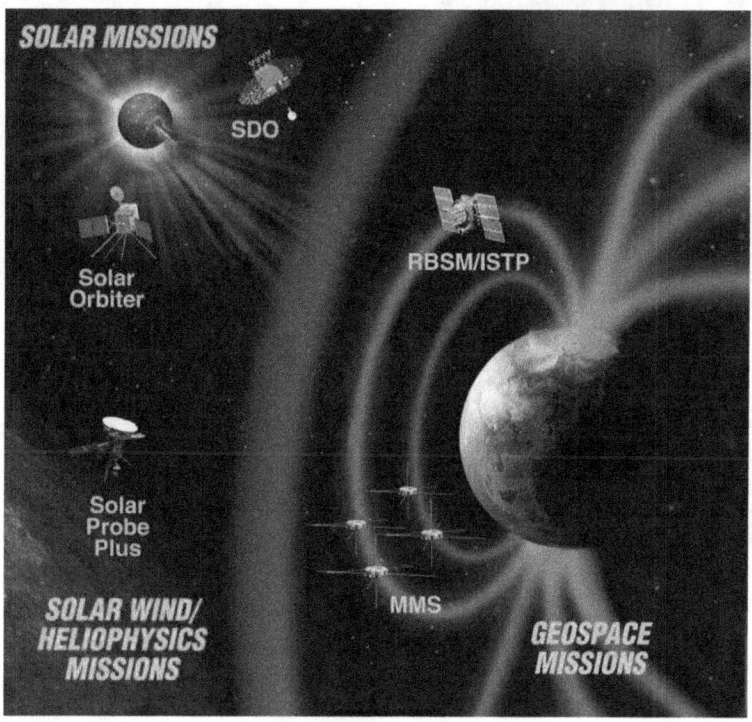

The next-generation HGO elements. Two missions will focus on the ionosphere, mesophere, and thermosphere regions; three missions will solve key questions regarding magnetic reconnection and charged-particle acceleration in the magnetosphere; four missions will measure the solar wind, determine its acceleration mechanism in the corona, and its evolution as it moves out into the heliosphere; and four missions will probe the outer layers of the Sun, its atmosphere, and the eruption of flares and coronal mass ejections.

GSFC Heliophysics Science Division 2009 Science Highlights

NASA heliophysics missions planned to fly beyond FY09:

Mission	Planned Launch	Project/Mission Scientist	Scientific Studies of:
SDO	2010 Feb	Dean Pesnell (GSFC)	Solar magnetic fields, corona
RBSP/ITSP	2012	Barry Mauk (APL)	Earth's radiation belts
MMS	2013	Mel Goldstein (GSFC)	Magnetosphere
Solar Orbiter	2017	Chris St. Cyr (GSFC)	High-resolution solar imaging
SP+	2018	Robert Decker (APL)	Origin of the solar wind

The nature and timing of the missions may change as a result of the NASA Heliophysics Roadmap study and the National Research Council (NRC) decadal survey. Appendix 4 provides some of the details of each of these missions.

GSFC Heliophysics Science Division 2009 Science Highlights

> Sharing NASA mission science and activities with the world provides critical enhancement in support of a robust heliophysics program

HELIOPHYSICS EDUCATION AND PUBLIC OUTREACH

Activities in formal and informal education, public outreach, and public affairs contribute greatly to the impact of the Division's scientific endeavors on current and future generations. Dr Holly Gilbert, Associate Director for Science, leads the effort to coordinate and propagate the most effective ventures toward our outreach goals to communicate heliophysics scientific discoveries and advances, raise public understanding of space weather impacts, and inspire the next generation of American space scientists.

HSD projects and programs also respond directly to NASA Education and SMD E/PO requirements, which include attention to partnerships, sustainability, a focus on customer needs, and diversity. Key to the success of these activities has been the involvement of many enthusiastic HSD scientists, whether directly interfacing with audiences, providing access to data, or simply confirming accuracy of product content.

The E/PO Leads for ACE, Hinode, SDO, SOHO, and STEREO, and the entire Sun-Earth Day team, reside at GSFC. A significant change in the HSD E/PO structure occurred when the Sun-Earth Connection Education Forum ended in September 2009. This 11-year partnership between GSFC and UCB facilitated heliophysics E/PO efforts across the country.

Division E/PO staff frequently provide presentations and workshops at educational conferences and other venues, sharing the science and interacting with these important audiences to learn how best to meet their needs. Coordinating efforts whenever possible, the teams work together as well as with strong partners outside NASA. Along with day-to-day work, the HSD E/PO group routinely contributes to community service, assisting each other with proposal development, on review panels, and with professional development presentations at community meetings. Several personal, project, and funding awards were received in the last year, including the AAS/SPD Popular Science Writing Award, an HSD Peer Award, and NASA grants covering several projects. SECEF received a NASA Group Achievement Award.

HSD team members staff the GSFC Integrated E/PO Strategy Group, Education Policy and Implementation Teams, and their working groups. They regularly interact and coordinate with the GSFC Education Chief and the new Assistant Director for Science Communication. Strategic coordination and collaboration within HSD, with other GSFC programs, throughout NASA, with universities, and with other organizations extend the impact and make optimum use of resources.

The following sections provide an overview and sampling of the FY09 HSD E/PO Program.

Advanced Student Research Programs

Through summer intern programs, such as the High School Internship Program and the Science and Engineering Student Intern (SESI) program in cooperation with CUA, HSD supported about 20 summer interns (high school and undergraduate), in addition to several graduate students through the Graduate Student Summer Program. HSD supported two students in the Cooperative Education Program, an important educational link between college-level academic study and full-time meaningful work experience, through a working agreement between GSFC and several educational institutions. The Graduate Student Researchers Program (GSRP) provides qualified graduate students, in residence at their home institutions, with fellowship support on research projects of mutual interest to the student and the GSFC mentor; in FY09 HSD supported eight GSRP students. The NASA Postdoctoral Program (NPP) provides talented postdoctoral scientists and engineers with valuable opportunities to engage in ongoing NASA research programs; HSD currently has 14 NPP staff members and several other postdoctoral scientists funded through science grants.

K-12 E/PO Programs

Sun-Earth Day (SED) has the widest reach of all Division programs. It is a well-coordinated collection of formal and informal E/PO programs, resources, and events under a unique yearly theme, highlighting the fundamentals of heliophysics, space weather research, and related missions. Events are hosted by educators, museums, amateur astronomers, community groups, and scientists at schools, parks, and science centers.

This year, SED was picked up in 47 countries including the US, with nearly 50,000 teachers and 1.4 million students participating. Eight NASA Centers including GSFC were involved. The program also develops podcasts and vodcasts, and as of this year provides updates on Facebook and Twitter. The SED website is rich in resources for every audience and contains an extensive multimedia section that includes virtual tours of ancient astronomy sites. Featured activities are added each year to enhance the theme and tell the interrelated story of the heliophysics missions.

Separate from SED, yet a significant component of it is the online Space Weather Action Center (SWAC). SWAC provides students with the opportunity to monitor the progress of a solar storm from its eruption on the Sun to the associated changes in Earth's magnetosphere – accessing and analyzing NASA satellite data in the process. Over 3000 unique downloads occurred during the first month after its update this year. Dozens of educators and over 100 students, including Native American tribal college staff members and students as well as teachers and administrators from the Bureau of Indian Education, were trained for SWAC, either through NASA's Distance Learning Network or via video conferencing. One of the program's developers gave a keynote address at the 9th International Federation for Information Processing World Conference on Computers in Education in Brazil.

The Space Weather Media Viewer provides access to near-real-time space weather data and contains illustrations and visualizations for classroom or public use. Its value to SED and heliophysics outreach in general is obvious. The major upgrade this year – with higher-resolution images, mission videos, easier downloads, and better descriptions of each multimedia product – was due to input and funding from Voyager, Ulysses, Hinode, THEMIS, STEREO, SOHO, IBEX, and ACE.

Elementary / Secondary Education

The SpaceMath@NASA website is a collection of problem sets that teach students about astronomy and space science by using mathematics and real-world problems. Problems are developed in collaboration with Hinode, STEREO, THEMIS, and SDO, among other missions, using mission science data, press releases and discoveries. The site's popularity is growing, and it is the top return on Google for "NASA + math".

SDO sponsors several formal education programs, including Day at GSFC, a field trip exposing students to a range of NASA career options that involves meeting scientists and engineers. The Ambassador in the Classroom program, a partnership with NASA's Aerospace Education Services Project, brings a scientist or E/PO professional into the classroom to share SDO-related science lessons. The AGU and AAS meetings have proved to be good venues for the K-12 Educator Reception, which provides local teachers with an invaluable opportunity to collect NASA classroom materials and interact one-on-one with scientists.

Informal Education

Family Science Night, a monthly two-hour intensive science program for middle school students and their families, is an HSD partnership with the Astrophysics Science Division and the GSFC Visitor Center. Evaluation after three years shows that participants have a positive change in attitude toward science, and that even one session improves the likelihood that families will engage in external science-related activities.

An E/PO team member frequently provides training in heliophysics and planetary science, and remote telescope operation to Girl Scouts, their leaders, and their master trainers, both at training sessions and at special events such as star parties, planetarium shows, "NASA Day," and astronomy club meetings. This year, our E/PO representative personally interacted with nearly 1500 scouts, leaders, and family members.

Public Outreach

These projects cover a wide range of activities. As part of the all-volunteer Goddard Ambassador program, Dr Gilbert conducts guided tours of GSFC and gives "Science on a Sphere" presentations to outside groups.

SDO uses the most popular social networking tools such as Facebook, Twitter, YouTube, Flickr, and Blogger to post mission status and science information to reach a wider, more diverse public and create stronger relationships with attentive audiences. After touring museums nationally, the Space Weather Center exhibit, developed in 2000 by a team of heliophysics E/PO specialists and scientists, was reinstalled at the GSFC Visitor Center and is being viewed by hundreds of students and members of the public annually. ACE's high-school-level *Cosmicopia* cosmic ray and heliospheric science website includes over 400 unique "Ask a Physicist" questions and answers; the last year saw a 10% increase in monthly accesses (averaging over 370K) and a 40% increase in unique users.

AstroZone (held at AAS), a.k.a. Exploration Station (at AGU), is a science open house to share NASA hands-on activities with local families, alongside local science exhibitors, that encourages involvement beyond the initial event. Co-developed by the SDO E/PO Lead, the program is a partnership with the Rochester Institute of Technology Center for Imaging Science Insight Lab with the support of the SDO Project Scientist and scientists from other divisions and institutions.

Public Affairs

Several HSD scientists have made appearances on national media, including Discovery, National Geographic, and History channels as well as local news. GSFC also has the Scientific Visualization Studio (SVS) producing superior quality images, animations, and data visualizations for a wide range of heliophysics communications and science activities, including press releases, live presentations, print publications, television, and video documentaries. SVS debuted the "Sentinels of the Heliosphere" film, which highlights NASA's HGO fleet, at the SIGGRAPH 2009 International Conference and Exhibition on Computer Graphics and Interactive Techniques, where it was nominated for an Academy Award – http://svs.gsfc.nasa.gov/goto?3595.

Other successful FY09 heliophysics stories include:

- "Sun Often 'Tears Out A Wall' In Earth's Solar Storm Shield" (12/16/08)
  http://www.nasa.gov/mission_pages/themis/news/themis_leaky_shield.html

- "How Low Can It Go? Sun Plunges into the Quietest Solar Minimum in a Century" (04/01/09)   http://www.nasa.gov/topics/solarsystem/features/solar_minimum09.html

GSFC Heliophysics Science Division 2009 Science Highlights

- "NASA's STEREO Spacecraft Reveals the Anatomy of Solar Storms" (04/14/09)
  http://www.nasa.gov/mission_pages/stereo/news/solarstorm3D.html

- "Tiny Flares Responsible for Outsized Heat of Sun's Atmosphere" (08/14/09)
  http://www.nasa.gov/topics/solarsystem/features/nanoflares.html

---

HELIOPHYSICS E/PO ON THE WEB

AstroZone -- http://sdo.gsfc.nasa.gov/E/PO/families/astrozone.php

Cosmicopia -- http://cosmicopia.gsfc.nasa.gov

Exploration Station -- http://sdo.gsfc.nasa.gov/E/PO/families/explore/explore.php

Family Science Night -- http://sdo.gsfc.nasa.gov/E/PO/families/fsn.php

Sun-Earth Day -- http://sunearthday.nasa.gov

SpaceMath@NASA -- http://spacemath.gsfc.nasa.gov

Space Weather Action Center -- http://sunearthday.nasa.gov/swac

Space Weather Media Viewer -- http://sunearth.gsfc.nasa.gov/spaceweather/

GSFC Heliophysics Science Division 2009 Science Highlights

The E/PO Team

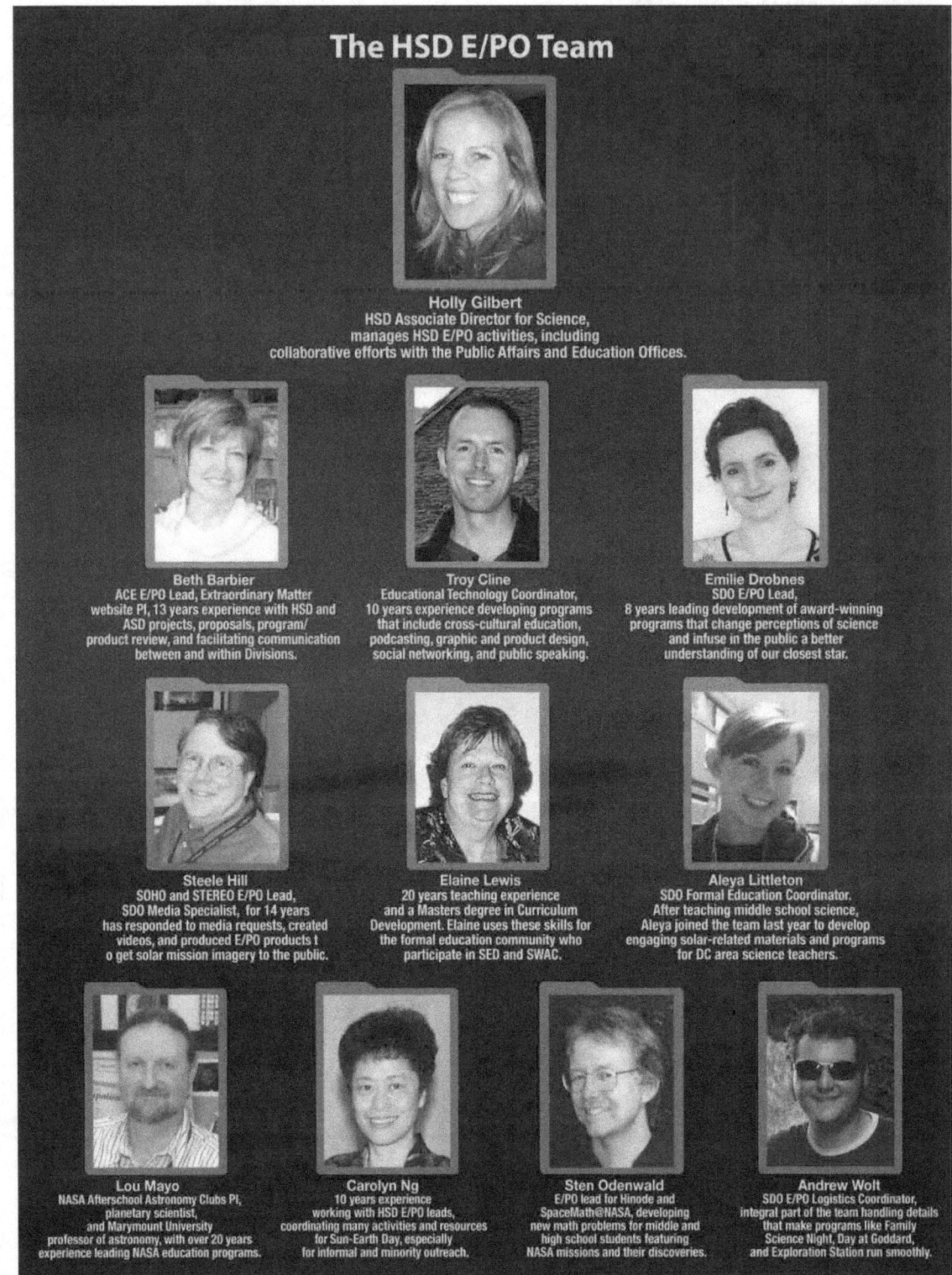

International Heliophysical Year (IHY)

Although IHY officially closed at the beginning of 2009, several activities continued throughout FY09 and beyond. These programs were selected because they required more time to complete (such as Whole Heliosphere Interval analysis, and scientific publications and sessions presenting IHY results), or because of the importance of continuing the activity to ensure long-term success. In particular, the instrument deployment and scientific collaboration programs with developing nations must continue if there is to be an impact on the researchers' participating nations.

A new UN program called the International Space Weather Initiative (ISWI) was established to continue legacy activities of the IHY UNBSS program. ISWI was initiated at the February 2009 IHY closing ceremony at the UN in Vienna. A steering committee has been established, and the first instrument coordination workshop took place on 2009 November 19-20 in Rabat, Morocco. Additionally, an ISWI space science school for graduate students is being planned for Autumn 2010 in Ethiopia.

The 100-page IHY final report was published and distributed by the UN.

A 370-page book on the IHY science, instrument development, outreach and history activities in 78 nations was published in 2009 by Springer Press. The book is available on amazon.com

GSFC Heliophysics Science Division 2009 Science Highlights

> The congruence of expertise and experience in HSD makes it a natural center for leadership in the design and implementation of science information systems.

SCIENCE INFORMATION SYSTEMS

NASA builds and flies heliophysics spacecraft and instruments to collect observational data and improve understanding of the heliophysical system and its detailed processes in a classic research wheel.

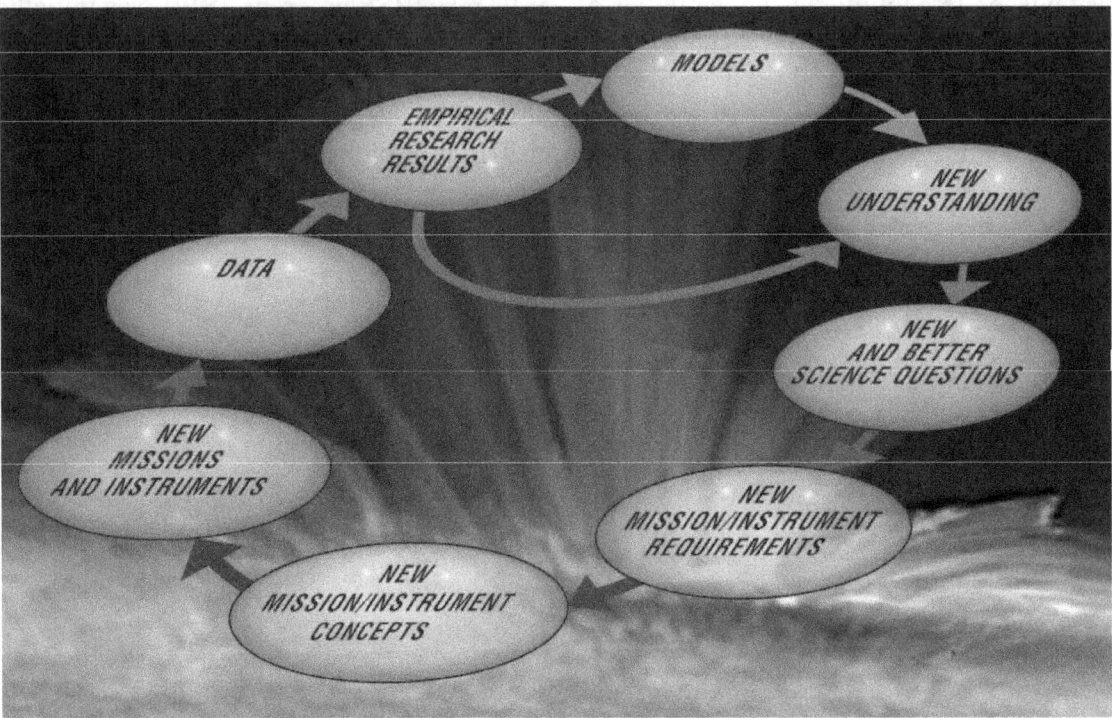

The heliophysics science challenge is to understand consistently the physics and coupling of this "system of systems" over many spatial and temporal scales. Science information systems are then key to drawing data and science both from individual science investigations and across the boundaries of the individual missions and instruments of the HGO. In addition, NASA's Heliophysics Data Policy (see links on http://hpde.gsfc.nasa.gov) demands open and useful access to NASA data by the broad national and international science research community to ensure full scientific return to the public on NASA's investment in science missions and investigations.

The congruence of instrument design, physics, modeling, data analysis, and information technology expertise and experience makes HSD a natural center for leadership in the design and implementation of science information systems (SIS), working in close partnership with the external science and technology communities. Categories for systems currently defined in the Heliophysics Data Policy include:

- Mission/Investigation Data Facilities
- Virtual Discipline Observatories (VxOs)
- Resident and Final Archives
- Science Data and Modeling Centers
- Deep Archive(s)

The evolving heliophysics data vision is a distributed set of mission and investigation data facilities closely tied to the scientists responsible for active processing of data. When coordinated with active archives and data centers connected by Virtual Discipline Observatory (VxO) data location, retrieval, and user services, it ensures easy consolidated access to usable data with views of the data customized to research discipline needs. In general, heliophysicsists expect to leverage existing capabilities and services wherever practical but to support the introduction of new technology to enable a more distributed architecture. The first versions of the VxOs are intended to follow a "small-box" model focused on finding fully described data relevant to research needs (e.g., selected by times with parameters in a given data range from a selection of instruments).

Mission and investigation data facilities under HSD direction include:

- Solar Data Analysis Center (SDAC) support for SOHO, solar instruments on STEREO, and data from TRACE, Hinode, and other missions
- STEREO Science Center (SSC)
- Critical support for SDO's distributed science archive and distribution system
- Operation of the Wind-Geotail (Polar-Wind-Geotail) ground data system
- Production of higher-level Wind (MFI, WAVES), Voyager (MAG and CRS), and STEREO (SWAVES) instrument products
- C/NOFS-CINDI data production

Two key HSD projects, the Space Physics Data Facility (SPDF) and SDAC, are now defined as active Final Archives of non-solar and solar imaging data, respectively, for NASA heliophysics science missions. They simultaneously serve as broad-based and heavily used current multimission data distribution and services projects.

SPDF also maintains the Common Data Format software, which is increasingly a standard self-describing, self-documenting format for non-solar heliophysics data products.

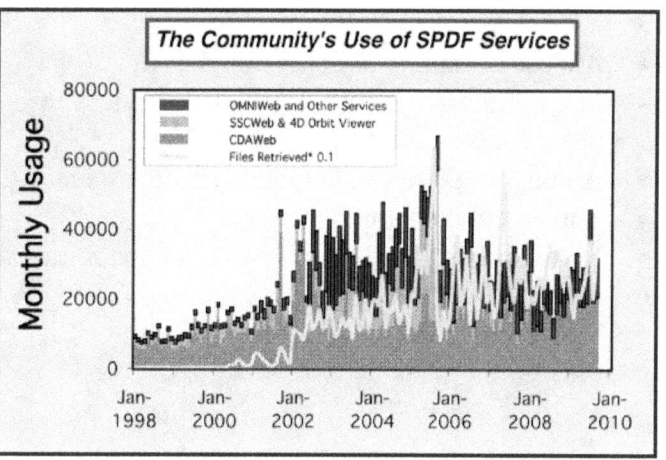

SPDF service usage, where an execution is a request to create a customized plot, listing, or output file.

GSFC Heliophysics Science Division 2009 Science Highlights

A primary center of activity for Virtual Observatories, HSD leads the newly formed Heliophysics Data and Modeling Consortium (HDMC) project for funding and coordinating the various VxOs and Resident Archives and provides direct leadership of the original set of VxOs selected by NASA:

- Virtual Solar Observatory (VSO)
- Virtual Heliospheric Observatory (VHO)
- Virtual Magnetospheric Observatory at GSFC (VMO/G)
- Virtual Ionospheric/Thermospheric/Mesospheric Observatory (VITMO) (as Co-Is)

and more special-interest VxOs focused to special requirements in:

- Virtual Energetic Particles Observatory (VEPO)
- Virtual Wave Observatory (VWO)
- Heliospheric Event List Manager (HELM)

HSD has active roles in the definition of the Space Physics Archive Search and Extract (SPASE) data model intended as a common language to allow queries among the VxOs and a foundation for the internal structure of several of them. In addition, the SPDF project is now supporting the Virtual Space Physics Observatory interface, an effort to comprehensively access heliophysics data holdings across these disciplines, leveraging both the present inventories of the individual VxOs and other information sources.

The overall goal is to make better data more readily accessible to and usable by researchers in a given discipline across the boundaries of specific missions, instruments, and times, using an optimal mix of existing capabilities with appropriate standards and new technology to improve data access, and using tighter coupling to the data-providing community to ensure that data are fully and properly described for correct and independent future research use. HSD is looking for:

- A closer coupling of data to models
- More extensive use of distributed services through which data can piped for specific processing or other added value
- An ongoing effort to define lower-cost science ground system design
- Implementation approaches
- More sophisticated on-board data capabilities that can lead to "sensorwebs" among active instruments.

GSFC Heliophysics Science Division 2009 Science Highlights

> The CCMC provides the tools and expertise to help support the heliophysics community in the development of the next generation of space weather models.

COMMUNITY COORDINATED MODELING CENTER

The Community Coordinated Modeling Center (CCMC) is a US interagency activity aiming at research in support of the generation of advanced space weather models. The CCMC consortium consists of NASA, NSF, NOAA, the USAF Weather Agency, Directorate of Weather, Space and Missiles System Center, AFRL, the Air Force Office of Scientific Research, and the ONR. CCMC's central facility is located at GSFC. The CCMC is supported primarily by NASA and NSF. At present, CCMC staff, consisting of space and computer scientists and Information Technology professionals, is 11 FTEs.

The first function of the CCMC is to provide a mechanism by which research models can be validated, tested, and improved for eventual use in space weather forecasting, such as is needed for NASA's Vision for Space Exploration. Models that have completed their development and passed metrics-based evaluations and science-based validations are being handed off to the NOAA and USAF forecasting centers for space weather applications. In this function, CCMC acts as an unbiased evaluator that bridges the gap between space science research and space weather applications.

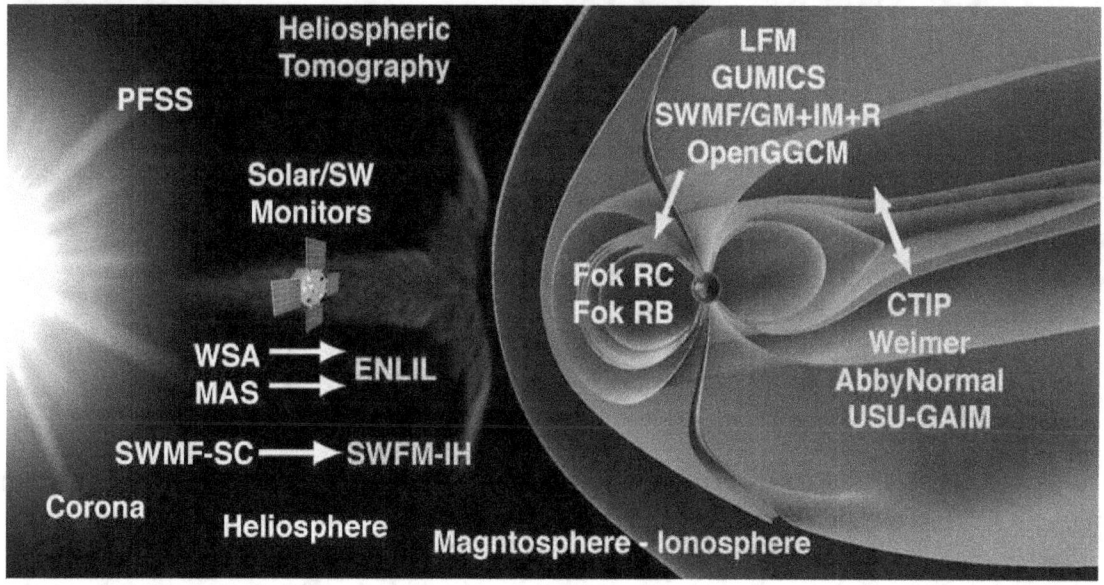

Overview of models at the CCMC

As a second, equally important function, the CCMC provides to space science researchers all over the world the use of space science models, even if the researchers are not model owners themselves. This service to the research community is implemented through the execution of model "runs-on-request" for specific events of interest to space science researchers at no cost to the requestor. Model output is made available to the science customer by means of tailored analysis tools and data dissemination in standard formats. Through this activity and the concurrent development of advanced visualization tools,

CCMC provides to the general science community unprecedented access to a large number of state-of-the-art research models. The continuously expanding model set includes models in all scientific domains from the solar corona to Earth's upper atmosphere.

Science Support

CCMC science services are provided through Web access (http://ccmc.gsfc.nasa.gov). Here, users can request calculations from more than 25 modern space science models, which are hosted at the CCMC through very positive collaborations with their owners.

After a run request, the user will be notified via e-mail when the calculation is complete. At this time, the run can be analyzed via tailored visualization tools, again via Web access. These tools have been continuously refined for almost 10 years, primarily in response to user feedback. Recently added capabilities include Poynting flux calculations, model outputs, field-line tracing along satellite trajectories, movie generation, and time line capabilities.

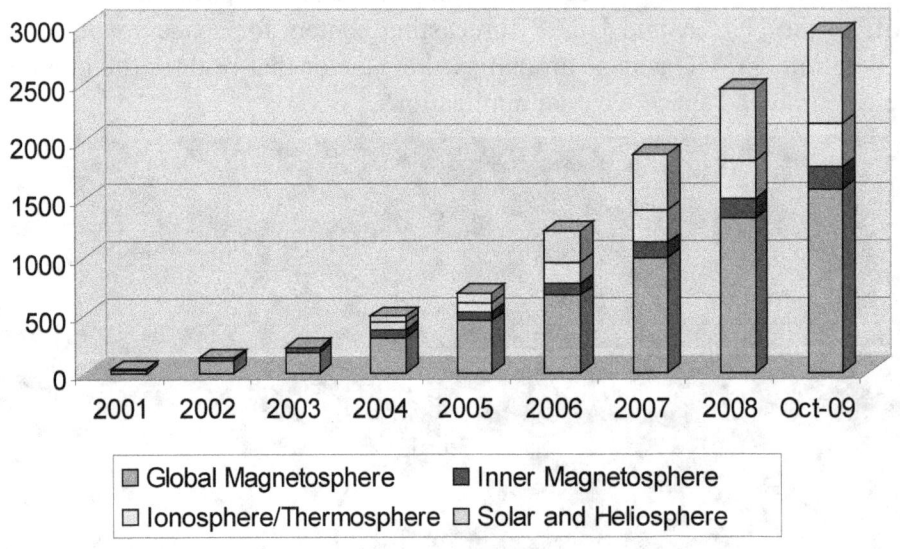

The utilization of CCMC Run-on-Request services continues to grow rapidly. CCMC run analysis services are also used heavily. In 2008, an average month saw 4,200 visitors, 19,200 visits, and 262,000 requests for pages. Similar statistics apply to 2009. Visualization requests from 870 users led to the creation of 34,500 visualization pages each month. These monthly averages continue to increase from year to year.

Furthermore, the CCMC staff continues to develop new means to support users in their scientific studies. In the realm of visualization, OpenDx-based visualization with enhanced 3D capabilities is now available for a number of different model outputs. With both OpenDx and IDL-based 3D visualizations, users can request output in Virtual Reality Modeling Language (VRML) form, which permits real-time, in-depth analysis of complex 3D structures.

GSFC Heliophysics Science Division 2009 Science Highlights

Further science services at the CCMC include the provision of the Model Web, where a large set of empirical or analytical models, such as the International Reference Ionosphere (IRI), are available for interested users for download or execution. CCMC is also supporting space science missions, for example STEREO and THEMIS. Science mission support includes background science information derived from routine runs and calculations in support of specific campaigns or specific science objectives.

As in the past, the future of CCMC services will be shaped by the needs of the science community. Community input is solicited formally and informally, through meetings, tailored workshops, and personal contacts.

Space Weather Modeling

The second focus of CCMC is bringing modern space science modeling to bear on the needs of space weather forecasting. NASA and forecasting agencies as well as need models of proven forecasting abilities. Operational models must be robust, have accuracies that are well understood, and be packaged in a manner that makes them easy to integrate into existing computational environments. The operational models must also be capable of executing reliably in real time. With these objectives in mind, the CCMC tests models

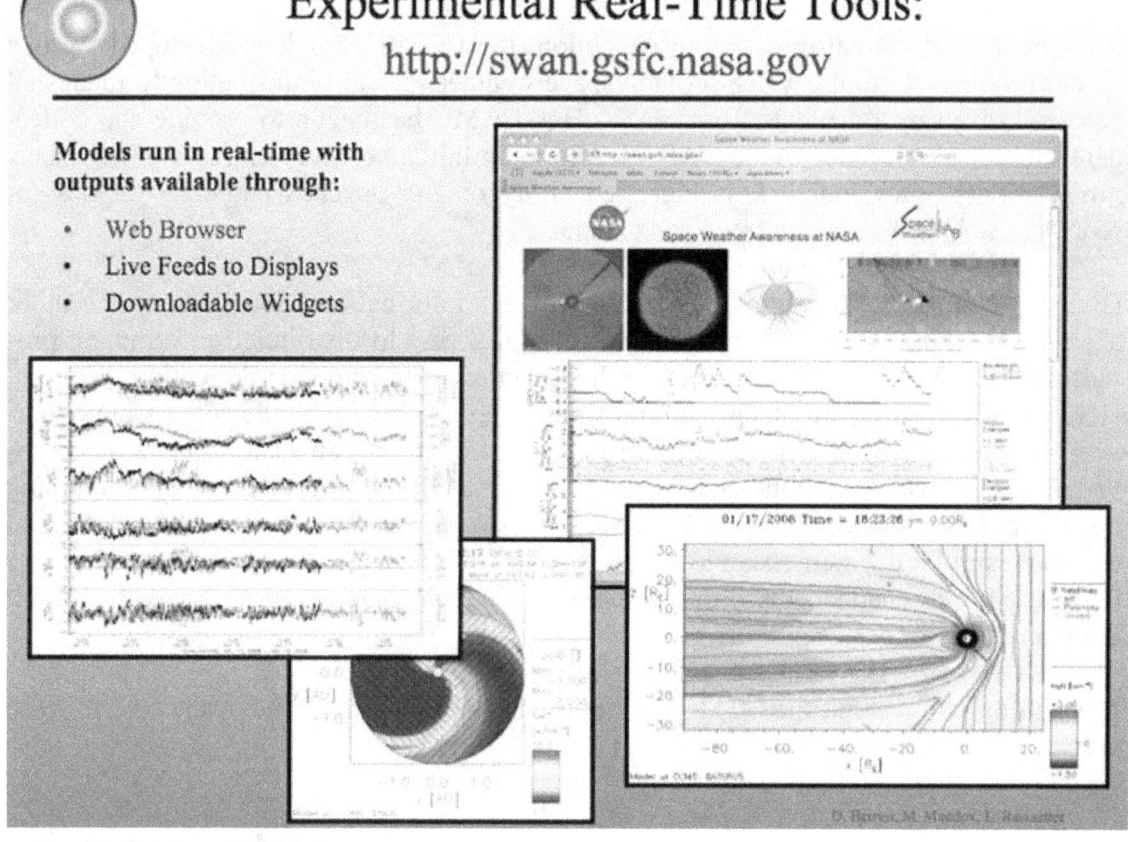

Sample plots from SWAN site

for accuracy, robustness, performance, and portability. Model accuracy is tested in two ways: through metric studies and through scientific validation studies. As a result, the CCMC can characterize the usefulness of these models for the forecasting role to which they would be applied. By testing how the models behave with a wide range of inputs, the CCMC can establish regimes in which the models appear robust and identify regimes that cause the models to fail. Through exposure to a range of compilers and hardware platforms, the CCMC improves the portability of these models. Finally, by measuring computational performance, the CCMC establishes the conditions needed for each model simulation to complete in real time or better.

Performing these types of analyses on models in a protected academic environment might produce less objective results. It is imperative that the models undergo tests in an environment that stresses them in the same manner as will occur at the operational agencies. Accordingly, the CCMC challenges all models by utilizing them in a quasi-operational setting, namely the Run-on-Request facility, as well as through a real-time simulation system. For the latter purpose, CCMC has established various real-time execution systems, starting with solar and solar wind modeling of photospheric magnetograms, and magnetospheric calculations driven by ACE data. The establishment of these systems has led to a wealth of expertise in the design and maintenance of real-time modeling. In collaboration with model owners, CCMC has created robust models of scientific and space weather utility.

As a byproduct of real-time model execution, the CCMC staff has developed various model products, or tools, which provide space weather-relevant information to interested parties. Driven by a mandate from NASA/HQ, CCMC has begun to provide and collect data sources of relevance to the NASA Exploration Initiative, and to NASA SSMO. This information is provided to any interested entity on the Space Weather Awareness at NASA (SWAN) Web site: http://swan.gsfc.nasa.gov.

Owing to the diversity of space weather-relevant information sources, space weather analysis and forecasting will increasingly rely on access to distributed information providers. The SWAN site is a first step in the direction of information collection. While it, along with further development activities supported by NASA's Office of Chief Engineer, addresses NASA's needs primarily, this information can also support other governmental as well as commercial space weather interests. CCMC therefore has ongoing collaborations with USAF, NOAA, international space weather interests, and commercial enterprises in the US. An example of the latter is the Electrical Power Research Institute (EPRI), a jointly funded agency that collaborates in the evaluation of space weather modeling products for power grid uses.

More information about the CCMC and its capabilities may be found at the CCMC Web site http://ccmc.gsfc.nasa.gov.

TECHNOLOGY DEVELOPMENT

GSFC and NASA support new technology in the HSD through three programs. The main source of support at GSFC is the GSFC Internal Research and Development (IRAD) program, which funds the development of new technology to support proposals to be submitted in response to future announcements of opportunity. The Technical Equipment program provides funding for purchases of advanced technology equipment and replacement of outdated equipment needed to support the development of new instrumentation. The Small Business Innovative Research (SBIR) program provides funding to develop new flight hardware and software concepts in the commercial sector in support of future flight opportunities.

The GSFC IRAD program runs on an annual basis for each fiscal year. An announcement of opportunity is released in early June describing the requirements of the opportunity for the subsequent fiscal year. Step-1 proposals are submitted in late June. A heliophysics proposal review panel led by Dr Sigwarth, the Chief Science Technologist for HSD, with support from Dr Khazanov meets in July to rank the proposals and recommend to the Center-wide integration panel those proposals that should progress to the second phase. Step-2 proposals are submitted in late July. The review panel meets again in early August to rank the Step-2 proposals and recommend funding to the Center-wide integration panel. Funding approval by the Center-wide integration panel is contingent on the available resources and the relative ranking of HSD proposals with respect to other focus areas within GSFC. In FY09, a total of 14 Step-1 proposals were submitted resulting in six projects funded by the GSFC IRAD program. An additional two projects were added midway through the year. For these winning projects, the FY09 IRAD program funded a total of 12.3 FTEs and an additional $570.8K for procurements in support of the selected IRAD proposals. In FY10, a total of 18 Step-1 proposals were submitted resulting in 11 projects funded by the GSFC IRAD program. For these winning projects, the FY10 IRAD program funded a total of 13.5 FTEs and an additional $528.7K for procurements in support of the selected IRAD proposals.

The new technologies supported by the GSFC IRAD program cover a broad range of heliophysics topics. These technologies include radiation-tolerant ion mass spectrometers, radiation-hardened extreme-environment ASICs; an image slicer for UV integral field spectroscopy; deployable radial imaging for velocity, electric field, and plasma density; a focal plane polarizer for solar and heliospheric imaging; evaluation of new micro-channel plate technologies; ultrathin windows for silicon-solid-state detector telescopes; a low-resource ionospheric sounder; a sparse sensor array for wave interferometry; TRL elevation of an advanced ASIC for heliophysics missions and instruments; and high-precision electric gate and miniature ion-mass spectrometers.

GSFC Heliophysics Science Division 2009 Science Highlights

Closeups of TTI, MINI-ME, and PISA integrated on the FASTSat spacecraft.

The teams working on three of the instruments selected for FY09 IRAD funding have included students from the US Naval Academy. Because of this cooperation, these three instruments – the Thermospheric Temperature Imager (TTI), the Plasma Impedance Spectrum Analyzer (PISA), and the Miniature Imager for Neutral Ionospheric atoms and Magnetospheric Electrons (MINI-ME) – have been briefed to the Department of the Navy and the Department of Defense (DoD) Space Experiment Review Boards (SERB) and selected to be included on the respective SERB lists. As a direct consequence of this inclusion on the SERB list, TTI, PISA, and MINI-ME will fly on the FASTSat spacecraft of the Department of Defense Space Test Program in May 2010.

GSFC Heliophysics Science Division 2009 Science Highlights

HSD LINE OF BUSINESS

Lines of Business (LOB) were established at GSFC in 2008 to enable the Center to integrate priorities across all research areas. The leadership team consists of the LOB Manager, E. Park (460); Chief Technologist, M. Johnson (550); Chief Scientist, J. Slavin (670); and New Business Manager, J. Breed (101). Essential scientific and technical support is provided by HSD Division Senior Staff (i.e., T. Moore, J. Sigwarth, and J. Davila). The research priorities are developed by the HSD scientists in close communication with the LOB. These priorities then guide the allocation of the available research and development funding. Each LOB establishes yearly goals and implementation plans focused primarily on near-term (3–5 years) new-business opportunities and their associated technology development requirements. The LOB goals and plans are reviewed by a Center integration team, then approved by Center management.

The HSD LOB has developed instrument roadmaps as a tool for correlating technology development, instrument development, mission development, and future competitive or directed opportunities for flight. The goals, implementation plan, and roadmaps are used in Center funded IRAD allocations, identification of technology gaps, and requests for B&P and any other available source funding for advanced work. In its two years of operation, the heliophysics LOB has been successful in developing a cohesive plan for heliophysics development at GSFC and increasing the funding received by the Division. Specific research and development tasks are selected on the basis of short proposals submitted by teams of Center scientists and engineers. The proposals are then evaluated on the technical merit and alignment with the approved LOB strategic priorities and instrument roadmaps.

FY09 LOB Internal Research and Development Priorities

1. Heliophysics Instrumentation
    a. FASTSat
    b. SP+
    c. Europa – Jupiter System Mission
    d. Next Generation Concepts for Sounding Rockets, Explorers, STP, LWS, and various Solar-Planetary Missions
2. NASA Space Weather Operational Requirements – Driven Situational Awareness and Forecast Technologies

FY09 LOB Internal Research and Development Achievements

A total of 11 IRAD proposals submitted by joint Code 500 and Code 600 teams were funded in FY09. Together they constitute a broad portfolio of new efforts in sensor, detector, electronics, and environmental modeling of new-technology development efforts designed to provide enabling capabilities for new flight-instrument and space-weather system implementation opportunities. Special attention is called to the IRAD-funded flight of three developmental instruments on the Department of Defense Space Test Program – MSFC FASTSat smallsat technology mission. The successful operation of these instruments on FASTSat will raise their Technical Readiness Levels to 6+, and allowing them to "graduate" from their development phases and be proposed for NASA science missions.

GSFC Heliophysics Science Division 2009 Science Highlights

APPENDIX 1: INDIVIDUAL SCIENTIFIC RESEARCH

The following section contains short summaries of the research being performed by scientists in HSD. The organization is divided up into four groups, or laboratories, with the civil servant complement shown in the detailed organization chart shown below.

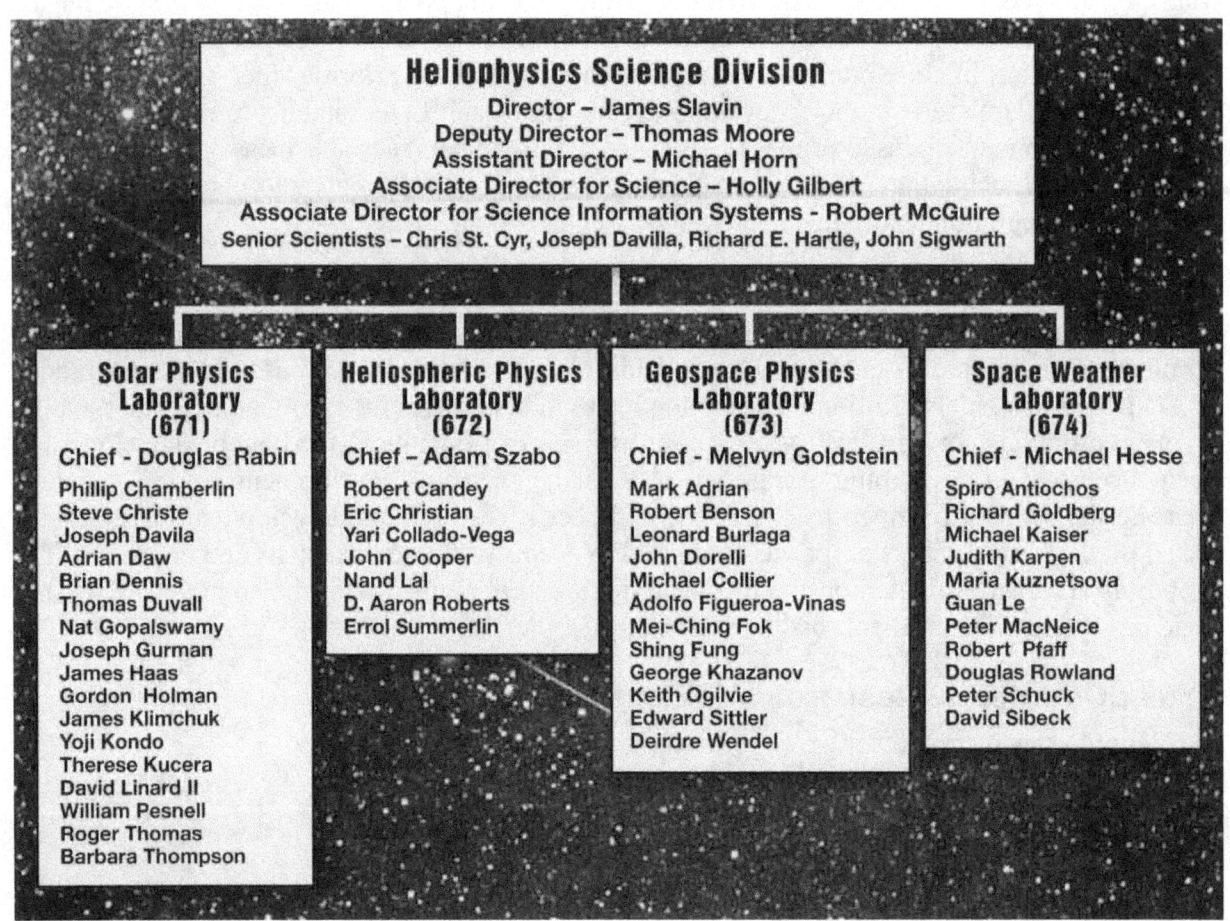

GSFC Heliophysics Science Division 2009 Science Highlights

Mark A. Adrian

Dr Adrian is a thermospheric-ionospheric-magnetospheric physicist who joined the Laboratory for Geospace Physics (Code 673) in November 2004. He serves as Deputy Project Scientist on the Magnetospheric Multiscale (MMS) satellite constellation mission and is the Instrument Scientist for the Dual Electron Spectrometer (DES) component of the MMS Fast Plasma Investigation (FPI). He received his BS (1989) and MS (1993) in physics from the University of Iowa and was awarded a PhD from the University of Alabama in Huntsville in 2000.

Dr Adrian supports ongoing research efforts in collaboration with MSFC, University of Arizona, University of Iowa, and Southwest Research Institute. Additionally, during FY09 he served on several NASA and NSF peer review panels. Locally, Dr Adrian plays dual roles for the Small Business Innovative Research (SBIR) program; both as a Deputy Subtopic Manager for Subtopic S1.12: Lunar Science Instruments and Technology and as a proposal reviewer. Dr Adrian also served as a manuscript referee to the *Journal of Geophysical Research* and *Space Science Reviews*.

In addition to his responsibilities on MMS, Dr Adrian continues to pursue a varied program of advancing both the development of novel plasma diagnostic instrumentation and basic research reaching from the thermosphere through the solar wind to the furthest extent of the heliosphere. During FY09 Dr Adrian lead an IRAD effort to prototype the Electron Concentration vs. Height from an Orbiting Electromagnetic Sounder (ECHOES): a simple, low-resource, space-based radio-sounding experiment designed to gather measurements of the altitude profile of electron density (N_e) for implementation on future heliophysics missions. In addition, he initiated an effort to implement and advance at GSFC a miniaturized electron spectrometer originally developed at MSFC. The Goddard Thermal Electron Capped Hemisphere Spectrometer (G-TECHS) is a miniature spectrometer designed specifically to measure the electron distribution of planetary atmospheres. Dr Adrian also conducts research into inner-magnetospheric coupling through efforts to understand the dynamical processes that generate and circulate plasma through Earth's plasmasphere.

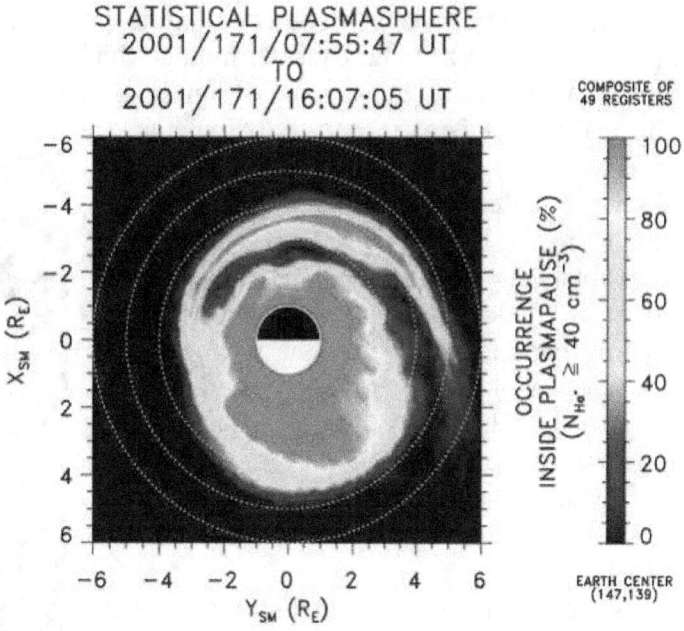

A statistical map of the occurrence and distribution of plasmaspheric He^+ during the recovery phase of a geomagnetic storm. The nightside of sector of the data is the presence of a low-probability region associated with the formation and evolution of an embedded plasmaspheric trough.

Vladamir Airapetian

Since 1995, Dr Airapetian has been an astrophysicist at GSFC (currently in Code 671) and a Research Associate Professor at George Mason University. He has a broad range of interests including the physics of solar coronal heating, MHD models of coronal streamers, physics of mega aurorae from extrasolar planets around young Sun-like stars, and MHD models of stellar winds from late-type evolved stars.

Dr Airapetian has been actively working in three major areas. First, he has developed a model that calculates 3D synthetic images of solar coronal loops obtained by extrapolating an observed photospheric magnetogram into the corona and nanoflare-driven impulsive heating. By determining which set of nanoflare parameters best reproduces actual observations, he hopes to constrain the properties of the heating and ultimately to reveal the physical mechanism. Second, Dr Airapetian has performed semi-empirical MHD simulations of an equatorially confined streamer belt by using observational constraints in 2.5D MHD modeling. Third, he has developed a 1D HD model of exoplanetary hot and dynamic mega aurorae driven by the collisional heating due to precipitating particles caused by the interaction of massive stellar winds with exoplanetary magnetospheres. Based on the model predictions, he suggests an observational technique to detect and characterize extrasolar giant planets.

Dr Airapetian is currently supporting the "Coronal Heating Origins of Solar Spectral Irradiance" project led by Dr J. Klimchuk and "Empirical Determination of Effective Heat Flux and Temperature Using Semi-Empirical 2D MHD Model of the Solar Corona and The Solar Wind" led by Dr Sittler.

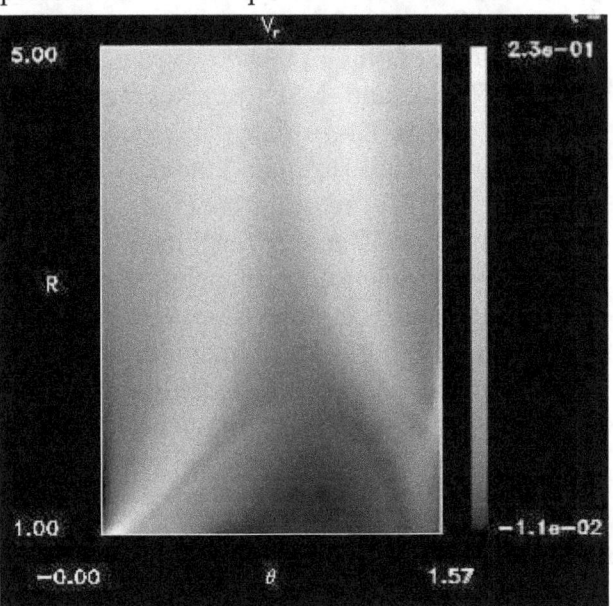

A 2D MHD simulation of an equatorial streamers (above) and a 3D synthetic image in 171Å of the AR NOAA 10938 (below)

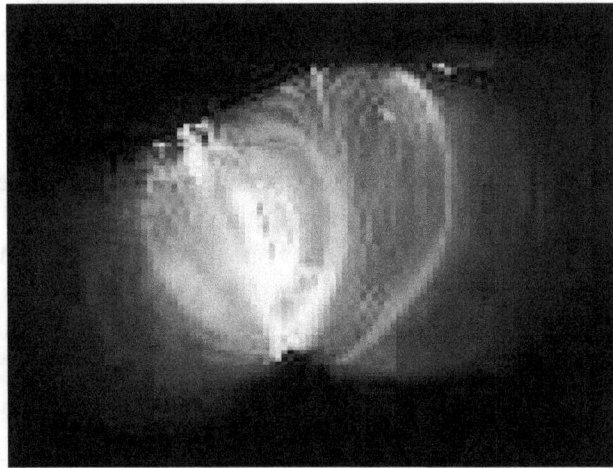

His research areas are
- 3D modeling of solar coronal loops driven by nanoflare heating
- Multidimensional MHD simulations of thermodynamics and kinematics of solar coronal streamers
- MHD and HD simulations of colliding stellar and exoplanetary winds driven by the wind-magnetosphere interactions
- MHD simulations of massive winds from evolved giants.

Sachiko Akiyama

Dr Akiyama is a postdoctoral researcher at CUA. She received her PhD in astronomical science from the Graduate University for Advanced Studies in March 2001. Her primary interest is solar eruptive events, such as flares, filament eruptions, and CMEs. She is also interested in the relationship between solar eruptions and related signatures of the interplanetary medium.

Dr Akiyama is working with Dr Gopalswamy and his group in solar-terrestrial physics. Her main task is data analysis and solar source identification of CMEs, magnetic clouds (MCs), solar energetic particles (SEPs), etc. She also conducts research with Dr Sterling, MSFC, regarding causal mechanisms of filament eruptions. They perform statistical studies of eruptive prominences in quiescent regions, focusing on the very slow rise phase before eruption.

Solar-terrestrial physics is one of the prominent topics for solar and geophysical scientists. Flares and CMEs with huge amounts of energy (about 10^{32} erg) have a great impact on the interplanetary medium and the environment of Earth. When strong geomagnetic disturbances associated with large CMEs reach the Earth, an aurora or a radio disturbance is sometimes observed at high latitudes.

Filament eruptions are one of the solar sources of CMEs. Though several theoretical models of their trigger have been proposed, the prime mechanism is still a matter of debate. In a study of the trajectories of 16 eruptive prominences using the SOHO/EIT data, all events show a slow-rising motion over a long time (>12 h) before the fast eruptive motion. The average speeds of the slow and fast rise phases are 1.2 and 46 km/s, respectively. The fast rising speed shows a weak correlation with the average magnetic field strength under the filament, but the slow rising speed does not. Of 16 events, 3 show several stepwise trajectories during the slow rise phase. The duration of the steps decreases with time.

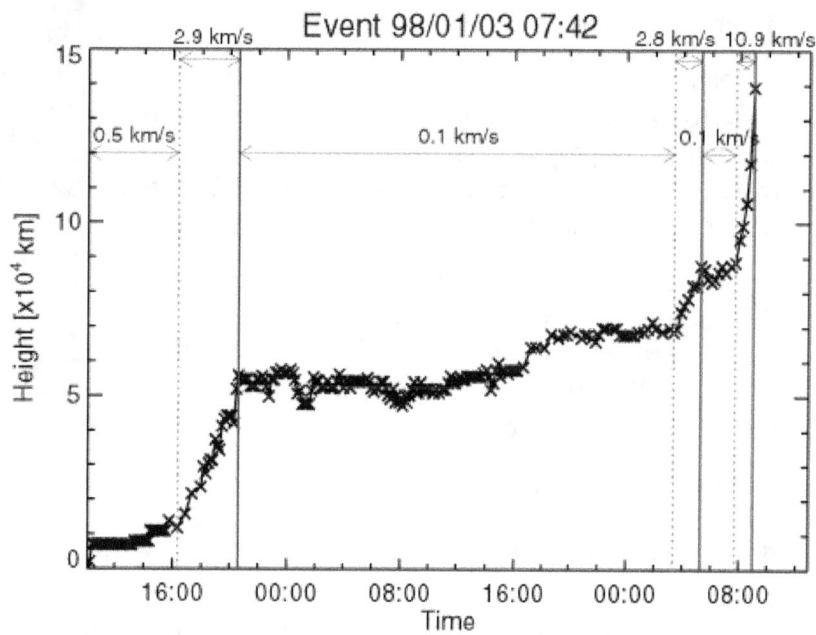

Height time-plot of the eruptive filament on 1998 January 3. Blue-dashed and red-dotted lines are the start and end time of the flat speed period.

GSFC Heliophysics Science Division 2009 Science Highlights

Spiro Antiochos

Dr Antiochos joined GSFC in January 2008 as a Research Astrophysicist in the Space Weather Laboratory. He has been an Adjunct Professor in the Department of Atmospheric, Oceanic, and Space Sciences, University of Michigan, where he was the Research Supervisor for two PhD students. He obtained his PhD at Stanford University in applied physics.

Dr Antiochos is Chair of the Living with a Star Targeted Research and Technology Steering Committee (TSC). He is responsible for the 2009 TSC report (http://lws-trt.gsfc.nasa.gov/trt_sc_report09.pdf). He served on the PhD Thesis Committee for Dr Justin Edmondson, U. Michigan. He also served on the External Advisory Committee of the Center for Solar Research and Big Bear Observatory, New Jersey Institute of Technology and produced the 2009 Advisory Committee report. He also served on the "Red Team" that reviewed the Solar Orbiter Phase-A Proposal to ESA.

He is the PI of a successful proposal to the LWS TR&T Program that integrates theory work for much of the Space Weather Lab. He was selected by NASA HQ to be Team Leader of the Focus Team on Kinetic-Global Coupling.

Dr Antiochos' main research theme is solar/heliospheric theory and modeling, with goals of understanding:

- The initiation of CMEs/eruptive flares, the primary drivers of the most destructive space weather. He has demonstrated that the breakout model produces the observed rotation in CMEs.

- Coronal jets, one of the most ubiquitous forms of solar activity and a major process coupling the corona and wind. He proposed and demonstrated a twist release model for jets and demonstrated that it successfully produces the observed velocities.

- The slow solar wind; its origin has long been one of the outstanding problems in solar/heliospheric physics. He proposed a new model for the wind (the S-web model) based on the topological constraints to the open-closed boundary derived in recent work.

- Magnetic reconnection, the fundamental process coupling plasma and field in the solar atmosphere and the process underlying most solar activity. He demonstrated that interchange reconnection maintains a smooth topology in the coronal field and does not lead to the mixture of open and closed field proposed by a number of heliospheric models.

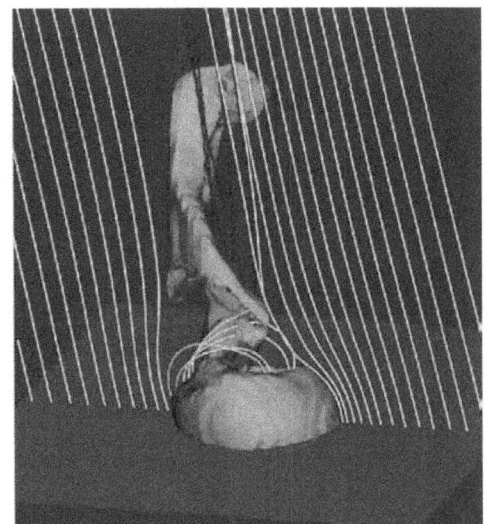

Simulation of twist release model for coronal jets. Open magnetic field lines are shown in white, initially closed in blue, and enhanced density in light blue.

GSFC Heliophysics Science Division 2009 Science Highlights

Manuel Luna Bennasar

Dr Bennasar is a solar physicist. He has conducting numerical investigations of solar prominences and coronal heating with Dr Karpen at GSFC since September 2009.

Dr Bennasar obtained his PhD at the University of Balearic Island, working on transverse oscillations of flux tube ensembles of the solar corona.

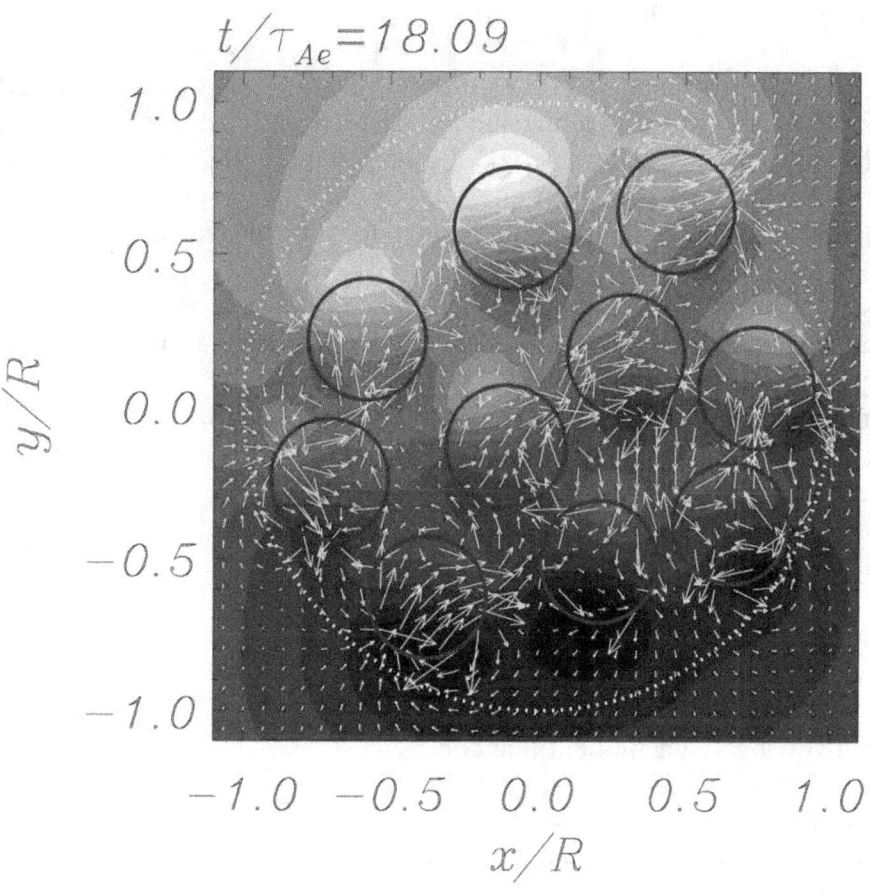

Transverse oscillation of a multi-stranded loop. The initial perturbation produces an ordered motion of the strands, but rapidly a non-organized motion appears.

Robert Benson

GSFC research includes plasma wave emissions, wave-particle interactions, solar-wind/magnetosphere/ionosphere coupling, and ionospheric electron density gradients.

Dr Benson has a BS (geophysics) and MS (physics) from the University of Minnesota (1956 & 1959) and PhD (geophysics) from the University of Alaska (1963). He was a scientific member of the first wintering-over party at the Amundsen-Scott South Pole Station (1957), Assistant Prof (teaching) in the Department of Astronomy at the University of Minnesota (1963/64 academic yr), and NASA/NRC Postdoctoral Resident Research Associate at GSFC. He joined GSFC as scientific civil servant in October 1965. He reviewed numerous papers for scientific journals and books, served on a PhD examining committee, and mentored high-school students from the National Space Club Scholars program.

He directed the GSFC International Satellites for Ionospheric Studies (ISIS) data restoration and preservation effort that produced digital topside ionograms from a select subset of the original 7-track analog telemetry tapes, which preserved a global sample of ionospheric topside-sounder data covering a time interval of more than one solar cycle. This data preservation effort enabled ionospheric topside-sounder data, dating back to 1966, to be made available in digital form.

One of Dr Benson's current research goals is to establish links between solar-wind and topside-ionospheric parameters based on the earlier demonstration of such a link between variations in the solar-wind quasi-invariant (QI – the ratio of the solar wind magnetic pressure to the solar wind ram pressure) and the magnetospheric plasma parameter f_{pe}/f_{ce} (ratio of the electron plasma frequency to the electron cyclotron frequency) as illustrated in the figure. Alouette/ISIS topside-sounder data, including recently produced digital ionograms, will be used to determine if solar-wind drivers directly modify the high-latitude topside ionosphere.

Recent research involving magnetic field line conjunctions between Explorer 45 and ISIS 2 indicated that plasmapause field lines are located near the minimum of the night time main (midlatitude) ionospheric electron-density trough at low altitudes (~600 km) but at higher altitudes they are located on the sharp low latitude trough boundary.

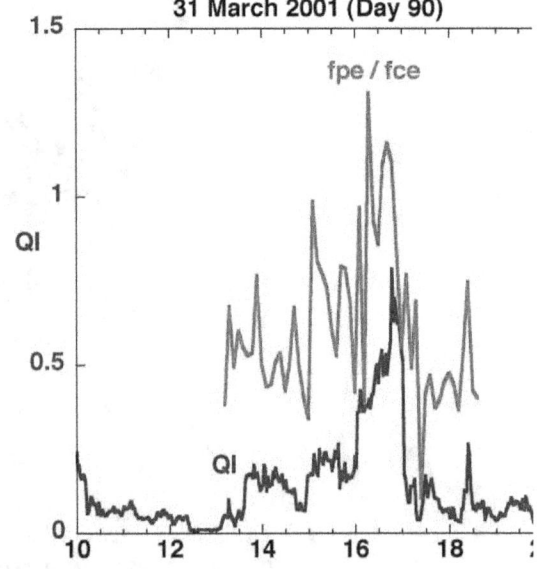

Comparison of solar-wind QI & IMAGE/RPI polar-cap f_{pe}/f_{ce} variations near 8 R_E with a 3-hr time shift [Osherovich et al., JGR, 2007].

GSFC Heliophysics Science Division 2009 Science Highlights

Anand Bhatia

Dr Bhatia has an emeritus appointment to work in atomic physics and solar physics. He has worked at GSFC for more than 46 years. He obtained his PhD from UMd.

Dr Bhatia makes theoretical atomic physics calculations. When he calculates new atomic data, they are added to the CHIANTI database, managed by NRL. These data sets are available freely and are being used by the solar and astrophysical community.

Scott Boardsen

Dr Boardsen has worked at GSFC since 1993. He received his PhD in physics from the University of Iowa in 1988, where he studied electrostatic ion cyclotron waves.

He is a member of the MESSENGER science team, and works on science support, analysis, and interpretation of magnetometer data taken by the MESSENGER spacecraft.

He studies plasma waves, plasma structures, and their interaction with particles in planetary magnetospheres. His current research focus is on the analysis and interpretation of ultra-low-frequency waves detected at Mercury by the magnetometer flown on the MESSENGER spacecraft. He is studying the role of these waves in energy transfer within plasmas, and as a diagnostic of magnetospheric processes and composition. He is also studying the interrelationship between Mercury's magnetosphere and heavy ions, mainly Na^+, created by ionization of Mercury's exosphere.

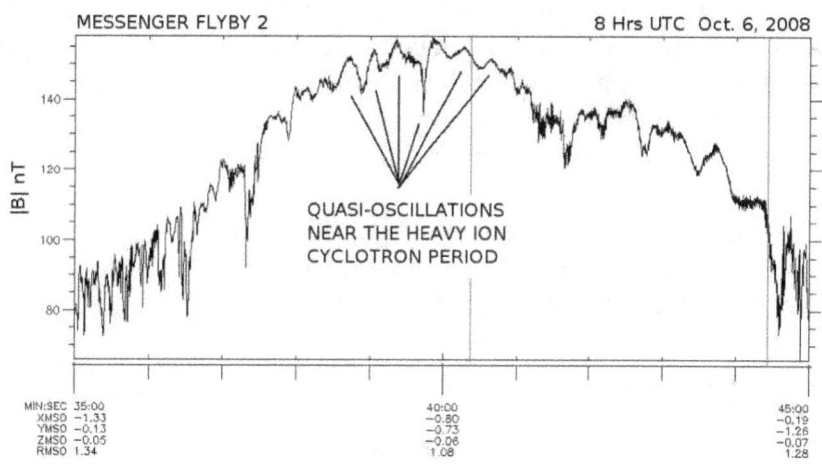

Quasi periodic oscillations observed during the 2nd MESSENGER flyby of Mercury, associated with strong southward IMF conditions. It is hypothesized that these oscillations are related to the Dungey cycle time in a small magnetosphere.

Dieter Bilitza

Dr Bilitza is an expert in ionospheric physics and the principal author of the International Reference Ionosphere (IRI). He is a Research Professor at George Mason University working on the SESDA contract in GSFC's Heliospheric Physics Laboratory. Dr Bilitza first came to GSFC as an AE GI in 1979 then returned in 1985, working first at the NSSDC and later at the SPDF. He received his Diploma and PhD in physics from the Albert-Ludwigs University in Freiburg, Germany.

Dr Bilitza is the chief support scientist at the Space Physics Data Facility (SPDF) that maintains the CDAWeb, SSCWeb, and OMNIweb services which are widely used in the science community. Recent science collaborations involved scientists from NASA/Langley, NOAA/NGDC, JHU/APL, Ohio State University, University of Colorado Boulder, University of Massachusetts Lowell, Rhodes University in South Africa, Institute of Atmospheric Physics in the Czech Republic, and German Geodetic Institute in Germany. Dr Bilitza was a member of the US delegation to the International Standardization Organization meeting in Berlin, Germany (2009 May 18-22).

Dr Bilitza's main research efforts in 2009 were directed towards improvements of the IRI model with special focus on storm-time effects on the E-region ionosphere and on the solar cycle variation of topside temperatures and densities.

He is working with TIMED GUVI and SABER team members and data to develop a new model that includes the contribution of auroral energetic electrons which is now being prepared for inclusion in IRI.

Dr Bilitza's modeling of solar cycle effects needs a large data volume and therefore requires the combination of ground and space data. A comprehensive analysis of simultaneous ground and space data for the topside electron temperature showed that there are still systematic discrepancies between the two techniques at low solar activities (= low electron densities). The figure on the right shows differences of up to 1000 K or about 30%.

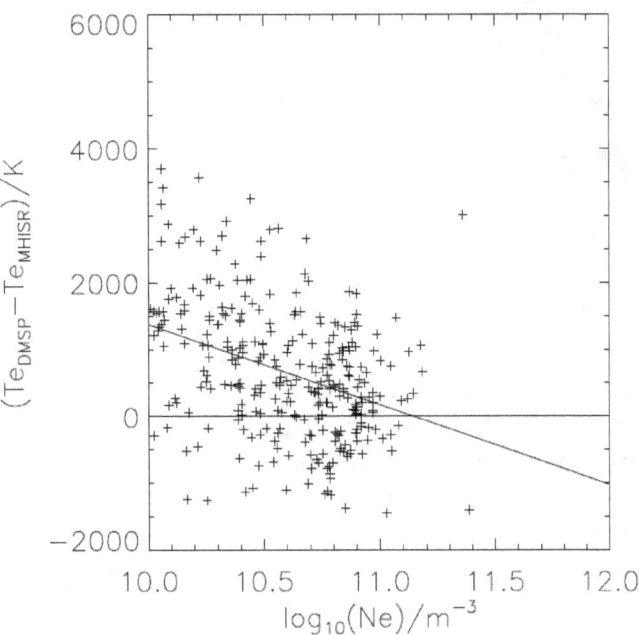

Difference between simultaneous measurements of topside electron temperature measured by the Millstone Hill radar and by the DMSP satellites

Stephen Bradshaw

Dr Bradshaw is an astrophysicist in the Solar Physics Group at GSFC and a Research Assistant Professor at George Mason University. Hailing from the U.K., he has a PhD from the Department of Applied Mathematics and Theoretical Physics at the University of Cambridge, and was a Science and Technology Facilities Council Post-doctoral Research Fellow at Imperial College London before arriving at GSFC in January 2009.

Dr Bradshaw develops analytical and numerical models for application to solar coronal loops. Through detailed comparisons between the predictions of theoretical models and observations made by satellites such as SOHO, TRACE, RHESSI, Hinode, and the forthcoming SDO, important insights into the physics governing these fundamental components of the Sun's atmosphere can be obtained. The ultimate aim is to understand the mechanisms responsible for heating the corona to million degree temperatures – one of the outstanding problems in modern astrophysics.

Recently Dr Bradshaw has developed an extremely efficient numerical solver for the system of time-dependent detailed-balance equations governing the ionization state of astrophysical plasmas. Both the dominant and minor ion populations are accurately treated. Non-equilibrium ionization effects are often neglected from observational and modeling studies due to their significant computational demands. The new code can calculate the ionization state for rapidly evolving plasma in just a few seconds on a standard desktop computer. Dr Bradshaw and others have shown that small-scale, impulsive heating events of the type generally thought to heat the solar corona could be entirely missed by observing instruments as a consequence of non-equilibrium ionization. Instruments must be designed specifically to detect the observable signatures of heating, which appropriate models can help to identify.

In ongoing work Dr Bradshaw has shown through analytical and numerical models that the bulk downflows observed in active regions by the EUV Imaging Spectrometer (EIS) on Hinode are consistent with a flux of enthalpy from the corona powering the transition region radiation. This marks a significant departure from previous ideas concerning the energy balance. The agreement between the models and observations provides substantial evidence in favor of impulsive heating if each coronal loop is composed of many sub-resolution strands that are heated and cooled independently.

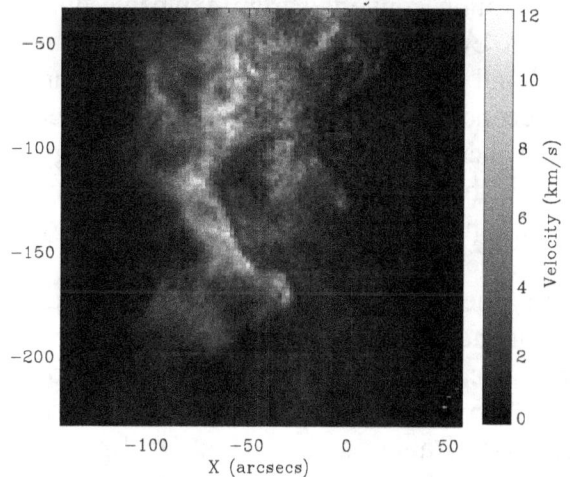

A theoretical velocity map showing the predicted distribution of bulk downflows and their magnitudes in an active region.

Ben Breech

Dr Breech joined GSFC in September 2008 as an NPP Fellow after completing PhD degrees in computer science and physics at the University of Delaware. He is currently part of the Laboratory for Geospace Physics.

Dr Breech studies turbulence in the solar wind, in both the super-Alfvénic and sub-Alfvénic coronal regimes. This work leads to and improves understanding of how the Sun interacts with the solar system as well as the celestial environment the Earth moves in. His current work focuses on developing models describing the transport of fluctuations in the corona.

Jeff W. Brosius

Dr Brosius received his BA in physics from Franklin and Marshall College, Lancaster, PA, in 1980, and his PhD in physics from the University of Delaware through the Bartol Research Institute, Newark, DE, in 1985. He has been with CUA since 2002.

His research during FY09 focused on flare dynamics based on rapid cadence (9.8 s) EUV spectra from SOHO's CDS, coordinated when possible with EIT, MDI, RHESSI, and/or TRACE. Investigation of a microflare in collaboration with Dr Holman revealed the same properties and behavior shown by larger flares undergoing gentle chromospheric evaporation driven by nonthermal electrons. The possibility that magnetic reconnection occurred low in the solar atmosphere, possibly even in the chromosphere, was presented for further investigation. A subsequent investigation of a GOES M1.5 flare revealed that chromospheric evaporation converted from explosive during the event's first impulsive burst seen in O V and Fe XIX emission, to gentle during its second. The change was likely due either to an increased absorption of beam energy during the gentle event because the beam passed through an atmosphere modified by the earlier explosive event, or to a weakening of the coronal magnetic field's ability to accelerate nonthermal particle beams (via reconnection) as the flare progressed, or both. In collaboration with Drs Wang, Thomas, Rabin, and Davila, Dr Brosius used the EUNIS-06 LW channel to calibrate CDS and the EUNIS-06 SW channel. A similar procedure, also based on insensitive ratios, will be used to cross-calibrate EUNIS-07 and Hinode EIS.

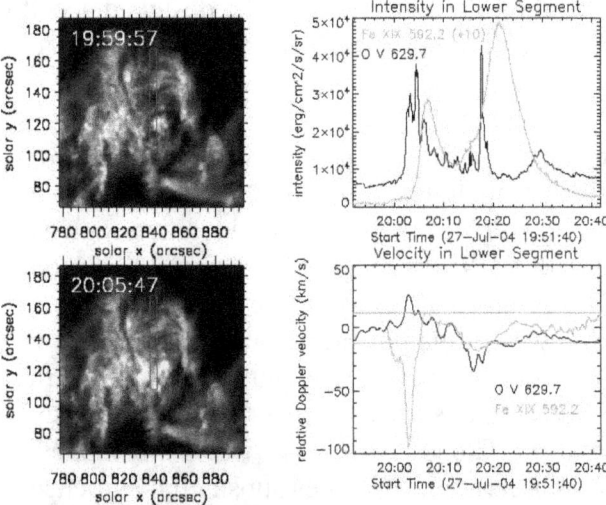

Left column: 120"×120" TRACE 195 Å images obtained early during a GOES M1.5 flare on 2004 July 27, with three 4"×20" CDS slit segments outlined in red.
Right column: light curves and relative Doppler velocities in O V and Fe XIX in the lower segment.

Leonard Burlaga

Dr Burlaga is an astrophysicist specializing in the structure and dynamics of the heliosphere and its interaction with the interstellar medium. He is a Co-I on the magnetic field and plasma instruments on Voyager 1 & 2, the SECCHI instrument on STEREO, the magnetic field instrument on ACE, the plasma instrument on WIND, and a theoretical team studying the dynamics of the distant heliosphere. He received his BS from the University of Chicago and MS and PhD from the University of Minnesota. He joined GSFC in 1968.

A principal result this year was the determination of the radial variation of the magnetic field strength in the inner heliosheath (the subsonic region beyond the termination shock) based on Voyager 1 observations. The magnetic field strength is increasing as Voyager 1 moves through the heliosheath from the termination shock toward the heliopause. In order to derive this result, it was necessary to correct for the decreasing magnetic field as a result of historically weak magnetic fields observed at 1 AU during the approach to the current solar minimum. This correction was possible because we can predict the decreasing magnetic field strength in the solar wind from observations at the Sun and near 1 AU by using Parker's spiral field model, which was verified by Voyager 1 between 1 AU and 94 AU.

Dr Burlaga presented an invited talk at the SHINE conference demonstrating the fundamental importance of nonextensive statistical mechanics in the dynamics of the heliosphere. He was also invited to present the Parker Lecture at the Fall AGU meeting.

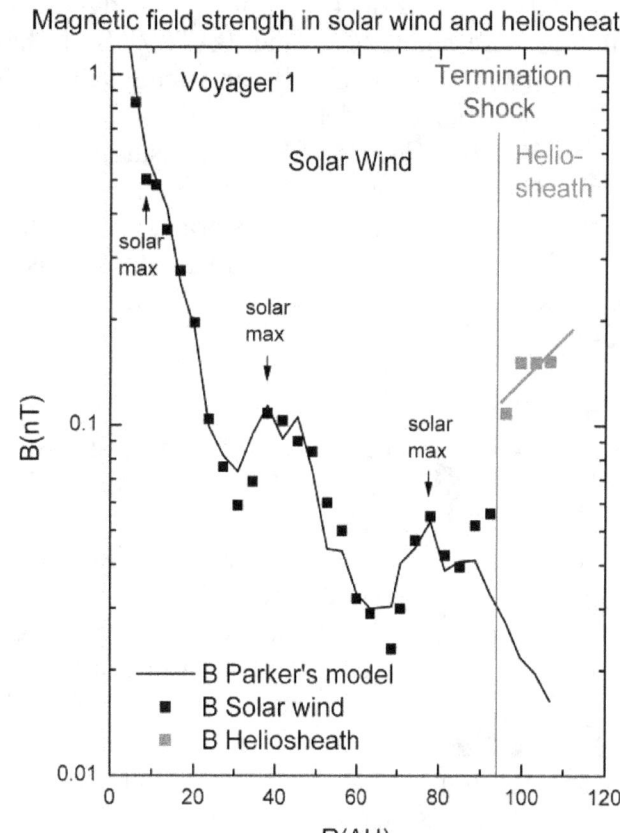

Radial variation of the magnetic field strength in the supersonic solar wind flow from 1 to 94 AU and in the heliosheath

Natalia Buzulukova

Dr Buzulukova is a third-year NASA NPP fellow. She is working with Mei-Ching Fok on modeling of the inner magnetosphere (ring current-plasmasphere-ionosphere coupling) and data analysis from the newly launched TWINS mission. She received her PhD from the Space Research Institute (IKI), Moscow, Russia, in 2003.

Dr Buzulukova works with the NASA CCMC on the project of ring current model (CRCM)-MHD model (BATSRUS) coupling. She is also involved in the TWINS project performing ENA emissions data analysis and data-model comparison.

Dr Buzulukova is working with data analysis and modeling of the Earth's inner magnetosphere. She is trying to extract and study the relationships between different processes and different domains in the inner magnetosphere, e.g., between the plasmapause location and ring current dynamics; ENA data analysis and ring current dynamics; storm-substorm relationship. Understanding how the inner magnetosphere evolves is critical for space weather applications; therefore her work is important for understanding the effects of solar activity on the Earth and its space environment.

In FY09, Dr Buzulukova finished studying one-way coupling between CRCM and BATSRUS. It has been shown that inclusion of self-consistent electric field in MHD-ring coupled models is critical for ring-current dynamics. With CRCM-BATSRUS, she studied a case with idealized substorm and development of injection in the inner magnetosphere. For real storms, the model reproduces the main features of storm ring current as seen by the IMAGE/HENA instrument. The paper was submitted to *JGR* in June 2009.

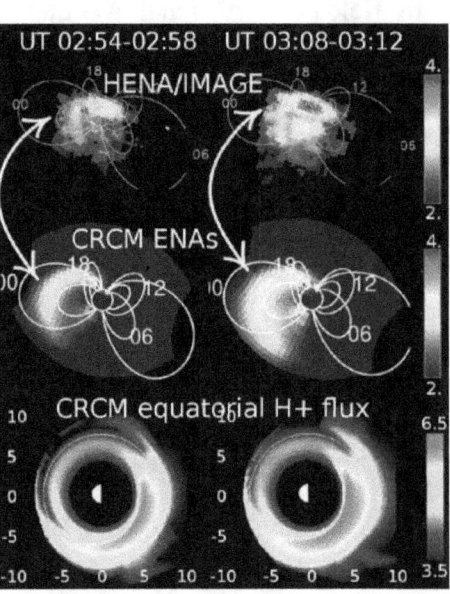

Ring current injection on August 12, 2000 as seen by IMAGE/HENA and reproduced by CRCM.

Robert Candey

Robert Candey arrived at GSFC in 1983 and has worked on the Dynamics Explorer, GOES, Search and Rescue (SARSAT), and CRRES and other chemical release campaigns, and is chief architect for the Space Physics Data Facility. He manages the science ground system for Wind and Geotail, and leads the new Heliophysics Event List Manager (HELM) VxO project and older xSonify sonification and Common Data Format (CDF) projects. He supports the VxOs, particularly VITMO and VWO. He led the Dynamics Explorer data restoration effort. He holds a BS in Optics Engineering from the University of Rochester and MS in Computer Science from JHU. He mentored two visually-impaired students and participates with the MMS E/PO effort. His areas of research include data access and analysis, long-term archiving, sonification, and science ground systems.

GSFC Heliophysics Science Division 2009 Science Highlights

Phillip Chamberlin

Dr Chamberlin joined Code 671 in July 2009. His research is focused primarily on measuring and modeling of the solar X-ray and ultraviolet irradiance. Dr Chamberlin leads the continued improvement of the Flare Irradiance Spectral Model (FISM), which not only improved the ability to empirically model the solar cycle and solar rotation irradiance variations in the VUV wavelengths (0.1-190 nm), but also added the ability to model the irradiance variations due to solar flares. He also continues to collaborate with researchers from the University of Michigan, NCAR's High Altitude Observatory, and Boston University to study the impact of solar flares on the atmospheres of Earth and Mars. Dr Chamberlin has also been involved with various tasks on sounding-rocket payloads, where the most recent launch in April 2008 led to the definitive measurement of the solar EUV irradiance during solar minimum conditions. He continues to be involved with the TIMED SEE data analysis, SDO EVE development, calibrations, and science analysis, as well as the development of the GOES-R EXIS instruments (XRS and EUVS), and is also a Co-I on the upcoming MAVEN mission working with the LPW-EUV instrument. Dr Chamberlin took on the role as a Deputy Project Scientist for SDO after coming to GSFC, and continues to develop new and improved instruments for the next generation of solar observing missions.

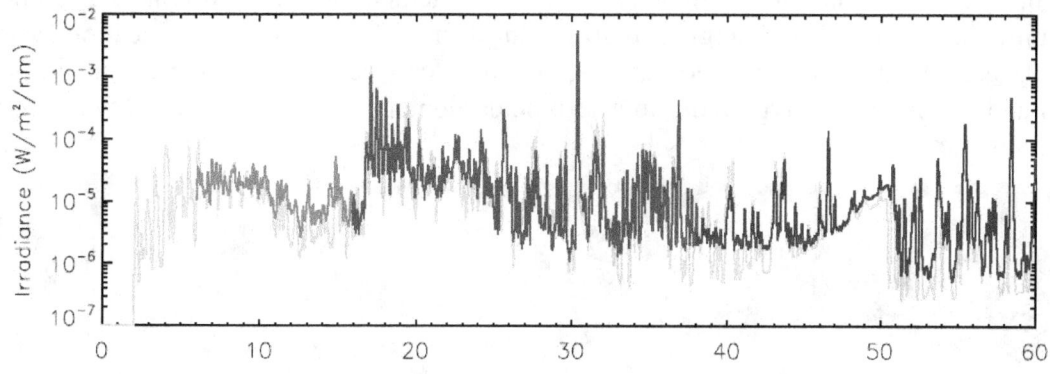

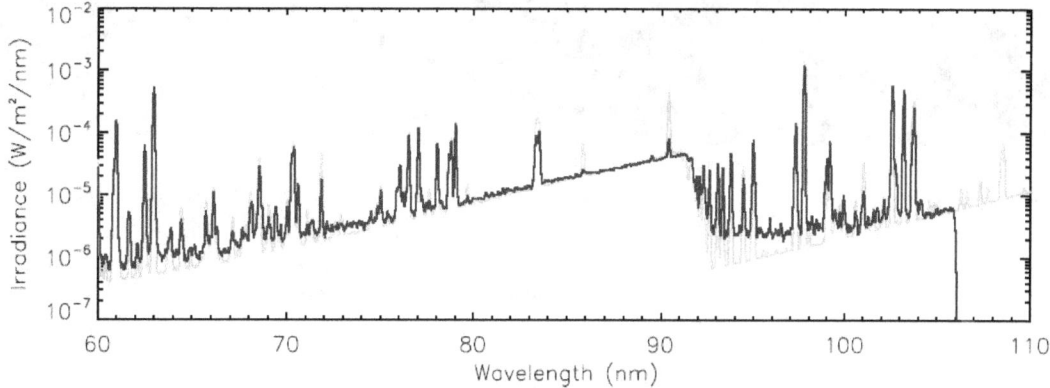

Measurements from the NASA 36.240 rocket flight from the three channels of EVE. MEGS A1 is given in red, MEGS A2 is given in blue, and MEGS B is shown in purple. The Warren [2005] reference spectrum is also shown for comparison in green.

Peter Chen

Dr Chen, in collaboration with D. Rabin (671) and M. Van Steenberg (610.4), is developing lightweight composite mirrors and advanced concepts for astronomy from the Moon. His group has developed a material ("lunar cement") that uses lunar regolith simulants to make a substance that is demonstrably harder than concrete. The fabrication process is simple, does not require water, and can take place under lunar conditions of vacuum and radiation. Preliminary tests suggest that the substance may be very stable.

This leads to a concept of fabricating large telescopes and radio interferometers on the Moon and then either deploying in situ or launching into deep space. On a much smaller scale, work is in progress to use the lunar cement to make replicas of Galileo's original telescope by using different lunar regolith simulants. One effort, undertaken in collaboration with colleagues at the University of Dallas, makes the Galilean telescopes using JSC-1A simulant. A parallel effort in collaboration with colleagues at the Beijing Astronomical Observatory makes use of the CAS-1 simulant developed in China.

Drs Chen, Rabin and Van Steenberg collaborated on the invention of a two-axis instrument pointing system using high-temperature superconductor (HTS) bearings. Unlike previous designs, this new configuration is simple and easy to implement. Most importantly, it can be scaled to accommodate instruments ranging in size from decimeters (laser communication systems) to meters (solar panels, communication dishes, optical telescopes, optical interferometers) to decameters and beyond (VLA-type radio interferometer elements). A working prototype is currently under construction.

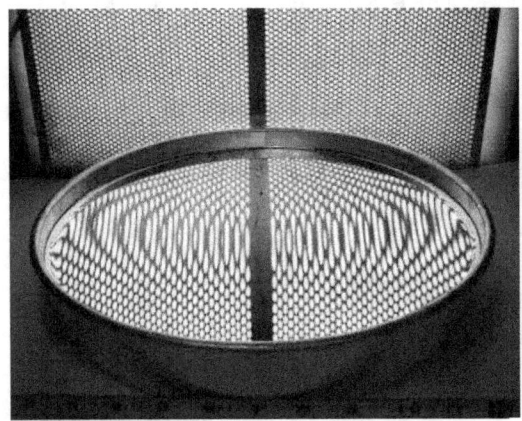

A prototype telescope mirror made using lunar cement. We conceptualize that very large mirrors and structures could be made in situ on the lunar surface.

A demonstration of the impact resistance of the lunar cement was carried out for a NHK Japan TV program filmed at GSFC. The sample (black solid rod) remained intact after being smashed with a rock hammer. The supporting cinder block broke apart.

Sheng-Hsien (Sean) Chen

Dr Chen is a research scientist at USRA working in the Geospace Physics Laboratory. Dr Chen is analyzing Polar, FAST, DMSP, Cluster, and THEMIS plasma and magnetic field data to study the interaction of solar wind, magnetospheric, and ionospheric plasmas at the ionospheric and magnetospheric boundaries. More recently Dr Chen has been working on the long-term solar cycle effect of the solar and geomagnetic activity to the ionospheric outflows in the polar cap regions and plasma heating processes associated with electromagnetic perturbations in the magnetosphere. Dr Chen has been working at GSFC since 1994, except for a couple years or so working in the IT industry. Dr Chen got his PhD in geophysics and space physics from UCLA in 1993.

Dr Chen is supporting Polar/TIDE and TWINS projects in the data analysis and scientific studies. Dr Chen has been a collaborator of projects at APL. Dr Chen has served in NSF and NASA proposal review processes.

It is well known that the ionospheric outflows of fully ionized atmospheric atoms and molecules contribute to the dynamics of the Earth's coupled magnetosphere-ionosphere-thermo-sphere system in an important and continuous way as the solar wind acts on the system. They join the circulation of plasmas in the magnetosphere and generate complex disturbances and instabilities to influence space weather in the magnetosphere and ionosphere below. However, the heating, ionization, and acceleration processes of neutral or weakly ionized gases of atmospheric origin are not well understood. Dr Chen is interested in using in situ measurements of fields and particles near the critical regions where the neutral gas is ionized and accelerated into the magnetosphere to quantify the possible dominant heating mechanisms and how these ions are further heated in the magnetospheric boundaries during geomagnetic storms.

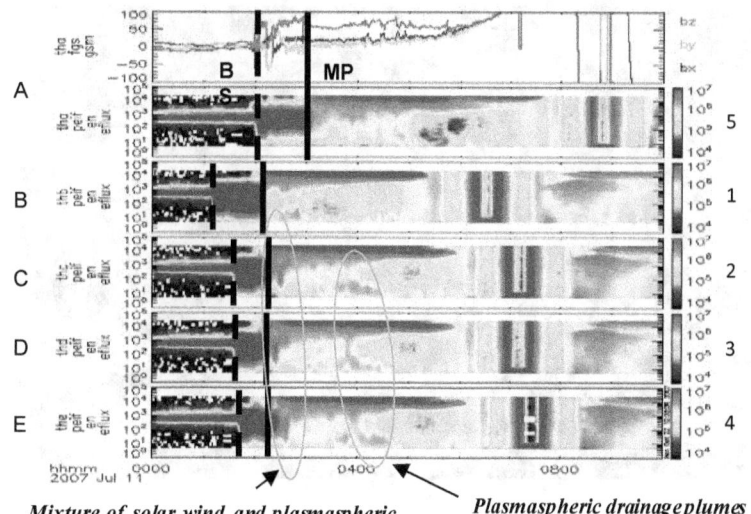

Mixture of solar wind and plasmaspheric plasmas just inside and outside the magnetopause

Plasmaspheric drainage plumes

THEMIS plasma and magnetic field measurements for the 2007 July 11 event. THEMIS was moving inbound radially near 14 MLT from upstream solar wind towards the inner magnetosphere, passing through bow shock, magnetosheath, magnetopause, magnetosphere, plasmasphere, around the Earth and going outbound near end of the interval.

Eric Christian

Dr Christian is working on a number of energetic particle detectors that span an energy range from 10 eV (IBEX) to tens of GeV (Super-TIGER). Most of the energetic particle instruments that he works on (ACE/CRIS, ACE/SIS, STEREO/IMPACT/HET, and IBEX) are launched and operational. Super-TIGER, a balloon-borne spectrometer, is currently under construction, and the EPI-HI instrument for SP+ is in design phase. His scientific interests are related primarily to the origin of energetic particles, including ENAs (IBEX), solar energetic particles (ACE/SIS, STEREO/IMPACT/HET, and SP+/EPI-HI), anomalous cosmic Rays (ACE/SIS) and galactic cosmic rays (ACE/CRIS and Super-TIGER). He has a PhD in astrophysics from Caltech, BA in honors physics and astronomy from U. Penn.

Dr Christian is Deputy Project Scientist for ACE and STEREO, Deputy Mission Scientist for IBEX, Standing Review Board member for RBSP, Guest Editor for two STEREO special issues of *Solar Physics*, and member of NASA Procurement Development Team for Web Enterprise Service Technologies. He is a science team member on multiple projects (ACE, IBEX, STEREO, and Super-TIGER) with many universities, primarily Caltech, JPL, Washington University in St. Louis, and SWRI. He also actively participates in E/PO activities for the HSD.

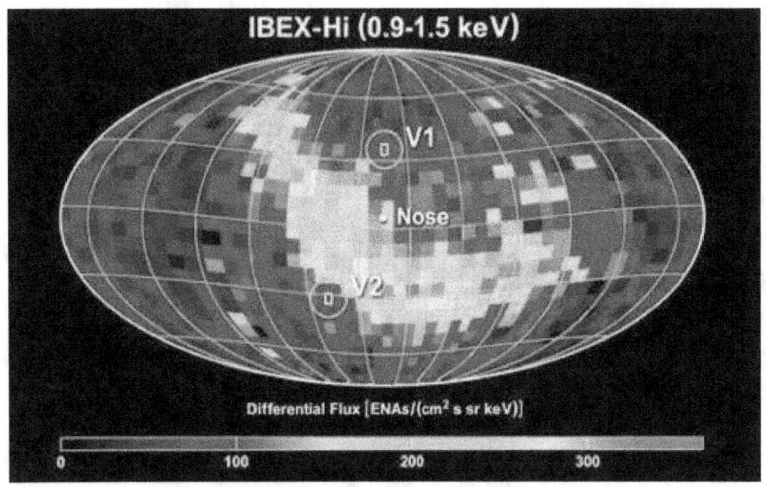

One of the IBEX All-Sky Maps showing the bright ribbon of ENA emission.

His primary research areas are:

1. Understanding energetic particles in the heliosphere including solar energetic particles, ENAs, anomalous cosmic rays, and galactic cosmic rays.

2. Instrumentation, data analysis, and technology development

3. IBEX results (part of a large team), Multi-Media Library (Co-I), and successful IRAD proposal.

GSFC Heliophysics Science Division 2009 Science Highlights

Yaireska Collado-Vega

Yari Collado-Vega is a third year physics PhD student at CUA. She is originally from Ponce, Puerto Rico, and has been living permanently in Maryland for over 2 years. She is part of the Graduate Cooperative Education Program (Co-op) at GSFC under the Heliospheric Physics Laboratory (Code 672) doing research on the solar wind interaction with the Earth's magnetosphere. She has been part of the GSFC family since August 2004. She received both her BS and MS in theoretical physics from the University of Puerto Rico at Mayagüez, Puerto Rico, in 2004 and 2007 respectively.

At GSFC she has been involved in E/PO activities and panels from the Science Exploration Directorate for Undergraduate and High School students to encourage them to pursue and to continue working in the scientific research.

Yari's current research is based on visualizing vortices caused by instabilities along the Earth's magnetopause, using MHD simulations: the Lyon-Fedder-Mobarry (LFM) code and the BATS-R-US code. Vortices are thought to be a source of energy and mass transport to the Earth's magnetosphere, but their mechanisms are not completely understood. The solar wind energetic particles entering the magnetosphere can cause serious damage to our technological systems.

The MHD simulations with fixed solar wind conditions yielded many vortices along the magnetopause with only a step function in the Interplanetary Magnetic Field (IMF). Comparison of these vortices with those that developed under dynamic solar wind conditions made possible the identification of curious characteristics of the vortices that have never been seen before. These characteristics can lead to the better understanding of the conditions under which vortices are mostly likely to develop, and how the conditions affect vortex evolution.

These results will be presented at the Fall AGU meeting in San Francisco, CA, and in the 9th International School of Space Simulations at the University of Versailles-Saint-Quentin-en-Yvelines, France.

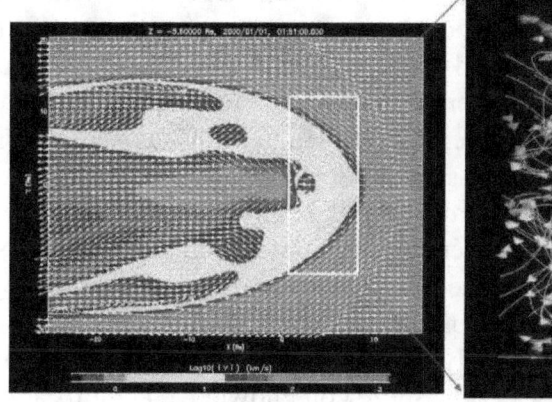

A 2D visualization of the Earth's magnetosphere using the results of the MHD simulations with an IDL-based tool. The colors represent the plasma flow velocity in a logarithmic scale and the arrows the flow direction. Dawn and Dusk sector vortices developed simultaneously for the fixed solar wind conditions. A blow up of these two vortices is shown in a 3D overview at the right. The vortex developed at the dawnside rotates clockwise, and the one developed at the duskside rotates counter-clockwise.

John Cooper

Dr Cooper is Chief Scientist for the Space Physics Data Facility, science team member for the Cassini Plasma Spectrometer (CAPS), and PI for four NASA-funded research projects in heliophysics and planetary science. He leads the Virtual Energetic Particle Observatory (VEPO), one of the newest of the heliophysics virtual observatories. He is a Co-I on several NASA and internal GSFC projects to develop advanced instrumentation for elemental and isotopic measurements of plasma ion composition, contributing input on the mass resolution requirements for design of the instrumentation in fulfillment of science objectives for future mission opportunities in solar, heliospheric, and planetary science. For the first 14 years at GSFC, he worked as a contractor, mostly recently with Raytheon on the Space Science Data Operations (SSDO) Project. Since 2005, he has worked as a NASA senior scientist in what is now the Heliospheric Physics Laboratory.

Enroute to GSFC, he graduated as a "Rambling Wreck" from the Georgia Institute of Technology with an undergraduate degree in physics, took an extended "Star Trek" vacation from science to serve as an officer in the USN aboard the USS. Enterprise, and then did his graduate studies and PhD research at the University of Chicago. Afterwards he did postdoctoral work in space physics at the Max-Planck-Institute Laboratory for Extraterrestrial Physics in Germany, CIT, and LSU.

As SPDF Chief Scientist he interacts with providers and users of NASA mission heliophysics data in the US and international space science communities and at GSFC advises SPDF management on requirements for evolution of SPDF data systems. In his role as PI for VEPO, he works more directly with these communities for on-line discovery and access of heliospheric energetic particle data as part of the emergent network of heliophysics virtual observatories. He is affiliated with the Cassini mission project as a team member for the Cassini Plasma Spectrometer, for which he received the NASA Group Achievement Award in April 2009. In his various roles as SPDF Chief Scientist and a NASA PI, he frequently collaborates with space physicists and planetary scientists at other institutions in the US and abroad. He participates in the Radiation and Education and Public Outreach working groups of the Europa Jupiter System Mission project managed by the JPL. Recently he led a community study group providing input to the NRC Planetary Science Decadal Survey on knowledge and mission applications of space weathering processes for planetary surfaces exposed to the space environment.

His principal research focuses on understanding how the heliospheric and plasma magnetospheric environment of plasma and energetic-particle radiation affect the chemistry and structure of planetary surfaces. This is important for interpretation of remote and in situ observations of these surfaces and for design of instrumentation on present and future missions to further explore these bodies. The Cassini-related principal work, of FY09 for publication, has covered a new radiolytic model for gas-driven cryovolcanism at Enceladus, on injection of Saturn magnetospheric oxygen into fullerenes expected to form in the upper atmosphere of Titan, and on measurements of Saturn plasma distribution parameters at the orbit of Titan.

Adrian Daw

Dr Daw joined Code 671 at GSFC as a research astrophysicist on 2009 August 31. This was preceded by a few years as a professor in the Department of Physics and Astronomy at Appalachian State University, a few years as a visiting scientist at the Smithsonian Astrophysical Observatory. Dr Daw received a PhD (2000) and AM in physics from Harvard University, and a BS from Yale University.

Dr Daw has ongoing collaborations with the University of Hawaii, on polarimetric imaging of NIR-visible coronal lines, as well as collaborations with Appalachian State University and Columbia University on measurements of atomic and molecular parameters for heliospheric ions.

He is currently probing the dynamics of the solar corona, as Deputy Mission Scientist for the Interface Region Imaging Spectrograph (IRIS), by working on cooling the detectors for the 2010 flight of the EUNIS sounding-rocket instrument to study variations on even shorter time scales than previously, and as Co-I on an IRAD project to develop an image slicer for UV integral-field spectroscopy. NIR-visible instrumentation was developed and used to observe the 2009 eclipse, but with some clouds, while analysis of the 2008 eclipse data continues with Dr Habbal and other collaborators.

Color overlay of the emission from highly ionized Fe lines (top panels) and combined with white light (bottom panels). For the 2008 eclipse observations (right column), Fe XI is shown in red, Fe XIII in blue, and Fe XIV in green. The 2006 data (left column) were obtained only in Fe XI (red) and Fe XIII (blue). These are the first such maps of the two-dimensional distribution of coronal electron temperature and ion charge state.

Brian Dennis

Dr Dennis was born in Lincolnshire, England and obtained a BSc and PhD from the University of Leeds. He emigrated to the US in 1964 to the University of Rochester, NY and moved to GSFC in 1966 as an NRC postdoc. He became a civil servant in 1969 and worked in the Solar Physics Laboratory on the observations and interpretation of X-rays, initially as part of the cosmic X-ray background, and since 1980, from solar flares.

He is RHESSI Mission Scientist and Co-I. He is working in collaboration with the PI team at the Space Sciences Laboratory of the University of California, Berkeley.

PI of the following projects:

- VxO for Heliophysics – Extending the VSO Incorporating Data Analysis Capabilities

- Low-cost Access to Space – Imaging X-ray Polarimeter for Solar Flares

- Fermi GI project – Fermi Solar Flare Observations

- GSFC Spontaneous IRAD project – X-ray Diffractive Optics

He has served on NASA heliophysics review panels.

His areas of research include:

- Understanding the energy release and particle acceleration processes in solar flares

- Developing new and advanced techniques for imaging, spectroscopy, and polarimetry of solar flare X-rays and γ-rays from solar flares and other astrophysical sources of interest

- Performing instrumentation, data analysis, data interpretation, theory and modeling, technology development

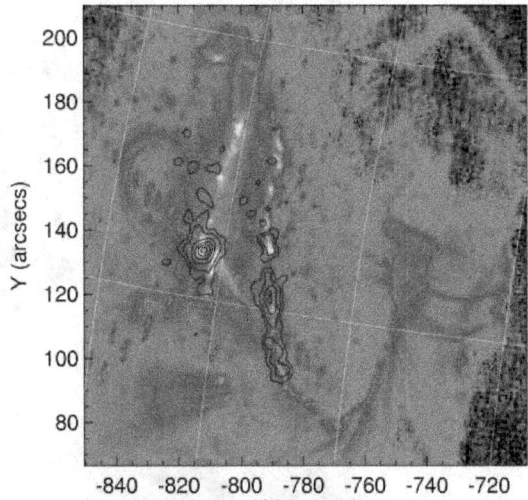

TRACE 171-Å image of flare on 30 July 2005 at 06:31:01 UT, with the RHESSI 25 – 50 keV blue contours showing the tight correlation between the EUV ribbons and the hard X-ray footpoint brightenings.

John Dorelli

Dr Dorelli joined the Geospace Physics Lab (Code 673) in March of 2009. His primary interest is in the physics of solar wind-magnetosphere coupling, but he has also worked on modeling of the solar corona as well as the physics of collisionless magnetic reconnection. During the last few years, he has used a combination of large-scale MHD simulations (spanning hundreds of R_E) and small-scale Hall MHD simulations (spanning tens of ion inertial lengths) to understand both the global topology and local geometry of magnetic reconnection at Earth's dayside magnetopause.

More recently, Dr Dorelli has been studying the topology and generation of flux transfer events (FTEs) to understand their role in triggering geomagnetic activity. In the last several months, he has become involved in analysis of the Cluster PEACE data as well as in modeling the response of the Fast Plasma Instrument (FPI) Dual Electron Spectrometer (DES) which is being developed at GSFC as part of NASA's Magnetospheric Multiscale Mission (MMS).

Dr Dorelli received his PhD from the University of Iowa in 1999. While at Iowa, he divided his time between developing and testing in-flight calibration algorithms for NASA's Polar-Hydra instrument (a suite of electrostatic analyzers designed to measure three-dimensional velocity distributions in Earth's magnetosphere) and modeling non-classical electron heat flow in the solar corona.

For the last several years, Dr Dorelli has been actively involved in the NSF's Geospace Environment Modeling program, serving as a Dayside Research Area coordinator as well as co-organizer of the Methods and Modules Focus Group. He is currently an Associate Editor for the *Journal of Geophysical Research*.

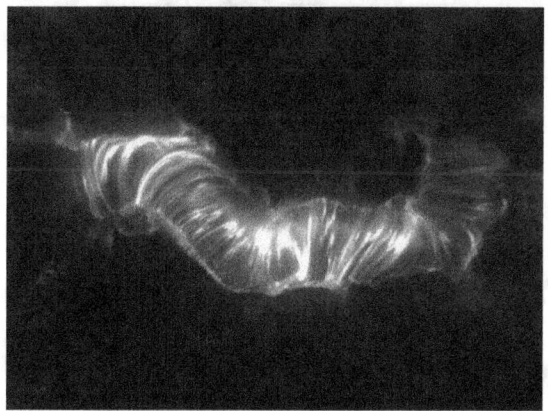
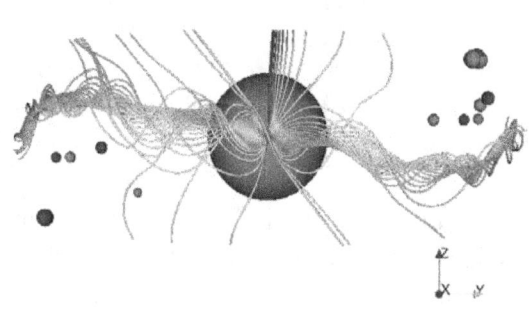

Flux rope formation appears to play a key role in generating storms on the Sun (solar flares). Recent computer simulations suggest that the same process may be occurring at Earth's dayside magnetopause and may be responsible for triggering magnetic storms in Earth's magnetosphere. Left-hand image from TRACE taken on 2000 July 14.

GSFC Heliophysics Science Division 2009 Science Highlights

Thomas L. Duvall, Jr.

Dr Duvall is an astrophysicist studying the Sun and has been doing so in the employ of GSFC for 30 years. His training is mainly in physics with degrees from Johns Hopkins (BA) and Stanford University (PhD)

He normally works remotely from GSFC at Stanford University with the instrument group that has developed the MDI instrument on SOHO and the HMI instrument for the upcoming SDO mission.

He served as co-chair of the NASA Sun-Earth Connection Roadmap Team in 2002 and on the NSO User's Committee from 1997 to 2000. He received the NASA Exceptional Scientific Achievement Medal in 1990. He has made five trips to Antarctica to do helioseismology research. As a reward for this onerous duty, the US Board of Geographic Names named Mt. Duvall at 73S22, 162E31 in 1996.

Dr Duvall is one of the pioneers of helioseismology and has been working in this area for 30 years. He is at least partly responsible for a number of discoveries, including the relationship commonly known as Duvall's law, the initial measurements of rotation and sound speed throughout much of the solar interior, and the asymmetry of solar oscillation spectral lines. In recent years he helped invent the technique known as time-distance helioseismology, in which travel times are measured between different surface locations. This new technique has led to interesting results, including the first measurements of meridional circulation in the solar interior and the measurement of flows and sound speed inhomogeneities below sunspots.

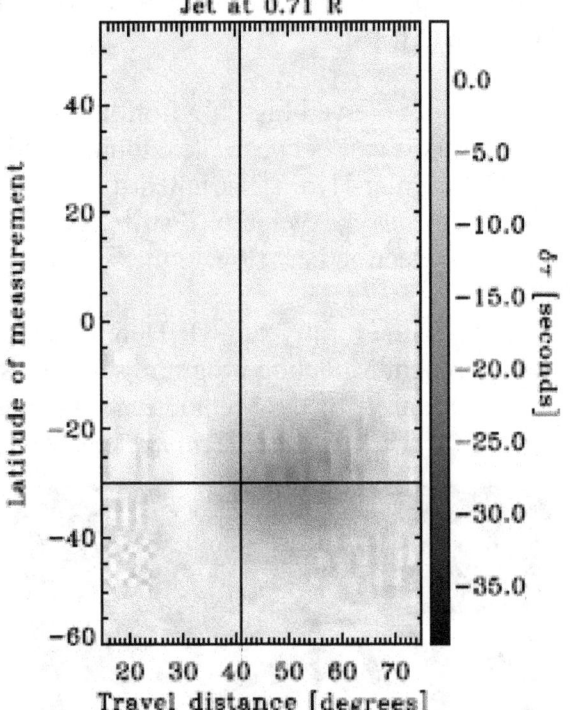

Travel time differences that would be observed from a jet simulated at a solar interior location r/R=.71.

The figure at right is from a simulation of interior flows and waves, in which a jet is inserted at 30° latitude and at radial location $r/R_s = 0.71$. The jet is actually small in extent, but the response to it is broadened by the large wavelength of sound waves in the solar interior.

Joseph Fainberg

Dr Fainberg has spent over 40 years in space research at GSFC and now serves in an emeritus status. He is a Co-I on the radio and plasma waves experiments on Wind, Ulysses, and STEREO. He has worked extensively on direction-finding techniques from both spinning and spin-stabilized spacecraft to determine the path and characteristics of type III radio bursts caused by energetic packets of electrons interacting with the decreasing density of the solar wind as they move from the Sun out to 1 AU and beyond. Type III bursts usually occur in groups, with single isolated events found less than 5% of the time. The characteristics of single bursts are usually described in terms of their rise time, their drift rate, and their decay rate. Along with colleagues Dr Osherovich and Dr MacDowall, he has reexamined a particularly important single event (1995 April 7) to see if the normalized amplitude-time observations at different frequencies obeyed a self-similar relation. One way of phrasing this is to ask if there is a parameter composed of time multiplied by a power of frequency such that the different observed frequencies expressed in this parameter will coincide.

This was the case for five frequencies of the 1995 April 7 event extending from 388 to 1040 kHz. The results are shown below with the observations in the first figure as a function of time. The second figure shows these observations after normalization (peak $S = 1$) expressed in self-similar space; this figure demonstrates that in this self-similar space all five frequencies line up in rise time, peak time, and decay rate. The red line in the second figure is a four-parameter function previously used to fit single type III bursts at individual frequencies but here used to successfully fit all five frequencies simultaneously.

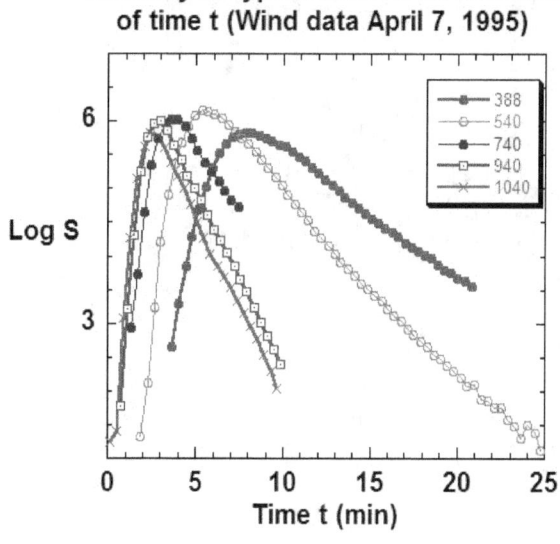

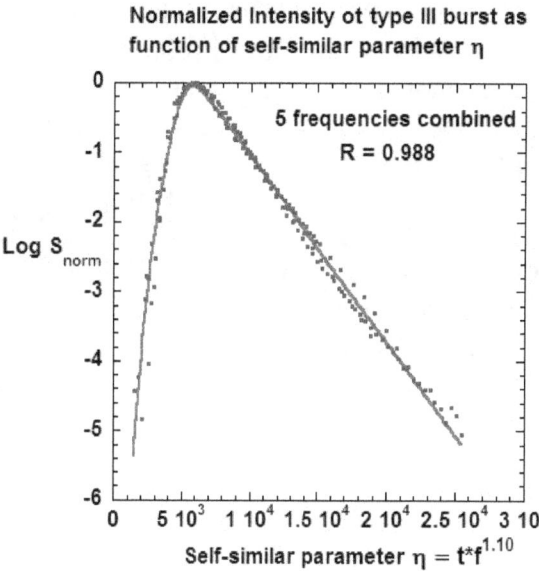

Artem Feofilov

Dr Feofilov is an atmospheric physicist working at GSFC since 2005. Currently he is affiliated with the Department of Physics of CUA at GSFC, Code 674. He received his PhD at St. Petersburg State University, St. Petersburg, Russia in 2001. He has been doing research aimed at better understanding fundamental processes governing the energetics, chemistry, dynamics, and transport of the mesosphere/lower thermosphere. He is also involved in work dealing with radiative transfer and non-LTE in planetary atmospheres.

In FY09 Dr Feofilov continued working on improving and validating the H_2O non-LTE model used for interpretation of the broadband 6.6-μm emission measured by the TIMED/SABER satellite radiometer. This work was done with A. Kutepov, R. Goldberg, and W. Pesnell from GSFC in collaboration with Hampton University of Virginia, GATS, Inc., an aerospace company, Atmospheric Physics Department of the Institute for Physics, St. Petersburg State University, Russia, and other institutes. The methodology for model validation was developed and implemented using the overlapping in space and time measurements of H_2O vapor by the ACE-FTS occultation instrument. Applied to SABER measurements, the updated model provides the H_2O VMR profiles that are consistent with other measurements and models.

This work has paved the way to processing an extensive SABER H_2O database. The retrieved H_2O profiles were used in the study performed together with S. Petelina from La Trobe University, Victoria, Australia, to establish the relationships between the polar mesospheric clouds (PMCs) and the mesospheric area parameters (pressure, temperature, water vapor concentrations) from nearly simultaneous common volume measurements performed by Odin/OSIRIS and TIMED/SABER satellite instruments in both hemispheres in 2002–2008. One of the results of this study demonstrating the correlation between the hydration of the "undercloud" area, freeze drying of the PMC area, and PMC brightness is shown in the figure. Together with M. Smith at GSFC and A. Kutepov at CUA/GSFC. Dr Feofilov also worked on the non-LTE analysis of broadband IR limb observations of the Martian atmosphere by the MGS/TES bolometer.

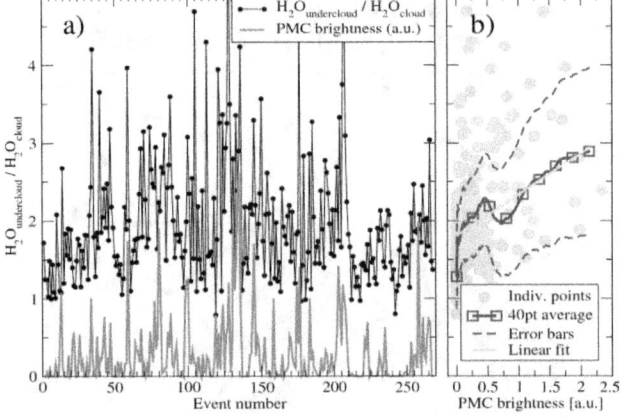

Correlation between the polar mesospheric cloudbrightness and the water vapor content in and below the cloud.

Mei-Ching Fok

Dr Fok is an astrophysicist in Geospace Physics Laboratory (673). She received her PhD in atmospheric and space sciences in 1993 at the University of Michigan. She did postdoctoral work on ring current and ENA simulation at MSFC before moving to GSFC in 1997. Dr Fok continued working on modeling the inner magnetosphere, ionosphere and coupling between plasma populations. She has developed two kinetic models: the Radiation Belt Environment (RBE) model and the Comprehensive Ring Current Model (CRCM). In the past year, Dr Fok has included wave induced cross diffusion in the RBE model (work in collaboration with Dr Zheng, UMCP/673). She also successfully reproduced the flux enhancements measured by the Akebono satellite during a substorm injection on 2008 September 4 (work in collaboration with Dr Nagai at Tokyo Institute of Technology and Dr Glocer, NPP/673). Dr Fok has been playing an active role in the TWINS mission as the project scientist of the mission. In addition to handling project related business, Dr Fok performed ring current-ENA simulation to help interpreting the ENA data and understanding the observable features in the ENA images (work in collaboration with Dr Buzulukova, NPP/673).

Dr Fok has applied her ENA analysis technique to nonmagnetized planets. She has worked with the Venus Express (VEX) team studying ENA images from VEX/ASPER-4, which were the first Venus ENA measurements ever. They found a strong tailward flow of hydrogen ENA tangential to the Venus limb. The figure shows the synthesized VEX H-ENA image taken from nightside (left panel) and the calculated image by Dr Fok. The observed ENA intensities are reproduced by the simulation within a factor of 2. It is concluded that the observed hydrogen ENAs originate from shocked solar wind protons that charge exchange with the Venus exosphere.

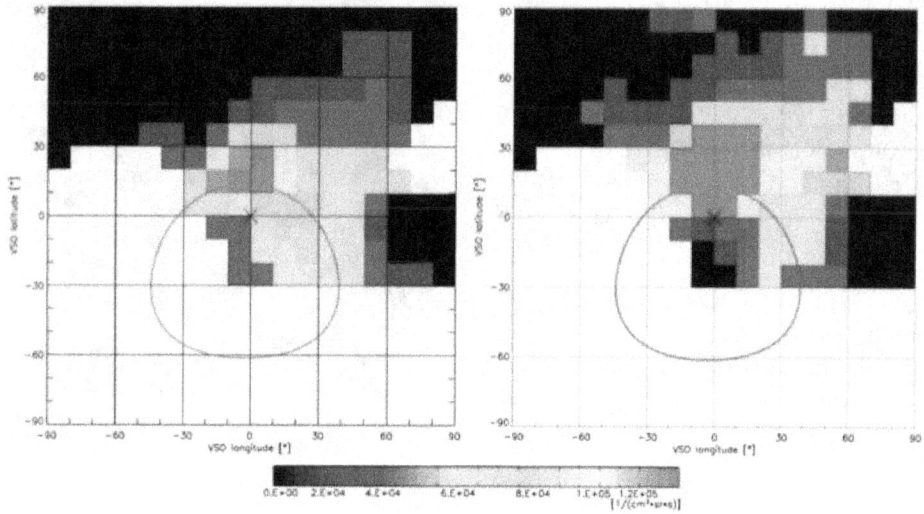

Left panel: synthesized VEX H-ENA image taken from 14 nightside intervals from 20 August to 3 September 2006. Right panel: model prediction of H-ENA flux convolved by instrument factors. Sun direction is at the middle of each image labeled by an 'X'.

Shing F. Fung

Dr Fung is a Space Scientist in the Geospace Physics Laboratory, Code 673. He has been a civil servant over 16 years.

Dr Fung is PI for the Heliophysics Virtual Wave Observatory (VWO <http://vwo.nasa.gov>), one of the latest VxOs selected to become a component of the NASA Heliophysics Data Environment (HPDE <http://hpde.gsfc.nasa.gov>). Over the past year, Dr Fung has led members of the VWO team to work with the SPASE Group Consortium http://spase-group.org to develop the SPASE data model for describing heliophysics data. Dr Fung was co-convener for a VxO special session in the fall AGU meeting held in December 2008. In addition, Dr Fung was a member of the Executive Committee for organizing the Conference on Modern Challenges in Nonlinear Plasma Physics (http://www.astro.auth.gr/~vlahos/kp/), held at the Sani Resort, Greece, 2009 June 15-19.

Dr Fung continues to investigate the large-scale plasmaspheric plume structure by analyzing Cluster and IMAGE crossings of plume structures. He is also studying the geophysical conditions under which a plasmaspheric plume may form. Working with colleagues at the University of Maryland College Park, Dr Fung and collaborators have also been analyzing Cluster data to investigate the role of ULF waves play in accelerating relativistic electrons in the inner magnetosphere. As Co-I on a Geospace SR&T project, Dr Fung and collaborators are investigating the direct influence of solar wind and IMF conditions on the ionosphere. In addition, working with a summer intern, Dr Fung is investigating the solar wind and IMF conditions that lead to the onset of extreme geomagnetic storms ($D_{st} < -250$ nT).

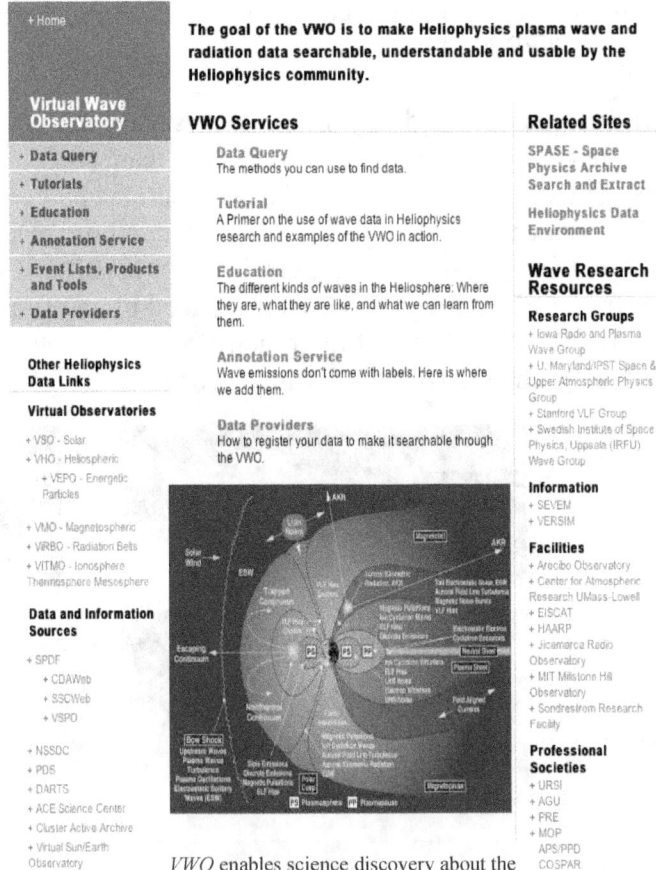

VWO enables science discovery about the Heliophysic wave environment

GSFC Heliophysics Science Division 2009 Science Highlights

Holly Gilbert

Dr Gilbert joined Code 670 in June 2008 as the Associate Director for Science. In her role, she manages the Education and Public Outreach (E/PO) activities of the Division and assists in the administrative duties involved with running the HSD. She also continues doing research on solar surface phenomena associated with CMEs. She did her undergraduate work in physics at the University of Colorado and obtained her PhD in theoretical astrophysics from the University of Oslo.

Dr Gilbert supports the Division office as a manager, but also spends a great deal of time participating in outreach activities. These activities span Public Affairs, including several media appearances (Discovery, National Geographic, and History Channels as well as the local news appearances), as well as meeting with small groups of 8th grade girls. She also regularly gives presentations, the most recent being a talk on Space Weather at the Howard University Space Weather Camp. As part of the GSFC Ambassador program, she gives guided tours of GSFC and makes Science-on-a-Sphere presentations at the GSFC Visitor Center to outside groups. She serves on several committees and groups on Center that involve the GSFC Education Office, and interacts regularly with potential interns, graduate students, and postdocs. She continues to serve the outside solar community at many levels. Specifically, she is an elected member of the SPD committee, a member of the SPD Hale & Harvey Prize Committee, a member of the SHINE steering committee, and the MLSO Users Committee.`

The primary focus of Dr Gilbert's research is determining the nature of prominence support, formation, and evolution and how this relates to CMEs. She has conducted detailed studies of the variation in mass of different types of prominences by addressing different mechanisms involved in mass loss. A search for a relationship between the nature of a prominence eruption and the pre-eruption mass variation is presently being conducted, as well as an examination of the basic CME properties (e.g., speed, mass, and energy) that are associated with the different types of prominence eruptions.

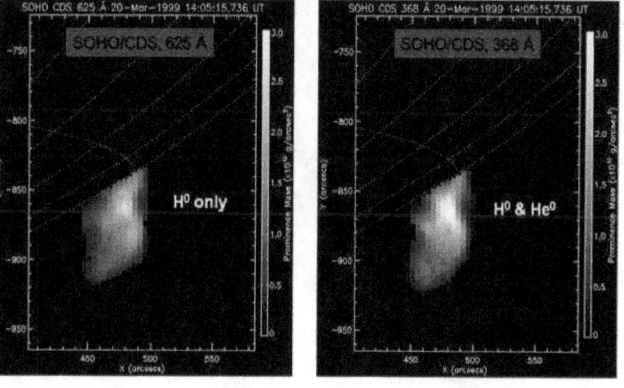

By looking at prominence absorption in a spectral line that ionizes H and He (368 Å) and one that ionizes H only (625 Å), we can compare the mass distribution of the two species

GSFC Heliophysics Science Division 2009 Science Highlights

Alex Glocer

Dr Glocer joined Code 670 as an NPP in September 2008 after completing his PhD in space and planetary physics at the University of Michigan. He has been doing research at GSFC primarily with Dr Mei-Ching Fok on modeling radiation belt electrons. Secondary projects include modeling magnetospheric composition and simulating Mercury's magnetosphere.

Community support functions for this past year include reviewing papers, and volunteering to lead student activities in Sun-Earth day.

Dr Glocer's primary research objective is to understand the sources of radiation belt electron enhancement and loss. Radial diffusion and wave particle are both known sources, but the importance of each processes varies from event to event. The reason for the variation is unknown, but is a topic of intense scientific and space weather interest. To study this problem he uses a wide range computational modeling techniques to bear, including the radiation belt environment model, and the BATS-R-US MHD model of the Magnetosphere. A paper on describing the modeling technique was recently accepted for publication in *J. Sol. Terr. Phys*.

Dr Glocer's secondary projects on modeling magnetospheric composition and simulating Mercury's magnetosphere have met with significant success in 2009, resulting in two papers in *JGR-space*, an invited talk at the Geospace Environment Modeling workshop, and a contributed presentation to the European Planetary Science Congress.

In 2009, Dr Glocer was honored to receive the Ralph Baldwin Award in Astrophysics and Space Science from the University of Michigan for his PhD thesis.

Richard Goldberg

Dr Goldberg is currently specializing in solar-terrestrial relations. He has flown more than 85 rockets with instrumentation designed to study the middle atmosphere, thermosphere, and ionosphere to learn how this region is affected by radiations and other energy sources both external and from sources closer to the Earth's surface. This has produced more than 140 publications. FY09 was spent concentrating on interpreting data from the SABER instrument aboard the TIMED satellite, and from the MaCWAVE Rocket Program. A major breakthrough leading to the extraction of water vapor from SABER has been made and is in press. He has worked at GSFC for more than 46 years, first for one year as a NAS/NRC Postdoctoral Research Fellow, and for the balance as a NASA employee. During 1989-91, he was on leave as a National Science Foundation Program Director of the Solar-Terrestrial Program. He has a BS (Rensselaer Polytechnic Institute) and a PhD (Pennsylvania State University) in physics.

As the GSFC Project Scientist for the TIMED Satellite, and with his activities as a rocket PI for many rocket programs, Dr Goldberg has worked with numerous agencies, companies, and universities, both nationally and internationally. He has served on various review panels for NASA and other agencies, and been involved on definition teams and study groups for various planning activities. He is also an adjunct professor with the Department of Electrical Engineering at Pennsylvania State University.

Dr Goldberg is interested in the mesosphere/lower-thermosphere region (MLT) which has been a difficult region to analyze and understand, largely because of the difficulty in making in situ measurements there other than with rocket soundings. Yet the MLT is a region of rapid change caused by the atmospheric dynamics coupled with numerous sources of highly variable energy sources. Currently scientists are attempting to explain the anomalously warm northern-mesopause region in the summer of 2002 and demonstrate the source for this heating to a warming of the stratosphere in the southern hemisphere using TIMED/SABER data.

Dr Goldberg is also a team member in the successful effort to develop an algorithm to extract mesospheric water vapor from SABER data, which is extremely important for understanding mesospheric processes and developing the climatology of the MLT. This year he has produced six publications and given six presentations.

GSFC Heliophysics Science Division 2009 Science Highlights

Melvyn Goldstein

A space plasma physicist, Dr Goldstein has been at GSFC since 1972, first as a National Research Council Postdoctoral Associate and, since 1974, as a member of what is now the Geospace Science Laboratory. His research focuses on a variety of nonlinear plasma processes that can be elucidated using data from the four Cluster spacecraft. In addition, Dr Goldstein has participated in large and complex simulations of the origin of magnetohydrodynamic turbulence in the solar wind. He also serves as the Project Scientist for the Magnetospheric Multiscale missions and as NASA's Project Scientist for the ESA/NASA Cluster mission.

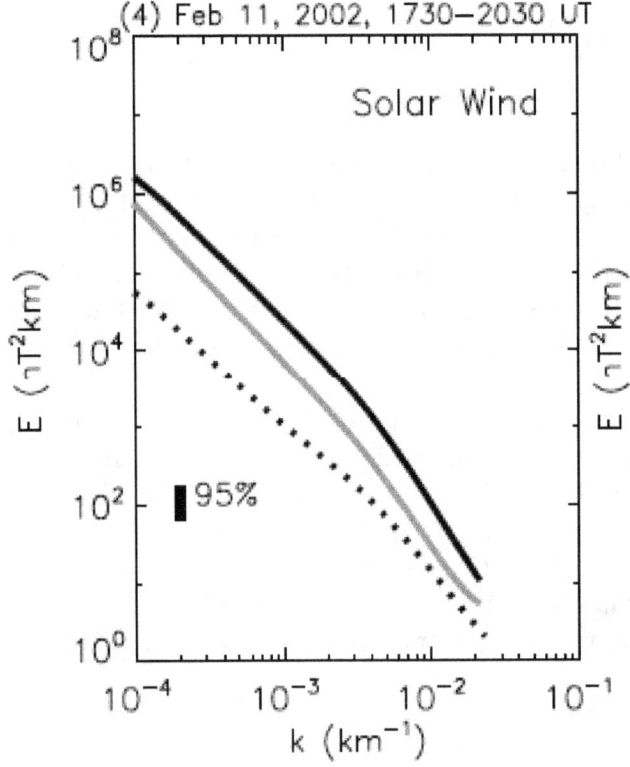

A cut of the wave number spectra for 2D (dark black), Alfvénic, (grey), and compressible (dotted) magnetic fluctuations in the solar wind as measured by the four Cluster spacecraft. The construction uses the "wave telescope" technique to determine the wave numbers directly.

Nat Gopalswamy

Dr Gopalswamy is a SOHO and Wind Team member, STEREO Co-I, lead of the CDAW Data Center and SOHO/LASCO CME Catalog. He has worked at GSFC since 1997, and for NASA since 2002. He has a BSc (1975) and MSc (1977) from the University of Madras; and he received his PhD from the Indian Institute of Science, Bangalore (1982).

Dr Gopalswamy runs the CDAW Data Center where the SOHO CME Catalog resides. He conducts CDAW as a community effort to focus on LWS topics. He completed his duties as International coordinator for IHY in Febrary 2009. He works with CUA on Co-op programs, mentoring two undergraduate students. He is associate editor of AGU's *GRL*, and one of the editors of *Sun and Geosphere*. He is president of IAU commission 49, organizing committee member of IAU Division II, and SCOSTEP Bureau member from IAU.

Dr Gopalswamy's research focuses on:

- How do CMEs and shocks evolve in the corona and interplanetary medium? Understanding this evolution is important in developing models to predict the arrival time of shocks and CMEs at Earth. CMEs and shocks are responsible for the two key space weather aspects: acceleration of solar energetic particles and production of geomagnetic storms.
- When and where do shocks release energetic particles near the Sun? This question is related to the formation of CME-driven shocks in the corona. Combining data on type II radio bursts (which indicate the presence of shocks near the Sun) observed from ground and space with the white-light CME data, the particle release height is determined.

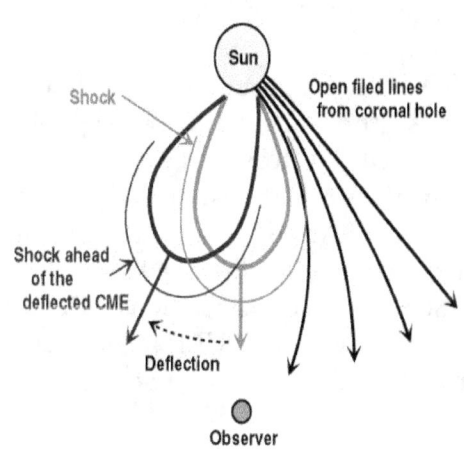

Illustration of CME deflection by the strong open field lines in a coronal hole

Dr Gopalswamy discovered EUV transient reflection from a coronal hole, which confirms the wave nature of the transient. CMEs were also found to be deflected by nearby coronal holes such that disk center CMEs behave like limb CMEs.

He has authored or co-authored 26 publications (first author in 15 of them) during FY09 and has presented 11 invited talks.

Joe Gurman

Dr Gurman is the US project scientist for SOHO, STEREO project scientist, TRACE mission scientist, and SDAC facility scientist.

Dr Gurman has worked at GSFC since 1979 and for NASA since 1985. AB in astronomy from Harvard College, 1972, MS in physics, University of Colorado, 1974, and PhD in astrophysics, University of Colorado, 1979.

In addition to being the US project scientist for SOHO, he leads the local team that operates and disseminates the data from the Extreme ultraviolet Imaging Telescope (EIT) on board. He will soon be ending three years chairing the Solar and Heliospheric MOWG. He is also the AAS Solar Physics Division Webmaster. He frequently instigates, reviews, and edits SOHO and STEREO outreach materials

Aside from senior review proposals, he works on the VSO, which the SDAC "manages." Statistics from the SDAC Senior Review proposal appear to the right.

The VSO is now engaged in establishing a data access server network for the Solar Dynamics Observatory (SDO) mission. The network use the existing VSO internals and NetDRMS database management software developed for the SDO AIA-HMI Joint Science and Operations Center (JSOC). The AIA and HMI instruments will produce well over 2 Tbyte of data per day when decompressed.

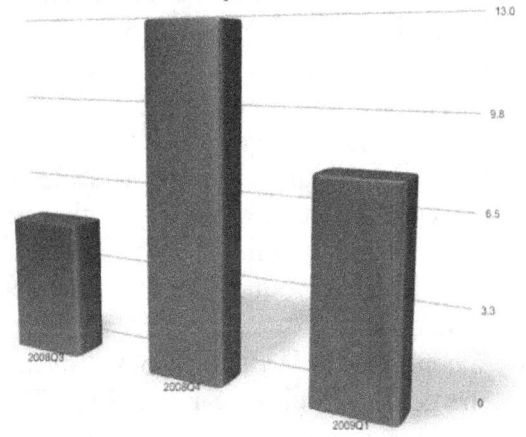

Total volume of data identified (and links provided) by VSO Web interface searches in recent quarters. At present, there is no way of knowing how much of the data was downloaded, since the data are obtained directly from the distributed data providers.

Michael Hesse

Dr Hesse is a space plasma physicist; he is Chief of the Space Weather Laboratory (Code 674), and is Director of the Community Coordinated Modeling Center.

His research focuses on the development and assessment of space weather modeling capabilities, and on basic research of the properties and dynamics of space plasmas.

In his role as Lead Co-I for Theory and Modeling for NASA's MMS mission, he develops new theories of magnetic reconnection, and he advises the MMS project on MMS measurement priorities.

As Director of the CCMC, he collaborates with governmental, academic, and commercial Space Weather interests across the globe. During 2009, Dr Hesse gave six invited talks on science and space weather topics. During the same period, he published seven papers in refereed journals.

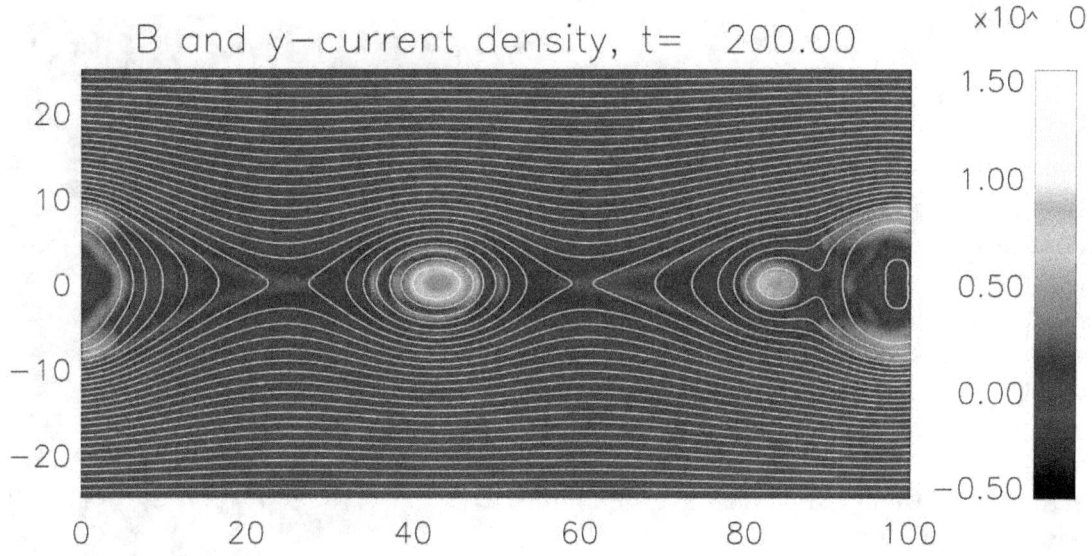

Magnetic field and current density during reconnection dynamics in a relativistic plasma. Modern research shows that nonrelativistic reconnection mechanisms carry over to relativistic plasmas.

Gordon Holman

Dr Holman is a solar physicist who works primarily on the analysis and interpretation of data from RHESSI. His scientific work focuses largely on obtaining a better understanding of the energy release and particle acceleration in solar flares. He has been an astrophysicist at GSFC since 1985. He received his PhD in astrophysics from the University of North Carolina, Chapel Hill, and a BS in physics from Florida State University.

Dr Holman is a Co-I on the RHESSI Mission. He leads the RHESSI theory effort at GSFC. He is PI on a NASA Heliophysics GI grant and Co-I on another. He served as Solar Flares Group Leader at the Solar Activity during the Onset of Solar Cycle 24 Workshop in Napa, California (December 2008), and participated in the 9th RHESSI Workshop in Genoa, Italy (September 2009). He also served on the Popular Writing Award Committee of the Solar Physics Division of the American Astronomical Society and serves as an adviser to the *Physics Today* book review section. He maintains or helps maintain several Websites, including the RHESSI Website and the Solar Flare Theory Educational Website. He helped mentor an undergraduate summer student from Trinity College in Dublin and is advising graduate student Yang Su from Purple Mountain Observatory in Nanjing, China.

Dr Holman's research has focused primarily on the physical processes that affect energetic electrons accelerated in solar flares and the X-ray emission they subsequently radiate. He is studying how X-ray images and spectra obtained with RHESSI reveal the evolution of these electrons in flares. He is also studying how the evolution of these electrons relates to the evolution of thermal plasma and magnetic fields in flares. He is currently examining the impact of nonuniform plasma ionization and return-current energy losses on flare hard X-ray spectra. He is also investigating the uncertainties in model parameters determined from multi-parameter fits to RHESSI spectra.

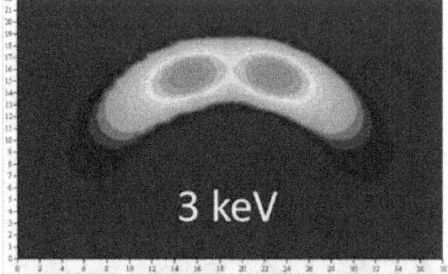
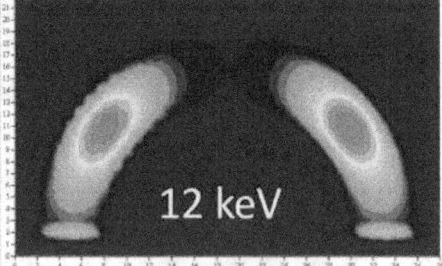

Model demonstrating how the X-rays emitted by nonthermal electrons streaming downward through a density gradient in a flare loop can differ in appearance at different photon energies. Red represents the highest surface brightness, and blue the lowest.

Kyoung-Joo Hwang

Dr Hwang has been using Cluster data to work on the electron diffusion region of reconnection, Kelvin-Helmholtz (KH) waves at the Earth's magnetopause, and dipolarization front (DF) event in the Earth's nightside magnetosphere. She contributed to detecting the implication of Cluster's traverse of the electron diffusion region in the reconnection site, from the electron distribution function showing a non-gyrotropy during one spin period.

She analyzed and reported well-developed KH waves at the flank magnetopause during southward IMF conditions, which is unusual from the observational point of view. The observation indicates that KH activities under southward IMF might have a more temporal or intermittent nature, which explains the preferential in situ detection of KH waves under northward IMF conditions. Along the boundary of KH waves, evidence for the reconnection phenomenon is also found.

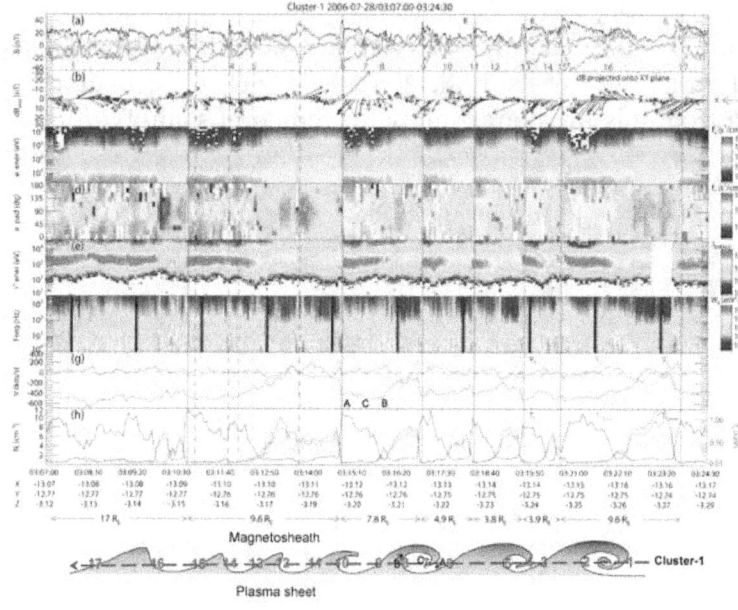

Cluster traverses of a series of KHI-generated structures. This observation is unusual in that well-developed KH waves are observed under southward IMF conditions, and they show temporally intermittent and irregular signatures within relatively short timescales

She also investigated a series of the depolarization front passing by Cluster in the near-midnight central current sheet. These DFs were presumably generated by bursty reconnection occurring tailward of Cluster from the observation of associated earthward fast flows. Each DF was followed by local magnetic disturbances that propagate mainly along the $-y_{GSM}$ direction, indicating a magnetic fluctuation generated by the impact of the DF. She will present these studies at 2009 AGU Fall meeting.

Suzanne Imber

Dr Imber came to GSFC in October 2008 as part of the UMBC/GEST program having completed her MSc at Imperial College, London, in 2005 followed by her PhD at the University of Leicester in the UK. She works primarily with Dr Slavin in studying the dynamics of the magnetotail at Earth but is also interested in Earth's field-aligned current systems and the location and rate of reconnection in the magnetosphere. Dr Imber has been a member of the Deputy Director's Council for Science since January 2009.

The primary focus of her research is to better understand the nature of reconnection in the magnetotail using in situ magnetic field and plasma data. She has been using data primarily from the THEMIS mission, which provides unique multi-point measurements of the near-Earth tail enabling the number of reconnection sites as well as the rate, the location and the spatial extent of the x-lines to be investigated. These data can then be combined with upstream solar wind data and ground-based auroral imagery and magnetometer data to investigate substorm dynamics.

She is also interested in the region 1 and 2 field-aligned current systems at Earth, and has been investigating their spatial variability by using data from the Space Technology 5 mission. She has also used data from this mission to study the temporal stability of theta aurora.

Jack Ireland

Dr Ireland is a scientist working for ADNET Systems, Inc. He has been at GSFC since August 2001. He obtained both his degrees, BSc in mathematics and physics, and PhD in physics, from the University of Glasgow in Scotland.

Dr Ireland is PI of the Helioviewer Project, a project to develop Web-based technologies for the browsing dissemination of heterogeneous solar and heliospheric data sets via intuitive interfaces. He was chair of LOC and SOC of the Fourth Solar Image Processing Workshop, 2008 October 26 – 30, Baltimore, MD. He is involved in SOHO-EIT and Hinode-SOT planning.

Dr Ireland's work in FY09 lay in four primary areas:

- He helped develop user-friendly Web interfaces for the exploration of solar datasets, catalogs and science, with application to the challenge of making SDO data easily browsable.

- He applied Bayesian probability techniques to the understanding of flare spectra from the RHESSI spacecraft. A Bayesian probability analysis allows one to directly calculate the odds that one spectral model explains the data over another, and to quote the probability that the measured value of a parameter lies within a given range.

- He has researched the wave propagation in the solar atmosphere, from which one may deduce the physical conditions of the solar atmosphere.

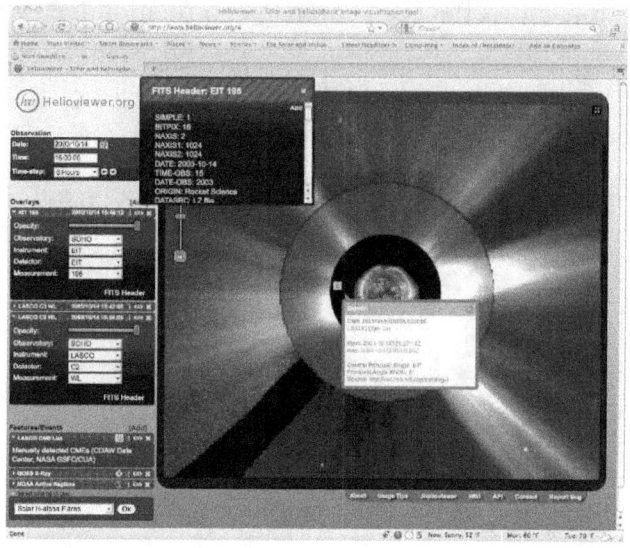

Screenshot of www.helioviewer.org, the Web-browser client of the Helioviewer Project. The tool lets you browse solar and heliospheric data by length-scale, time-scale, wavelength, instrument, and solar feature.

- The final research area concerned the detection of CME events in STEREO/COR1 beacon data.

He expects to continue working in these research areas in the coming year. Dr Ireland also supervised two summer students this year and undertook SOHO-EIT and Hinode-SOT planning duties. He also won an Outstanding Achievement award from ADNET Systems, Inc, and a NASA HGI award "Surveying active region oscillations".

Shaela Jones

Shaela is a PhD student at the University of Maryland. Prior to coming to Maryland she received a BS in Physics from the University of Florida. She has been working with Dr Davila for the past six years.

Currently she is conducting dissertation research measuring the rotation rate of the corona by using the STEREO COR1 coronagraphs. Previous measurements of rotation rates have indicated that the outer corona appears to rotate much more rigidly than the photospere.

If true this would suggest the presence of a region of shear between materials rotating at different rates and could have important consequences for both coronal heating and the initiation of CMEs. However, no studies have definitively located such a region and problems have been identified with earlier coronagraph studies of rotation rate.

She is attempting to measure rotation rates in the COR1 field of view, using both single- and multi-spacecraft observations. Shaela hopes to quantify the degree of influence of projection effects on coronagraph rotation measurements, and to use the multi-spacecraft capability to address the question of whether smaller, shorter-lived coronal features are rotating at a different rate than surrounding large-scale features.

Hyewon Jung

Dr Jung is a solar physicist at CUA. She joined Code 671 in January 2009. She received her PhD degree at Seoul National University, South Korea, in 2008. Her thesis was about magnetic helicity transfer through the solar surface.

She is working with Dr Gopalswamy and his group. In this group, she performs data analysis and interpretation, especially by measuring magnetic helicity. She studies the links between the properties of CMEs (or corresponding MCs) and the properties of their source active regions. Also She is interested in studying the characteristics of CME-productive active regions. Her recent study is about the relationship between magnetic helicity and CME speed in source active regions.

Dr Jung studied linking the speed of CMEs to the magnetic helicity in the source active regions. The motivation comes from the fact that the CME speed may depend on the active-region free magnetic energy, which in turn may be represented by the helicity, a proxy for the nonpotentiality. After selecting a set of active regions from solar cycle 23, she measured their helicity. Using EUV and magnetogram data from SOHO, she extrapolated the photospheric magnetic field to the corona to obtain the coronal helicity that fits a EUV image before each CME eruption. The CME speeds used here correspond to the average speed within the SOHO coronagraph field of view. She found that magnetic helicity is positively correlated with the speed of the CME.

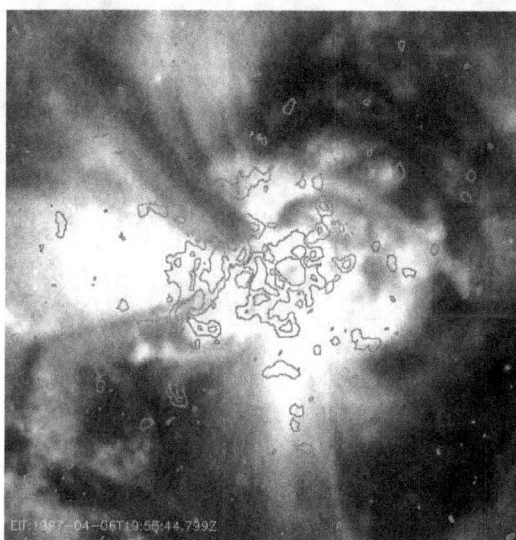 

Magnetogram contour (MDI/SOHO) of AR 8027 superposed on the coronal EUV image (EIT/SOHO) (left), and the constructed coronal magnetic structure (right) whose magnetic helicity has been calculated to be 2.54×10^{42} Mx^2 From this active region a CME occurred whose speed was 878 km/s.

Judy Karpen

Dr Karpen has been a research astrophysicist in the HSD Space Weather Laboratory since July 2008. Her primary research interests include analytical and numerical modeling of dynamic solar and heliospheric phenomena, and applications of plasma physics and magnetohydrodynamics to solar and heliospheric activity. Degrees: PhD in astronomy, University of Maryland (1980); BS in Physics, University of Michigan (1974).

This past year, Dr Karpen has performed 2.5D simulations of driven cancellation of unsheared and sheared magnetic flux, to explore the poorly-understood process of flux cancellation and its potential role in filament channel formation. She is also a member of the LWS TR&T Focused Science Topic team aimed at understanding how flares accelerate particles near the Sun and how they contribute to large SEP events; her contribution is to derive estimates of velocity and magnetic fluctuations from 3D MHD breakout simulations that will be used as input for existing models of particle acceleration in turbulent regions above and below the flare reconnection site. Most recently, she has begun to explore the effects of finite resistivity on the breakout mechanism for CME initiation, with 2.5D MHD modeling.

Dr Karpen is a member of the AAS/SPD Prize Committee. She has served on numerous NASA, NSF, and NAS advisory committees and review panels. She also represents the HSD on the SEMD Women's Science Forum.

Her research includes:

- Gaining insight into fundamental physical processes governing solar and heliospheric activity, such as magnetic reconnection and MHD instabilities.

- Developing and implementing 1D hydrodynamic and 2.5D/3D MHD models of key solar phenomena, including flux cancellation, CMEs/eruptive flares, prominences, and jets, to successfully explain and predict observations from NASA space missions and ground-based instruments.

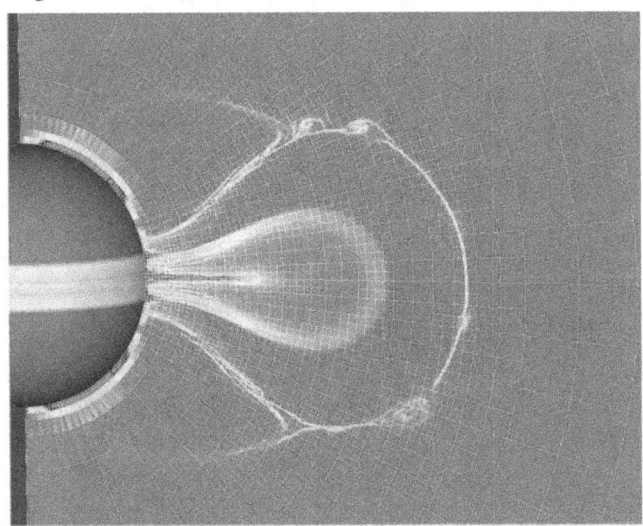

MHD simulation of a breakout CME shortly before eruption (t=90,000 s), showing the current density (color shading) and adaptively refined grid (white lines indicate block boundaries, where each block contains 8x8 cells). The ARMS code, a state-of-the-art 3D MHD code with adaptive regridding developed by C. R. DeVore (NRL), was used for this challenging calculation with six levels of refinement.

Larry Kepko

Dr Kepko is a magnetospheric scientist who recently joined the Space Weather Laboratory (Code 674). He is interested primarily in magnetospheric substorms, but retains in interest in periodic solar wind number density structures. His current focus is to understand the dynamics of the auroral zone around the time of substorm onset. It has been observed for decades that auroral substorm onset begins on the equatorward edge of the nighttime aurora, which maps to the near-geosynchronous region. However, it is thought that magnetic reconnection in the midtail plasma sheet occurs prior to auroral onset, and the lack of any ionospheric signature mapping to this region has been puzzling.

This past year Dr Kepko has studied 630.0-nm auroral emissions for a particular event during the THEMIS tail season. He has found in these images a precursor to white-light auroral onset, which has been the traditional marker of substorm onset. This observation has the potential to fill an important observational gap and further suggests that the 630.0-nm emissions may be a sensitive indicator of plasma-sheet dynamics.

Dr Kepko is a member of the NSF Geospace Environment Modeling (GEM) steering committee, a GEM Focus Group leader (Modes of Magnetospheric Transport), IAGA 2011 Division III.2 Lead Convener (Magnetospheric Substorms and Tail Physics), and chair of the NASA Geospace Management Operations Working Group (G/MOWG).

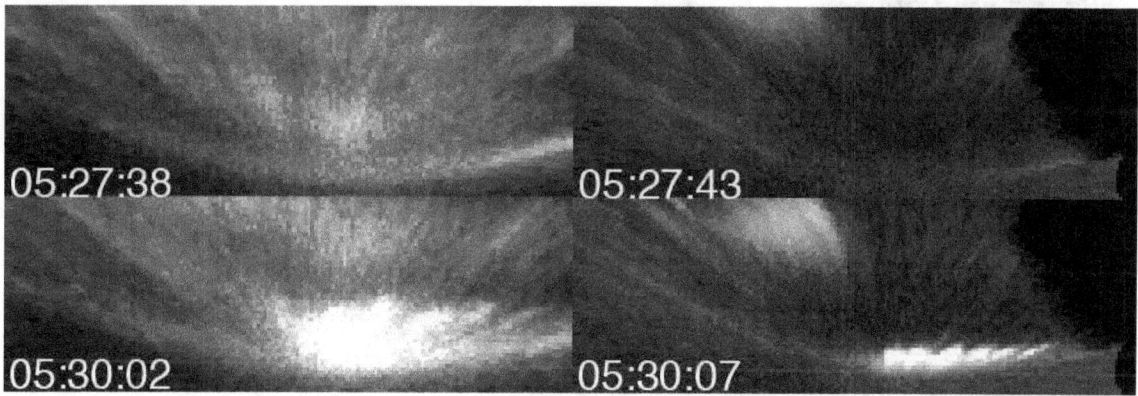

Images from the NORSTAR (630.0 nm) and THEMIS white-light imagers during a magnetospheric substorm. Substorm onset is seen in the lower right image. For several minutes prior to auroral onset the 630.0-nm images showed a diffuse, equatorward moving auroral form. When this form reached the equatorward edge, auroral onset commenced.

George Khazanov

Dr Khazanov has a PhD in physics and mathematics from the Irkutsk State University, Irkutsk, Union of Soviet Socialist Republic (USSR). Prior to joining NASA, Khazanov was Full Tenured Professor of Physics at the University of Alaska Fairbanks. Dr Khazanov has extensive experience in space plasma physics and simulation of geophysical plasmas.

His specific research areas include: analysis of hot plasma interactions with the thermal space plasma with special emphasis on hot-plasma instabilities, investigation of current-produced magnetic field effects on current collection by a tether system, space plasma energization and transport; kinetic theory of superthermal electrons in the ionosphere and plasmasphere; hydrodynamic and kinetic theory of space plasma in the presence of wave activity; theoretical investigation and numerical modeling of ionosphere-plasmasphere interactions; and waves and beam-induced plasma instabilities in the ionosphere. Dr Khazanov supervised and directed more than 30 MS and 15 PhD graduates. He is author or coauthor of five books and about 250 peer-reviewed publications.

FY09 research area: Electromagnetic ion-cyclotron waves generation and coupling in the Earth's magnetosphere. This is important for the following applications: ring current precipitation, plasmasphere energy deposition, and radiation belts MeV electrons scattering.

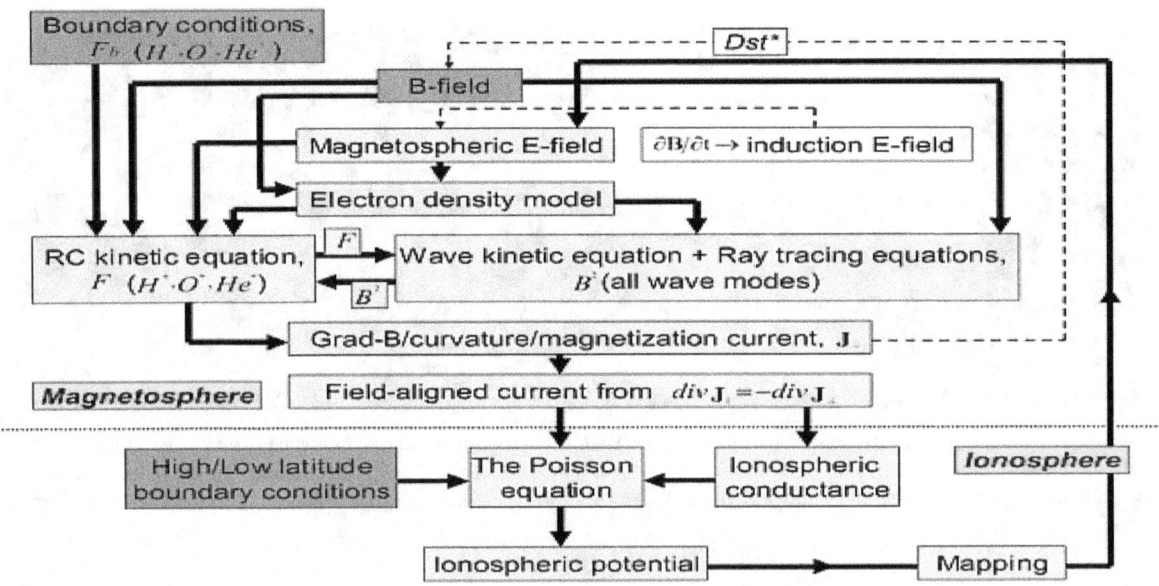

The block diagram of the RC, EMIC waves, plasmasphere, and ionosphere coupling in our model.

Gary Kilper

Dr Kilper started at GSFC in May 2009, via a NASA Postdoctoral Program (NPP) fellowship, which is administered by Oak Ridge Associated Universities. He primarily analyzes observations of solar prominences and the Sun's atmosphere, from several observatories and instruments.

Dr Kilper received a BS with Honors in physics and a BS in mathematics from the University of Chicago in 2004, and an MS in 2007 and a PhD in 2009 in astrophysics from Rice University.

He sometimes works in conjunction with the STEREO/SECCHI group at GSFC, and is currently collaborating on research and observations with colleagues at MSFC, NCAR/HAO, and the National Observatory of Japan.

His current research goals are to understand the process of prominence eruptions, including their role in CMEs and space weather, and to find a way to predict the eruptions to improve space weather forecasting, which is very important for manned space missions, satellites in orbit, and several industries on Earth. More broadly, he wants to understand the magnetic fields and material in the solar atmosphere, and how it relates to the surface below and to interplanetary space above.

The research is done by directing the observations of ground- and space-based instruments, then intricately analyzing that data and describing the physics behind the trends found in the data. Several instruments are used from observatories, including SOHO, STEREO, Hinode, MLSO, and BBSO. From the past year, we have found that the prominence's mass and composition changes in relation to activity level and in the build-up to prominence eruptions and CMEs.

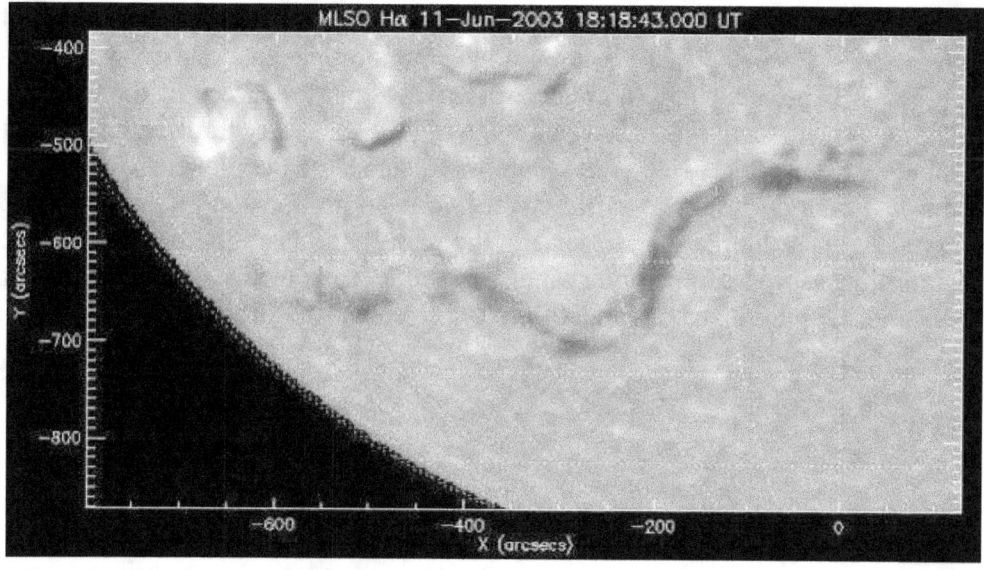

Example of an eruption on 2003 June 11

Joe King

Dr King supports efforts in heliospheric data accessibility and related information management.

He joined GSFC in February 1967 as an NAS/NRC Postdoc in the Laboratory for Theoretical Physics. He was a civil sevant until 2003 and acted as scientist and manager, at the National Space Science Data Center. Since 2003 he is a part-time contractor. He obtained his PhD in space physics at Boston College in 1966.

Dr King supports the Virtual Space Physics Observatory, Virtual Energetic Particles Observatory, Space Physics Archive Search Retrieve (SPASE), IMP 8 magnetic field data reprocessing, and NSSDC legacy data prioritization. He previously was the IMP 8 Project Scientist

He mainly builds multisource, value-added data products and guide development of interfaces thereto. In 2009, they extended the parameter set in, and the time span of the High Resolution OMNI data set, we extended the functionality of the interfaces to this and to related data sets, and created a new interface, called OMNIWeb-Plus, that integrates over multiple, pre-existing interfaces.

Yoji Kondo

Dr Kondo is Co-I of the Kepler Mission and is involved in the pipeline processing of the light from some one hundred thousand stars within the Kepler field.

The Kepler Mission was launched on March 6 and is performing nominally. According to the current understanding of the planetary formation, we expect to find about 60 Earth-like stars within the habitable zones of the G-type mother stars by the end of the 4-year mission. The Kepler field covers some 100 square degrees of the sky in a region of Cygnus. The detection is based on observing the planetary transits; since the transit must be repeated at least twice to evaluate the distance of the planet from the mother star, the mission period must necessarily cover two or more transits. Planetary detections thus far have been based largely on observing the perturbations on the primary star due to the orbital motions of the companion planets, hence it tends to favor detections of massive planets – i.e., Jupiter-sized or super-Jupiter-sized planets – orbiting near the primary.

GSFC Heliophysics Science Division 2009 Science Highlights

Jim Klimchuk

Dr Klimchuk has been an astrophysicist in the Solar Physics Lab at GSFC for 2 years, after having spent 14 years at the Naval Research Lab and 7 years at Stanford University before that. He received a BA degree from Kalamazoo College and a PhD from the University of Colorado.

Dr Klimchuk has served on many panels and committees for NASA and NSF and has held several leadership positions in professional organizations. During 2009, he was President of IAU Commission 10 (Solar Activity) and Vice-President of Division 2 (Sun and Heliosphere). He was a member of the AAS Solar Physics Division Committee, after having completed terms as Chair and Vice-Chair. He served on the AAS Committee on Astronomy and Public Policy, the Editorial Board of the journal Solar Physics, the Solar-C Science Definition Team, and the GSFC Deputy Director's Council on Science. He also led the NASA LWS Focus Team on Solar Origins of Irradiance Variation. Dr Klimchuk is an honorary fellow of the Royal Astronomical Society.

Dr Klimchuk's research during 2009 was mostly concerned with the heating and thermal structure of the solar corona, especially coronal loops. He used a combination of observations and numerical simulations to infer the fundamental properties of the heating. An important breakthrough was the detection of extremely hot (> 4 MK) plasma during times when the Sun was not flaring. The very faint emission is widespread in active regions and is the strongest evidence yet that the corona is heated by a myriad of tiny impulsive energy bursts called nanoflares. Dr Klimchuk used MHD simulations to show that the nanoflares are not due to classical turbulence, as many people had proposed, but instead are likely to be caused by the secondary instability of electric current sheets. He discussed his results in 5 invited and 6 contributed presentations over the course of the year, including a keynote address at the Hinode II meeting. He authored or coauthored 8 published papers and 2 that are in press.

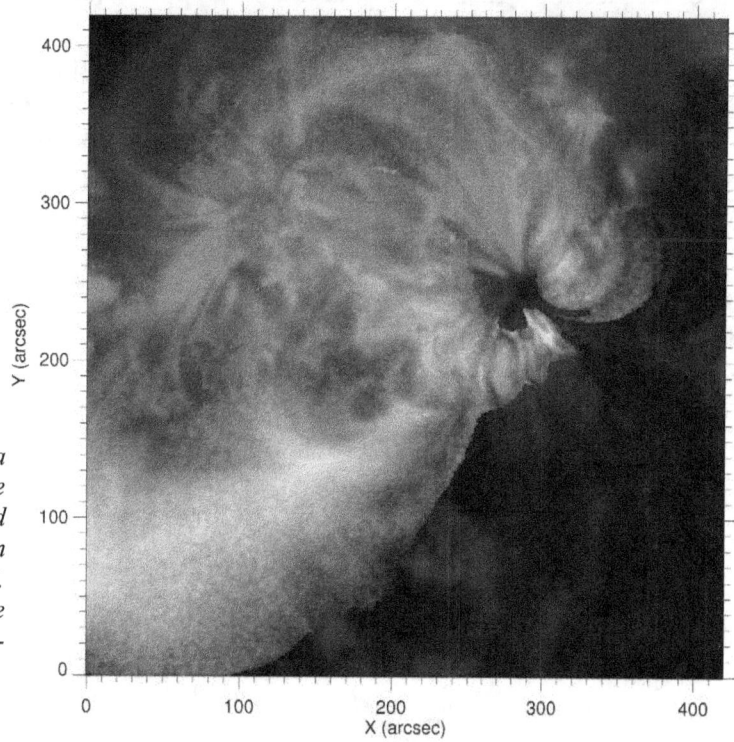

False-color temperature map of a solar active region observed close to Sun center. Blue regions and possible also green regions contain plasma near 10 million degrees K. Measurements were made with the X-Ray Telescope on the Hinode mission.

Andriy Koval

Dr Koval is an assistant research scientist at Goddard Earth Sciences and Technology Center (GEST) at UMBC from August 2009. He was a fellow at NASA postdoctoral program from August 2006 until August 2009. He received a PhD at Charles University in Prague (Czech Republic) in 2006 and BSc and MSc at Sumy State University (Ukraine) in 2000 and 2001.

His main scientific interests are:

- Properties of shock waves in the solar wind and magnetosheath.

- Development of an improved technique for shock parameter determination from single-spacecraft measurements by taking into account measurements of temperature anisotropy and alpha-particle parameters and by simultaneous determination of shock normal direction and propagation speed.

- Estimation of shock global shapes from multi-spacecraft observations to find deviation from the widely used planar assumption.

Based on the introduced new approaches to the shock parameter determination, a program with a graphical user interface was created using Interactive Data Language. The program serves for fast determination of shock parameters and allows user to investigate and compare different methods and approaches in order to determine shock parameters more accurately. It supports measurements of several currently available spacecraft and provides a simple method for adding a new spacecraft.

Dr Koval developed an improved calibration technique for the Wind spacecraft Magnetic Field Investigation to allow the production of 11^{22} vector/s high-time-resolution data. In the current stage, the improved technique results in significantly decreased spin noise in the data as compared to the original calibration technique.

Maxim Kramar

The pre- and post-CME coronal electron density structure has been studied and compared it with the potential field source surface (PFSS) models for the corresponded periods. Two CME cases belonging to different types have been considered. One case consists of a slow CME on 2008 June 1, probably originated relatively high in the corona, and the second case consists of two fast CMEs on 2007 December 31 and 2008 January 2 having source regions close to the Sun's surface in the same active region. The reconstructions of the coronal electron density in the range from 1.5 to 4 R_s were performed using a tomography technique based on COR1/STEREO data.

For the first case of slow CME, it was found that:

- The potential magnetic field configuration in the CME initiation region before the CME does not agree with the coronal density structure while after the CME the agreement between the field and density is much better. This could be manifestation of that that the field was non-potential before the CME and after the CME the field relaxes towards a more potential state.

- The dimming caused by the slow CME is not due to rotation of the corona and a line-of-sight (LOS) effect but a streamer blowout effect took place.

Therese A. Kucera

Dr Kucera is a solar physicist, who joined Code 671 as an NRC post-doctoral fellow in 1993-1995, and then worked in the branch as a contractor until 2001 when she became a civil servant. She received her BA in physics from Carleton College and MS and PhD degrees in astrophysical, planetary and atmospheric sciences from the University of Colorado, Boulder. She has served as the Deputy Project Scientist for SOHO and as STEREO E/PO lead.

She is currently Deputy Project Scientist for STEREO and was a Guest Editor for the STEREO Special Issue of *Solar Physics* published in 2009. She has just returned from a detail to NASA Headquarters as the Heliophysics Solar Discipline Scientist.

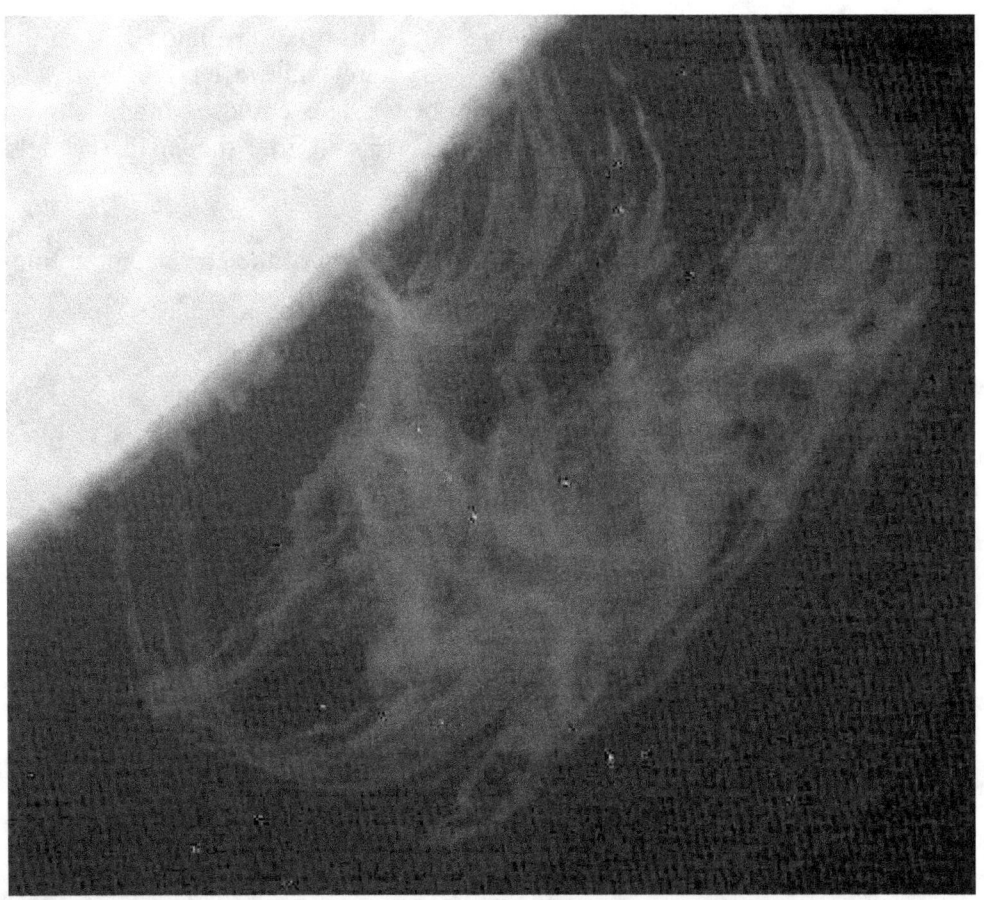

Prominence observed with TRACE 1216 Å band, 1998 August 9

Dr Kucera has been researching the solar atmosphere with special emphasis on solar prominences and ultraviolet spectroscopy. Other interests include active regions and coronal cavities; she is currently a member of the International Space Science Institute's Prominence Cavity Team.

GSFC Heliophysics Science Division 2009 Science Highlights

Alexander Kutepov

Dr Kutepov is an atmospheric physicist, PhD in physics (Candidate of Sciences) at the Leningrad State University, currently senior research associate at the Department of Physics of CUA, and working at GSFC, Code 674, since 2003.

He collaborates with the GSFC Planetary Systems Laboratory (Code 693) as PI on the NASA MDP project "Retrieval of temperatures of Martian atmosphere in the altitude region 60–100 km from the MGS/TES bolometer infrared limb radiances"; and with the Department of Atmospheric and Planetary Sciences, Hampton University as the TIMED/SABER science team member.

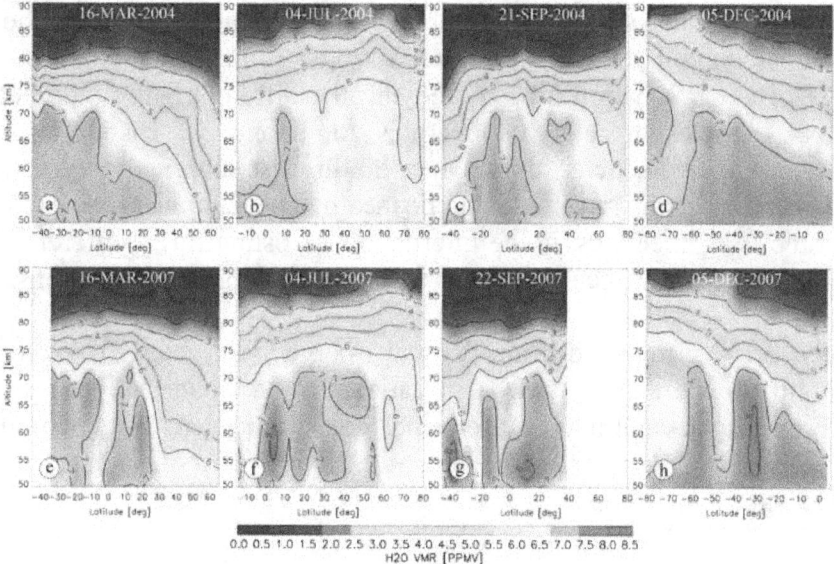

Water vapor volume mixing ratio distributions for eight seasonal turning points in 2004 and 2007 obtained from the SABER data with the revised non-LTE model.

Dr Kutepov continued working together with M. Smith at GSFC and Feofilov at CUA and GSFC on the non-LTE analysis of broadband IR (5.5–100-μm) limb observations of the Martian atmosphere by the MGS/TES bolometer aimed at retrieving pressures and temperatures in the middle and upper Martian atmosphere (60–95 km) from the MGS/TES observations. Together with A. Feofilov, R. Goldberg, and W. Pesnell (both at GSFC), he continued working on the water vapor density retrievals in Earth's mesosphere and lower thermosphere (MLT) from the IR broadband TIMED/SABER observations of the Earth's limb emissions. In this work an update was suggested to the rate coefficients of the most important vibration-vibrational (V-V) and vibration-translational (V-T) intermolecular energy processes that affect the $H_2O(\nu_2)$ level populations in MLT relying on the ACE-FTS satellite solar occultation water vapor density measurements in MLT. It was demonstrated that applying the updated H_2O non-LTE model to the SABER radiances provides retrieved H_2O densities in the 50–85-km region consistent with climatological data and model predictions.

Nand Lal

Dr Lal is a member of the Voyager Cosmic Ray Subsystem (CRS) team. He is also a member of the Virtual Energetic Particle Observatory (VEPO) team. He serves as the Technical Officer for Applied Information Systems Program grants. He is also the Division Computer Security Official.

He received his MSc in physics from the University of Delhi, and a PhD in theoretical physics from Cornell University. Dr Lal came to GSFC in 1972. His work here has involved design and development of data processing and analysis systems for cosmic ray, γ–ray and X-ray experiments, as well as management and support of information technology programs.

Both Voyager spacecraft are now in the heliosheath. Voyager 1 crossed the solar wind termination shock on 2004 December 16 and Voyager 2 crossed it several times on 2007 August 30-31. The CRS team has been studying the different particle populations (Termination shock particles at the lowest energies, the anomalous cosmic rays at intermediate energies and the galactic cosmic rays at the highest energies) observed by CRS as the two Voyager spacecraft traverse the heliosheath. The team has reported on these observations at the AGU meetings, in the International Cosmic Ray Conference and the Institute of Geophysics and Planetary Physics as well as in journal publications.

We have been actively involved in the Virtual Energetic Particle Observatory efforts to improve usability of energetic particle observations by the broader science community. In addition to preparing several new datasets and their descriptions in the common SPASE format, we have developed a prototype for interactively browsing related energetic particle data products.

Derek Lamb

After defending his thesis at the University of Colorado, Dr Lamb (CUA) started at GSFC in November. He is analyzing solar CMEs as observed by the STEREO-SECCHI/COR1 coronagraphs. Some of these ejections show a particularly slow liftoff, and he is analyzing these events with multiple instruments from multiple observing positions to determine whether these events are fundamentally different from their more common, much faster counterparts.

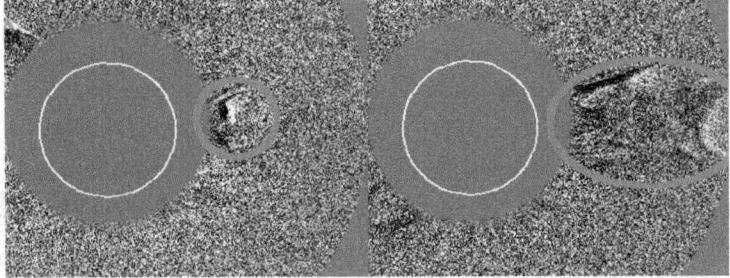

Running difference images of the corona as observed by STEREO/COR1. (left) Red circle highlights the initial appearance of a CME. (right) Red ellipse highlights significant structure still visible over 8 hours later.

Guan Le

Guan Le is a magnetospheric physicist in Space Weather Laboratory (Code 674). She is the science lead for the Space Technology 5 (ST5) mission and has been studying field-aligned currents (FACs) using multi-points magnetic field measurements from ST5. Her research focuses on the temporal variations of FACs at short time scales, as well as the ionospheric closure currents of FACs. She is a science Co-I of Vector Electric Field Instrument on C/NOFS mission. Her research focuses on the ionospheric currents using the C/NOFS magnetic field data. She is a detailee in NASA HQ Heliophysics Division. She is the Deputy Project Scientist for MMS and Project Scientist for Geotail. She is a member of AGU SPA Section executive committee and serves as the editor of SPA Website and co-editor of *SPA Newsletter*.

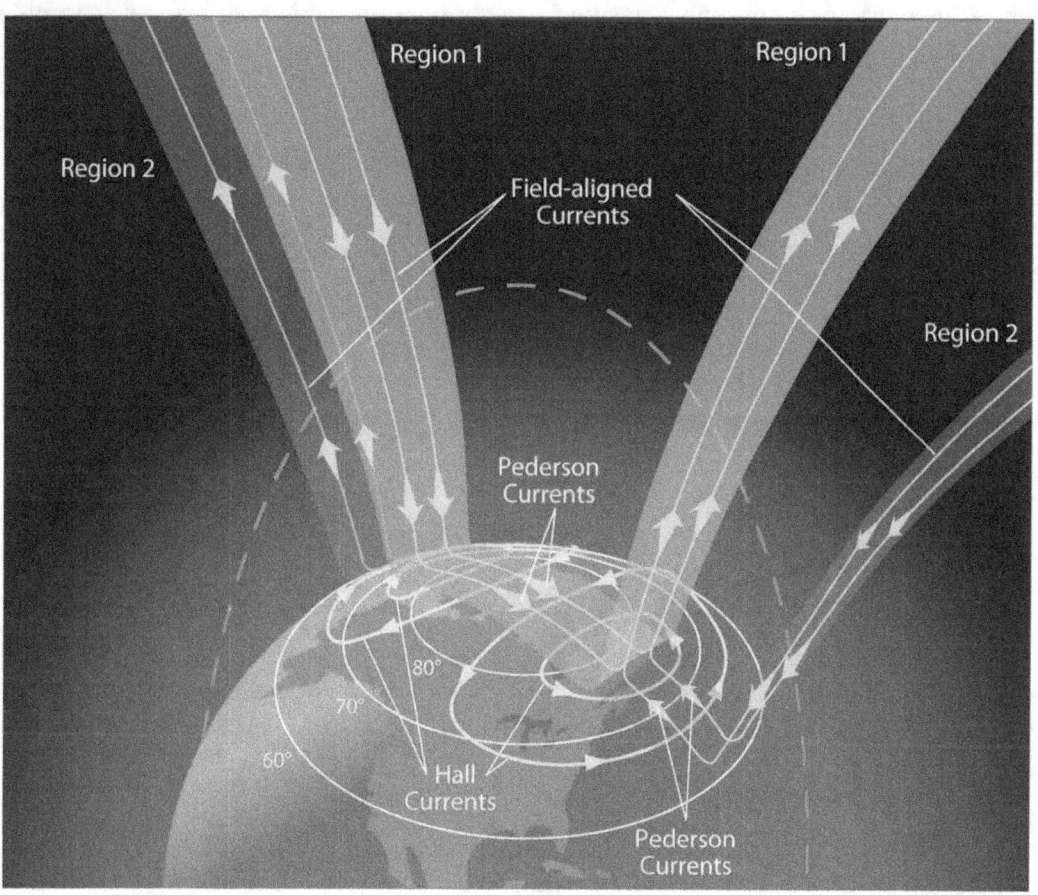

A schematic of combined field-aligned currents (FACs) and ionospheric current systems. In the ionosphere, Region 1 (R1) and Region 2 (R2) FACs are closed by Pederson currents flow between the ionospheric footprints of R1 and R2 FACs. The unbalanced portion of R1 and R2 currents are closed by cross-polar cap Pederson currents. Hall currents are not part of the closure currents for FACs. Also known as auroral electrojets in the auroral zone, they close in the ionosphere by flowing sunward over the polar cap to form current loops. ST5, a constellation of three micro-satellites, provides first multi-point magnetic field measurements of field-aligned currents at low-altitudes. The ST5 data enable us to quantify the imbalance of R1 and R2 currents and the cross-polar cap Pedersen currents.

Robert Leamon

Dr Leamon moved to GSFC in September 2003, after a post-doctoral appointment at MSU. His BSc is from Imperial College, London, and his PhD is from the University of Delaware. Although he is in the Solar Physics Laboratory (Code 671.1), his research interests extend further (both literally and metaphorically) than just the Sun.

He works primarily in data analysis and observation; the overarching theme of his post-dissertation research career to date has been the correlation of interplanetary phenomena with their solar sources. This is true on both large scales with CMEs and their interplanetary manifestations, and the small-scale tracking and forecasting of the quiet solar wind.

Much of his work in the past year has been on applications of MRoI. This is a simple realization of the magnetic environment and reflects the radial distance required to balance the magnetic field contained in any pixel in a magnetogram. While it contains no directional information, it can be thought of as a measure of magnetic partitioning of the plasma into open and closed regions. When the MRoI is small, the magnetic field is "closed" locally, while if the MRoI is large, the magnetic field at the center is largely unbalanced and the magnetic environment is effectively "open." High MRoI can result from either a flux concentration (e.g., an active region), or from a large predominantly unipolar region (e.g., a coronal hole). It has implications on small scales (e.g., blue shifted flows as observed at Transition Region heights by SUMER), and also on the largest scales, as in where the source of Earth-Directed solar wind.

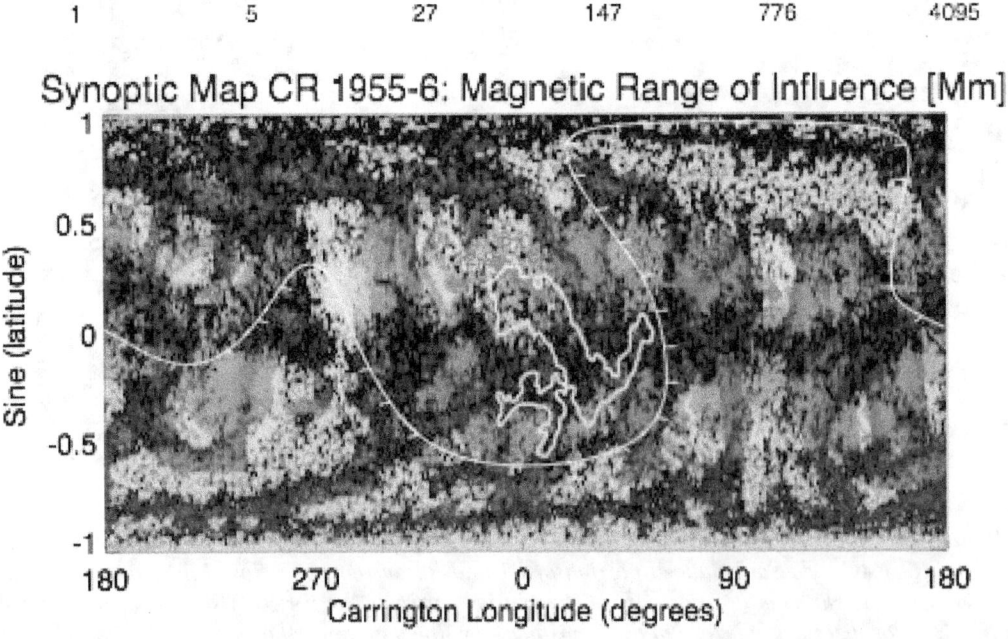

MRoI map for synoptic maps for Carrington Rotations 1955-6 (half of each), showing the motion of the foot-point of earth-directed solar wind does not progress smoothly across the photosphere – rather, it jumps from one patch of (relatively) high MRoI to the next. When there is no close region of high MRoI, the foot-point stays put for several days, but when there is no real dominant patch, the foot-point moves rapidly. The time that the footpoint jumps into the yellow coronal hole contour matches the start of fast wind observed in situ.

GSFC Heliophysics Science Division 2009 Science Highlights

Alexander S. Lipatov

Dr Lipatov (Senior Research Scientist, GEST Center UMBC) joined the GSFC (Code 673) in Jan. 2007 as the Senior Visiting Research Fellow. He received an MS in gas dynamics and thermodynamics (1969), and a PhD in plasma physics (1972) from Moscow Institute for Physics and Technology (State Univ.), Russia. He also received a PhD in theoretical and mathematical physics (1988) from Space Research Institute (IKI) Russian Academy of Sciences (RAS), and the title "Professor of Space Physics" from the Russian Ministry for Higher Education (1995). He worked in IKI RAS from 1972 as Scientist, Senior Scientist, Lead Scientist/Professor and after Jan. 1997 in the Dialogue-Science, A.A. Dorodnitsyn Computing Center RAS as a Lead Scientist/Professor. From 1973 he also occupied the Assistant, Dozent, Deputy Head of Basic Speciality (Chair) "Space Physics" (including Space Physics, Astrophysics and Computational Physics), and Professor positions in the Dept. of Problems of Physics and Energetics, Moscow Institute for Physics and Technology. He also occupied the long term visiting professor positions from 1992: Univ. of Maryland at College Park; Max-Planck Inst. for Extraterrestrial Physics, Berlin; Bartol Res. Inst., Univ. of Delaware, Newark; Max-Planck Inst. for Solar System Research, Katlenburg-Lindau; Inst. for Theoretical Physics, Technical University, Braunschweig; Univ. of Michigan, Ann Arbor; Univ. of Alberta, Edmonton.

Currently he works on the 3D hybrid kinetic simulation of the plasma environment dynamics near SP+ spacecraft, Titan, Europa and the Moon in collaboration with Dr Sittler, Dr Hartle, and Dr Cooper. The results of these simulations will serve as an expert system for SP+ spacecraft design, for a future mission to Europa and for an interpretation of the CAPS data during the Cassini mission near Titan.

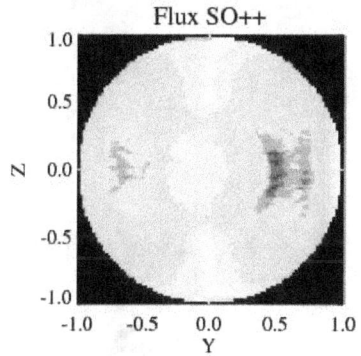
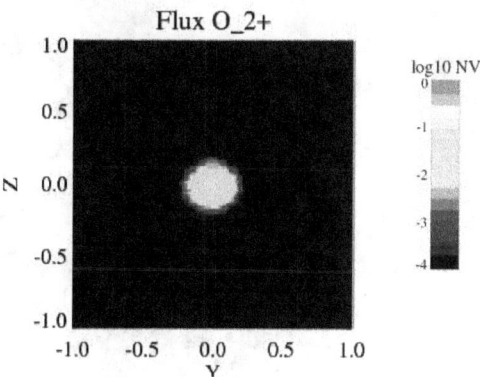

Example of fluxes of SO^{++} (left) and O_2^+ (right) ions at the surface of Europa

Peter MacNeice

Dr MacNeice works on the development and application of solar and heliospheric models, for both research and forecasting applications. He has been studying the accuracy of the models of the coronal and heliospheric magnetic fields and plasma, which are run at the CCMC and developing specific tests for their validation. He is also leading an effort to develop a 3D model for the slow quasi-static evolution of solar active region fields. This effort encompasses potential, nonlinear force-free and MHD codes with adaptive mesh refinement. It will also include a utility for synthesizing time series of magnetograms for use by the models. This suite of models is being developed for use by the research community as an LWS strategic capability. The magnetogram synthesis work is also supported by AISRP. Dr MacNeice has also developed a 2.5D model for the initiation and propagation of CMEs in the presence of a realistic solar wind. This model is available to users through the CCMC's Website.

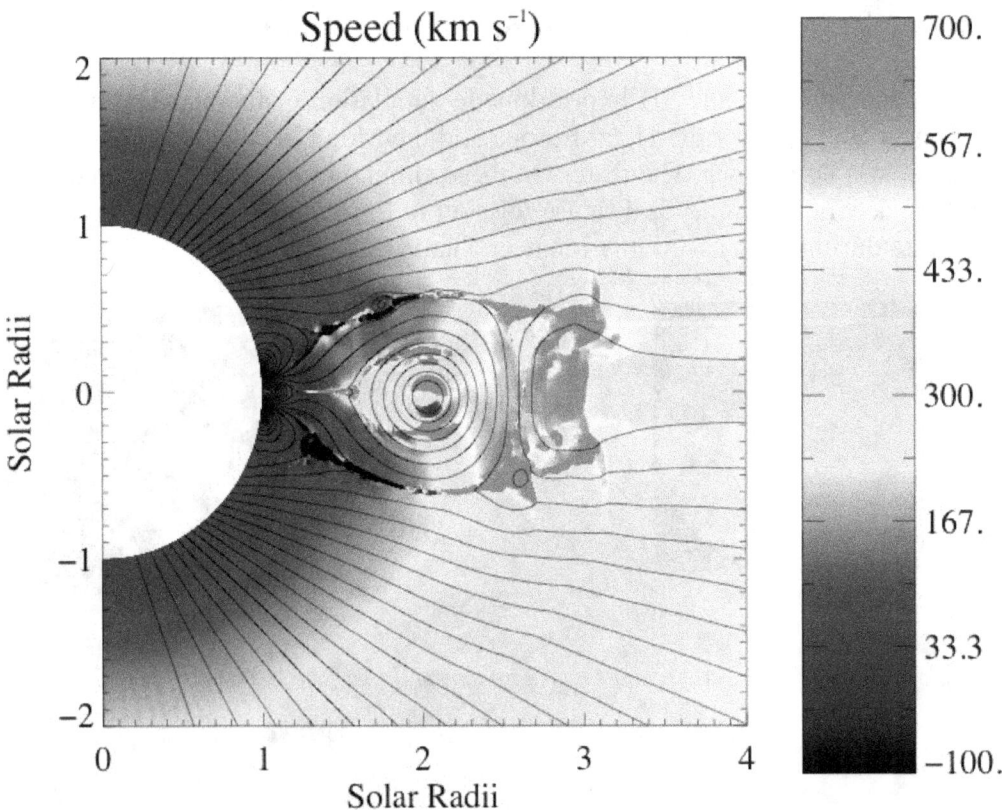

Pertti Mäkelä

Dr Mäkelä is a postdoc researcher currently at the Solar Physics Laboratory (671). He has been working at GSFC since January 2007. He obtained his MSc in physics and PhD in space physics at the University of Turku, Finland. In Finland, he worked at the Space Research Laboratory in the Department of Physics, where he was involved mainly in the data analysis of the Energetic and Relativistic Nuclei and Electron instrument on SOHO.

He helps with the CDAW Data Center (http://cdaw.gsfc.nasa.gov) that maintains the extensively used SOHO/LASCO CME list and also other CME related data sets.

While working here at GSFC with Dr Gopalswamy and the other members of his group, Dr Mäkelä has been able not only to further his experience on SEPs but also to extend it into optical and radio observations of the associated solar events. His main field of research concerns particle acceleration at the Sun. He is therefore working mostly on the analysis of SEP and CME observations by both spaceborne and ground-based instruments. As energetic particles and CMEs transport the effects of solar activity to Earth, the subject of his research is important in the field of space weather, even though he is not studying the effects of space weather itself.

In 2009, Dr Mäkelä research concerned particle acceleration during CME-driven radio-quiet (RQ) shocks. The RQ shocks are defined as interplanetary (IP) shocks that are associated with CMEs without observed type II radio burst in the metric or decameter-hectometric wavelength range, i.e., no coronal shock. The study is based on the observations by SOHO, Wind, and ACE spacecraft and ground-based radio observatories, and it extends the previous studies of the group on RQ shocks and CMEs.

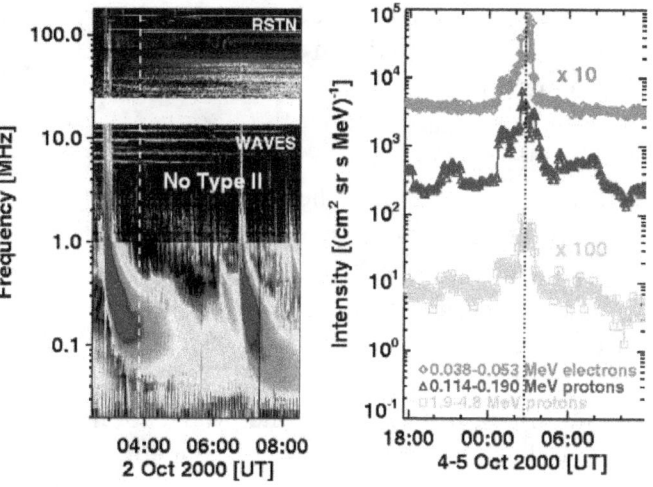

Radio emission (left) and particle flux (right) associated with the 5 October 2000 radio-quiet shock.

Results indicate that shock acceleration efficiency of electrons and ions in the solar corona and IP space are not strictly correlated. IP shocks can accelerate particles, but less efficiently, even though they do not produce observable type II radio bursts, as evidenced by the Figure showing the lack of type II radio burst in the observations of the 03:50 UT 2 October 2000 CME by the Learmonth radio observatory and Wind/WAVES and the associated proton and electron enhancement observed by the ACE/EPAM instrument during the passage of the 2000 October 5 RQ shock (dotted vertical line).

GSFC Heliophysics Science Division 2009 Science Highlights

Leila Mays

Leila Mays began at GSFC in August 2009 as a NASA Postdoctoral Program (NPP) Fellow working with Dr St. Cyr and Dr Sibeck. In 2004 she received a BS with high honors in physics and astronomy from the University of Maryland College Park. She graduated with a PhD in physics at the University of Texas at Austin with supervisor Wendell Horton and expects to finish in August 2009.

Leila Mays' NPP research of the Solar-Terrestrial chain of events in the inner heliosphere involves using data from various spacecraft and models to understand:

- The propagation of interplanetary disturbances in the solar wind (e.g., CMEs, shocks, and density fluctuations);
- The interaction of the disturbances with the dayside magnetosphere;
- The magnetospheric response. Spacecraft observations by THEMIS in the magnetosphere will be used to validate WINDMI, a model for the solar wind-driven magnetosphere-ionosphere system which characterizes geomagnetic storm and substorm activity.

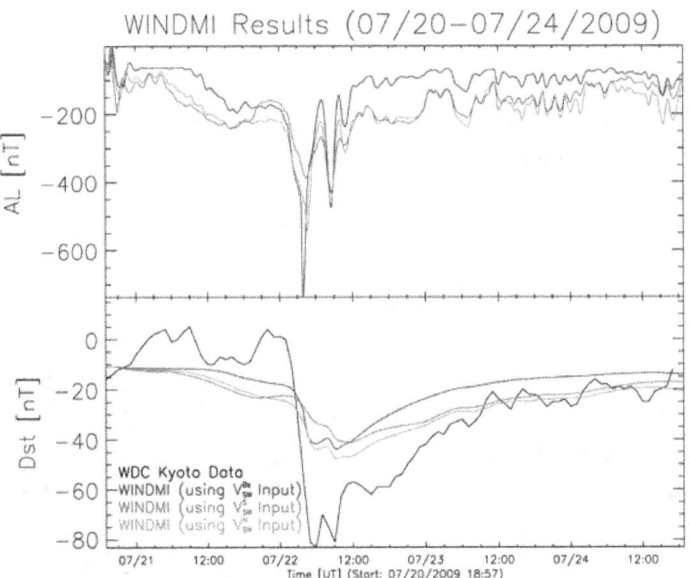

WINDMI model results of the AL and Dst Indices for 20-24 July 2009 shown in color for different solar wind input drivers. WDC Kyoto Quicklook Dst data is shown in black.

It is important to understand the propagation of interplanetary disturbances from the Sun because the solar wind drives geomagnetic activity and solar wind spacecraft data is used as input for many magnetosphere models. The magnetospheric response aspect of this work is relevant to the operational forecasting of Space Weather. WINDMI and other model runs will be performed using the public resources at the CCMC.

Robert McGuire

Dr McGuire is currently the Associate Director for Science Information Systems in HSD. In this role, he coordinates science data activities across the Division and represents the Division on various Directorate and Center groups. He also leads and directs the Space Physics Data Facility (SPDF) project as its Project Scientist. SPDF develops and operates multi-mission and active archive data and display services (e.g. CDAWeb and OMNIWeb), orbit planning and display services (e.g. SSCWeb and the 4D Orbit Viewer), and supports maintenance and use of the Common Data Format (CDF) standard, now used by many current NASA missions.

SPDF is also now one of two designated heliophysics active Final Archives ensuring the long-term preservation and ongoing accessibility of important heliophysics data from past, present and future NASA missions.

Bob has science interests in interplanetary particle composition and acceleration, as well as his work in science data systems and archives. He is a Co-I on several current heliophysics Virtual discipline Observatory (VxO) investigations.

Jan Merka

Dr Merka has been an Associate Research Scientist at the University of Maryland, Baltimore County, since 2006. Located at NASA/GSFC since 2001, he works on several space physics research projects. He is the PI of the Virtual Magnetospheric Observatory. Dr Merka got his PhD degree in plasma physics from the Charles University, Prague, Czech Republic. He provides essential support to HSD in development of Web and database systems for the emerging heliophysics virtual observatories.

Dr Merka's primary area of research is the interaction of the solar wind with the Earth's magnetosphere, in particular the solar wind plasma effects on the magnetosphere shape and plasma entry inside. Understanding solar wind plasma entry into the magnetosphere is important for developing better insight into magnetospheric processes and how their affect space exploration and ground systems.

He leads the development of the Virtual Magnetospheric Observatory (VMO) and collaborates with the Virtual Heliospheric Observatory (VHO) team to create unified, easy-to-use portals for heliophysics data discovery and retrieval.

Ryan Milligan

Dr Milligan has been a solar physicist as part of the RHESSI team at GSFC for 3 years. He is originally from Northern Ireland, where he received both his BSc degree and PhD from Queen's University Belfast.

As part of ongoing collaborations with Trinity College Dublin, this summer Dr Milligan hosted a student who worked on the analysis of early-impulsive flares using RHESSI data. He continues to collaborate with colleagues at Queen's University Belfast on EUV spectroscopy, and with Hinode/EIS team members at NRL in Washington, DC and MSSL in the UK, on the development of observing sequences optimal for studying solar flares. He is also one of the Max Millennium Chief Observers who monitor and report on solar activity, and often volunteers as a guest scientist for SDO's education and public outreach program.

Dr Milligan's research focuses primarily on the physics of energy release and transport associated with solar flares. By combining observations from RHESSI with those of other space-based instruments such as Hinode and STEREO we can diagnose both the cause and the effect of the energy release. In a recent publication, he showed that the solar chromosphere responds dynamically to a beam of injected high-energy electrons during a flare's impulsive phase. The velocity at which this heated (or evaporated) material rises was found to be directly related to its temperature. This work was presented at the annual Solar Physics Division meeting in Boulder, CO, and as part of the RHESSI workshop hosted in Genoa, Italy. He is currently using a combination of RHESSI and STEREO data to investigate the mechanism responsible for accelerating the electrons.

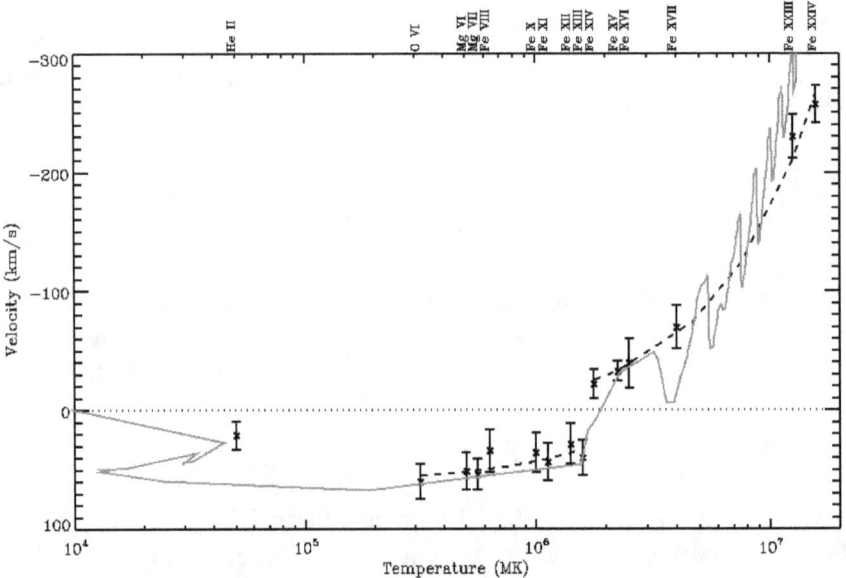

Plot of evaporation velocity as a function of temperature for a broad range of emission lines using data Hinode/EIS. Negative velocities are toward the observer. The solid red line is the same relationship from a recent numerical model.

Tom E. Moore

Dr Moore is the deputy director of the HSD, lead investigator for the MMS Fast Plasma Instrument, Co-I for IBEX low energy instrument, and PI for a Living With a Star project called "Storm Time Plasma Redistribution." He has been at GSFC since May 1997. He was originally from New Hampshire, and earned a BS and MAT from UNH. He taught middle and high school, and then returned to the University of Colorado in Boulder for a PhD in astrogeophysics, returning to UNH as a postdoc. He joined NASA in 1983 at MSFC to work on the NASA Dynamics Explorer mission and SpaceLabs.

At GSFC, Dr Moore has worked on the Polar mission (as PI, and briefly as Project Scientist), the IMAGE mission (as Project Scientist and Lead Investigator), the Magnetospheric Constellation mission (as Project Study Scientist), and other research tasks related to the ablative effects of solar wind plasmas on Earth's atmosphere. These roles involved collaborations with the UAH, UTD, SwRI, CETP, UMich., LMATC, ISAS, CESR, and many others, as well as numerous reviewing assignments and participation in groups such as the LWS TR&T Steering Committee, the GEM Working Group on M-I Coupling, the SP+ STDT, and service as the AGU/SPA secretary for Magnetospheric Physics.

Dr Moore's research interests focus on the coupling between the atmospheres of the Sun, planets, small bodies, and the interstellar medium. These effects span the origin, evolution, and fate of our solar system.

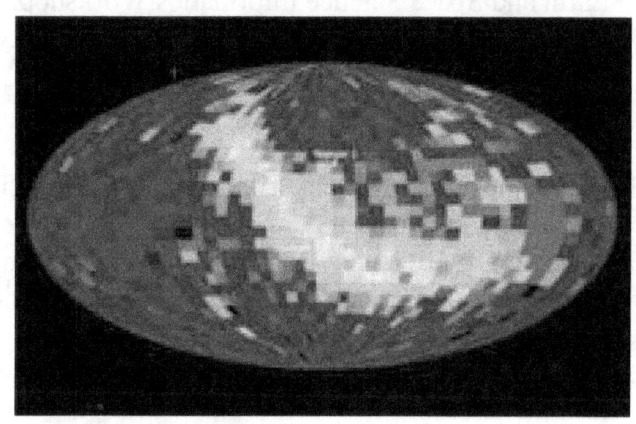

The heliopause "ribbon" of bright emission of ~1 keV neutral atoms, as observed by IBEX.

In FY09, his research activities included studies of the ENA emission from the heliopause region as observed by the IBEX mission, inference of interstellar magnetic field from those observations, entry pathways of solar wind plasma into the magnetosphere, and mechanisms for the conversion of bulk directed motion into random thermal energy by plasma-gas wave-particle interactions in the auroral ionosphere.

Tom Narock

Tom Narock is a member of the Heliospheric Physics Laboratory and has been at GSFC since 2001. His work focuses on data access and data integration, which lead to improvements in heliophysics research. In recent years this work has centered on the emerging "Virtual Observatories" that provide uniform search, retrieval, and visualization capabilities to disparate heliospheric and magnetospheric missions. Tom earned a BS in astronomy from the University of Maryland, College Park, and an MS in physics from Johns Hopkins University. The technical challenges related to heliophysics research have led him to the field of Information Systems (IS). Currently, Tom is pursuing a PhD at UMd, Baltimore County, in IS.

His duties include:
- Virtual Heliospheric Observatory – Co-I – lead IT development.
- Virtual Energetic Particle Observatory – Co-I – technical advisory
- Virtual Waves Observatory – Co-I – technical advisory
- American Geophysical Union, Earth and Space Science Informatics Focus Group, Executive Committee
- Earth and Space Science Informatics Workshop, Organizer and Program Committee Chair, University of Maryland, Baltimore County, 2009 August 3-5
- Special Guest Editor, *Earth Science Informatics*, journal special issue related to above workshop

The Semantic Web is an extension of the World Wide Web in which the meaning of information and services is defined. This enables the possibility for computers to understand and more effectively satisfy the requests of human users. Tom's research focuses on applying this emerging technology to heliophysics data and research problems. Having intelligent computers and software applications enables more efficient access to data and more effective ways of conducting space science research.

Research efforts this year have included the completion of a prototype data integration system that utilizes the aforementioned technologies. The system is being evaluated on actual space science research problems and the results are indicating that we can now do in minutes what had previously taken months of heliophysics researchers' time.

Teresa Nieves-Chinchilla

Dr Nieves-Chinchilla completed her PhD at Alcala University in Madrid, on the study of the evolution of the magnetic cloud in the interplanetary medium. She was developing a non-force-free model (with expansion and deformation in the cross-section) and implementing an algorithm to fit Wind data.

She joined the NASA Postdoctoral Program in May of 2006 and she has been working with the Dr Vinas. In the first part of her postdoc, she was working using Wind/SWE data in the line of investigation of the properties of electrons in magnetic clouds. In the last year in the NPP program, she worked with high-time-resolution data from Cluster/PEACE to study the kinetic aspect of the electron distribution function. She emphasizes the Strahl electron component to understand the properties, the sources, and the role of processes that regulate instabilities. These studies have been submitted to peer review.

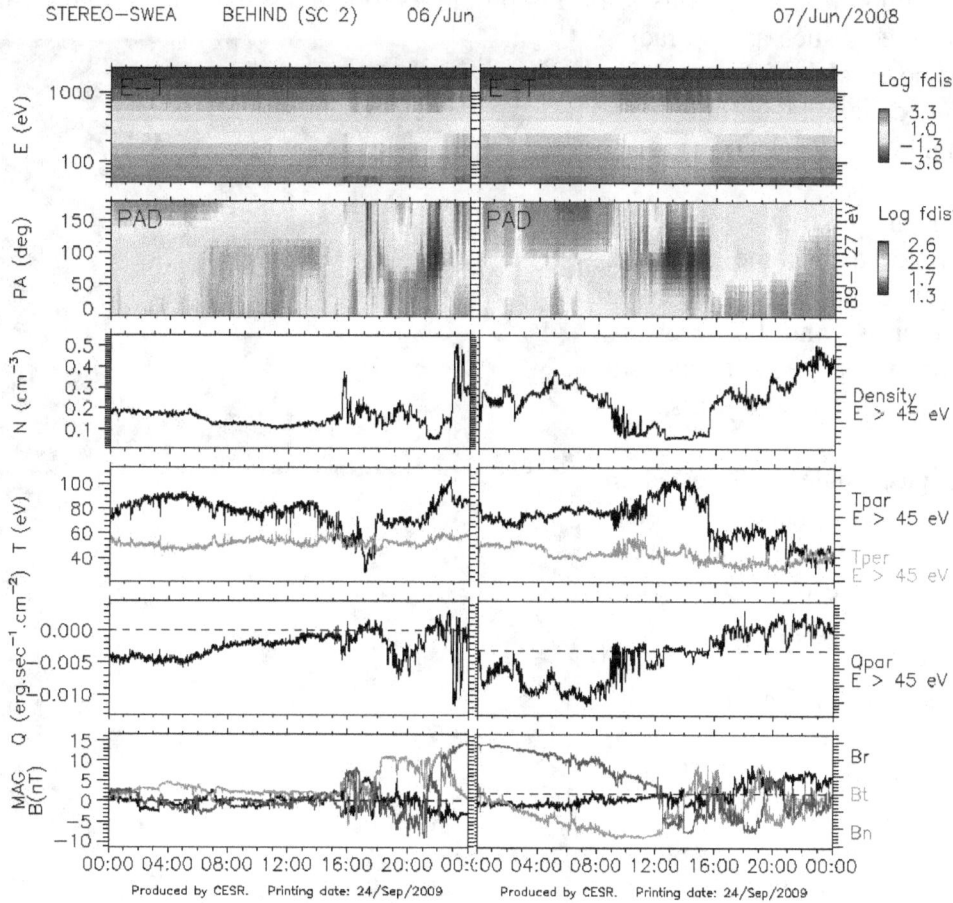

This year, as a CUA research associate, Dr Nieves-Chinchilla started analyzing STEREO IMPACT and SECCHI to develop a global model of CMEs in the interplanetary medium.

Sten Odenwald

Upon receiving his PhD in astronomy at Harvard University in 1982, Dr Odenwald worked extensively in areas related to galactic and extragalactic infrared astronomy, including work with IRAS and COBE. He also conducted a number of research programs in extragalactic astronomy using the radio telescope facilities at Kitt Peak, Greenbank and the VLA. Currently he is heavily involved in developing mathematics education resources for NASA missions in collaborations with a number of SMD education programs. He also conducts NSF-funded research on the cosmic infrared background, as well as the impact of space weather to various technology systems.

Over the years, Dr Odenwald has served on review panels for NASA Astrophysics and Earth Science, as well as NSF Astrophysics research grant review panels. He works extensively with the HSD Sun-Earth Connection Education Forum to help coordinate and develop opportunities for mission E/PO programs to increase their leveraging using a variety of media such as TV, radio, magazine article-writing, website content development and vodcasting.

During the last five years, Dr Odenwald has become interested in the historical impact of severe solar storms, and how society and the news media reacted to them over the last 200 years.

He is passionately interested in supporting US mathematics education goals, by developing programs and resources such as SpaceMath@NASA.

Dr Odenwald collaborates with many E/PO programs, such as those of Hinode, STEREO, LRO, Terra, THEMIS, and SDO to create mathematics resources that support student mathematics education using mission science data, press releases and discoveries.

Space Math @ NASA books and "weekly" problems are used by thousands of teachers across the country.

GSFC Heliophysics Science Division 2009 Science Highlights

Leon Ofman

Professor Ofman is working on observationally driven theoretical research in solar and space physics. He has been working at GSFC since 1992. He obtained his PhD at the University of Texas, Austin, TX, USA and his BSc and MSc at Tel Aviv University.

He administers several postdoctoral researchers who work on STEREO/Cor1 and Hinode/EIS at GSFC. He is a Research Associate Professor at the Department of Physics, CUA, and Visiting Associate Professor, Tel Aviv University.

He has served on NASA review panels, and as a reviewer for leading journals in the field. He refereed the PhD thesis and defense of Dr Rui Pinto, Dynamic Modeling of the Corona and the Solar Wind at the Paris Observatory, Meudon.

Professor Ofman's main research goal is to understand the transport of energy in the solar corona, heating and acceleration of the solar wind, and to interpret observations in terms of the underlying physical processes. These goals are important for the study of Sun-Earth connection, and the characterization of the heliospheric physical conditions.

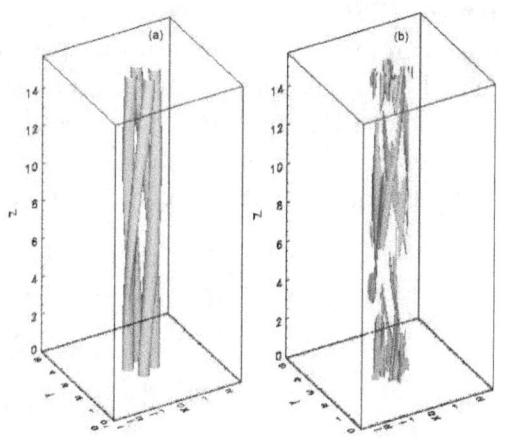

Results of 3D MHD model of oscillating twisted coronal loop. (a) Density structure. (b) Current j^2

The main methods of Professor Ofman's work are based on numerical models of the solar coronal and interplanetary plasma. He used 3D MHD modeling to study the twisting and the oscillation of solar coronal loops. The studies led to better understanding of wave energy trapping and dissipation in active regions. Professor Ofman also studied interplanetary shocks and heating of multi-ion solar wind plasma using 1D and 2D hybrid and multi-fluid models. The studies improved our understanding of solar wind heating, and of kinetic effects in shocks seen with STEREO plasma instruments.

In FY09 Professor Ofman authored and co-authored 6 papers that were published in refereed journals, one proceedings paper, and 19 meeting presentations. Seven other papers were in various stages of publication. He was awarded a new NASA grant, and continued working on several NASA grants.

Keith Ogilvie

Dr Ogilvie has been at GSFC since 1967; he was originally an NRC fellow. He received both BSc and PhD in nuclear physics at the University of Edinburgh. He is working on observations of phenomena in the interplanetary medium, especially plasma and solar related; the reduction and interpretation of plasma data; and development of plasma instrumentation. He is working on the launch and interpretation of data from DSCOVR, which is expected to be launched in 2012. He was awarded a NASA Distinguished Service Medal in 2009. He produced a paper on "Extreme Rarefactions in the Solar Wind" is in preparation. He is organizing and leading a series of talks in the "Project Scientist Forum."

Vladimir Osherovich

Dr Osherovich (CUA) has been working at Code 673 in collaboration with Dr Fainberg (Code 673) on the relation between helium abundance in the solar wind and sunspot numbers (SSN) – both as yearly average values. The solar cycle variation of He has been studied from the beginning of the space age. However, the quantitative relation between He and SSN has not previously been established. Their research has found a linear relation between He and SSN with a correlation coefficient of 0.969 after finding that an unexpected time shift of 162 days has to be taken into account (with maximum of He after solar maximum for each cycle). The data used were from NSSDC.

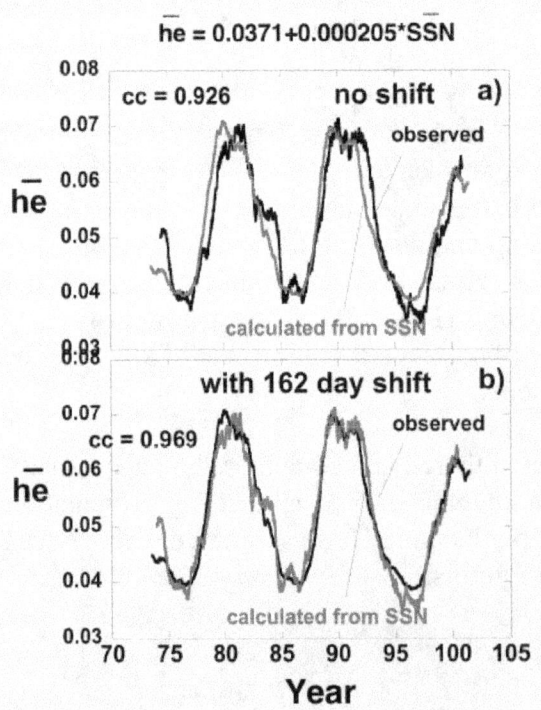

Judit Pap

Dr Pap got her PhD in astrophysics at the Eotvos University, Budapest, Hungary, in 1986. She worked in the Kiepenheuer Institute for Solar Physics in Freiburg, Germany, The World Radiation Center in Davos, Switzerland, and also in the Observatory in Nice as visiting scientist. She came to the US in 1988, and started working at University of Colorado, in January 1989. She moved to the Jet Propulsion Laboratory as a staff member in 1991, and from 1995 she was a research astronomer at the Department of Physics and Astronomy, till November 2001. Dr Pap joined GSFC in September 2000 as a visitor from UCLA, and as a contractor for GSFC. She was working for GEST/University of Maryland Baltimore County in October 2008 when she was transferred to University of Maryland College Park under the contract with CRESST.

Dr Pap is working on solar irradiance variations, their relation to solar activity, evolution of active regions and the effect of irradiance variations (both bolometric and at various wavelengths from X-ray to IR) on the Earth's atmospheric and climate system. Her research relates to both Sun-climate relations and space weather studies.

Dr Pap is a Co-I on SOHO/VIRGO and MDI, on SDO/HMI, and the French PICARD experiment. She collaborates with scientists of UCLA, JPL, National Solar Observatory at Kitt Peak, University of Washington, Yale University, San Fernando Observatory at California State University, LASP/University of Colorado, World Radiation Center at Davos, Switzerland, CNRS/CNES in France, University of Graz in Austria, Heliophysical Observatory of the Hungarian Academy in Debrecen, Hungary, and of course GSFC.

At GSFC Dr Pap is working on analyzing MDI, SOLIS, and SFO images to extract solar features to explain irradiance variations, and to establish their contribution to the observed changes at various wavelengths. She is also involved in space weather issues via studying the evolution of active regions. This research also relates to explaining the problem of the missing energy in the sunspot-related irradiance dips. She is working on irradiance modeling and irradiance variability and its effect on the Earth. She was participating in implementing a small radiometer on commercial satellite platform, and this work is still in progress in collaboration with JPL.

Dr Pap's research includes trying to understand to what extent solar variability and to what extent global solar changes contribute to irradiance research. Since the Sun is the major natural driver of the Earth's climate system, understanding its variability is important for solar physics, climate research, and society. She also works on the question of the spectral distribution of TSI variations from 200 nm to 2 micron, which is an important task since various parts of the incoming solar flux are absorbed by various layers of the Earth's atmosphere, establishing the chemical and dynamic processes there.

Dean Pesnell

During 2009 Dr Pesnell continued working towards the launch of the Solar Dynamics Observatory as Project Scientist. The observatory was moved to Florida and is preparing for launch.

FY09 was also a busy year for research and outreach. Shea Hess Webber returned as a graduate student to continue the analysis of the area of the polar coronal holes. The prediction of solar cycle 24 and the discussion of the current solar minimum continued to occupy a prominent spot in his research. *Solar Physics* accepted a paper analyzing a well-known prediction method, and talks describing solar minimum were given at several venues. Both efforts are related to the long-term manifestations of the solar dynamo in observations. Peter Williams was renewed as a postdoc to study observable changes in the convective velocity spectrum. Dr Pesnell was also a mentor for Nishu Karna (an undergraduate student from Nepal) under the Heliophysics Student Intern Program during the summer of 2009. Collaborative work on the interpretation of SABER data (especially the water vapor measurements) continued with Dr Feofilov, Dr Kutepov, and Dr Goldberg (all 674).

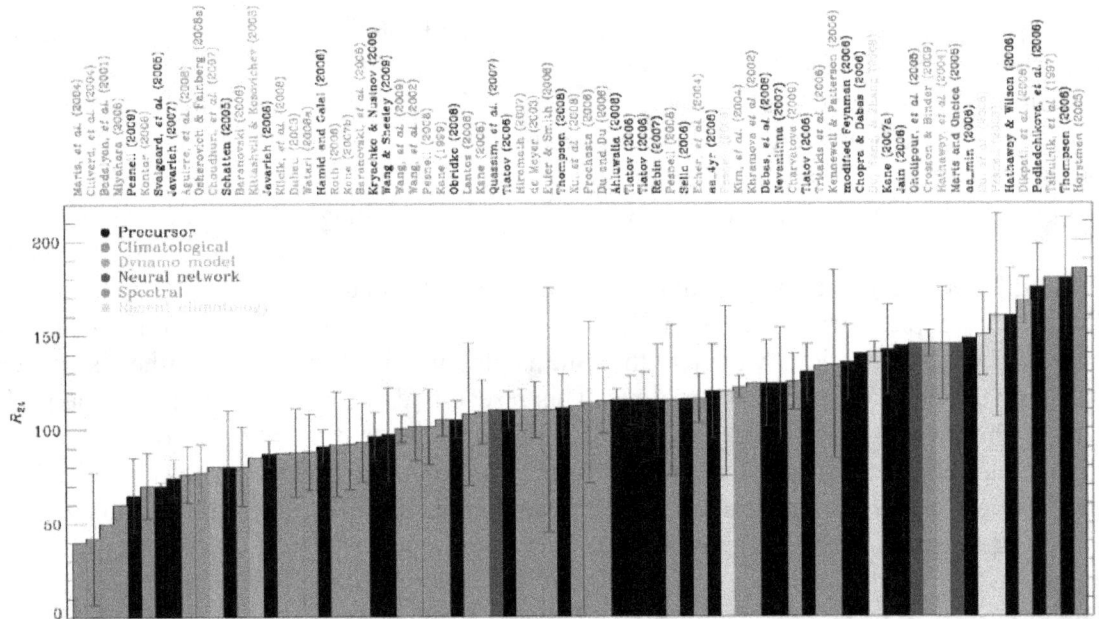

Seventy-five predictions of the amplitude of solar cycle 24, plotted in order of increasing predicted maximum for Cycle 24. The prediction categories are color-coded as described in the legend. The predictions in the earlier version had a distribution almost centered on the long-term average amplitude while the current predictions are now have an excess of low-amplitude values.

GSFC Heliophysics Science Division 2009 Science Highlights

Robert Pfaff

Joining the GSFC in 1985, Dr Robert F. Pfaff, Jr. is a Space Scientist in the Space Weather Laboratory in HSD. Dr Pfaff received his PhD from Cornell University in 1985.

Dr Pfaff has served as Project Scientist for the FAST satellite since 1990, Project Scientist for the Sounding Rocket program since 1994, and NASA Project Scientist for the Air Force C/NOFS satellite since 2003.

As Study Scientist, Dr Pfaff led the Ionospheric Mappers planning for NASA's Living with a Star (LWS) program, and later served on the Geospace Mission Definition Team for LWS. He has served on the Geospace Electrodynamics Connections (GEC) definition team, as well as on the Magnetospheric Management Operations Working Group (MOWG) at NASA HQ.

Dr Pfaff's experimental electric field research has involved advancing electric field double probe research techniques, including the fabrication of low-noise detectors, burst memories and on-board signal processing such as for the C/NOFS satellite and numerous sounding rockets, new low-cost boom systems, and miniature spherical payloads with dual sphere, vector double probes, Langmuir probes, and magnetometers.

Dr Pfaff is the PI of the Vector Electric Field Investigation (VEFI) on the C/NOFS Air Force satellite, launched in 2008, that includes a vector DC and AC electric field detector, magnetometer, Langmuir probe, lightning detector (developed with the University of Washington) and burst memory. Within NASA's sounding rocket program, Dr Pfaff has provided electric field, magnetic field, and plasma density measurements for 42 rocket missions to date. All of the hardware has been delivered on time and on budget and all experiments have worked exceptionally well.

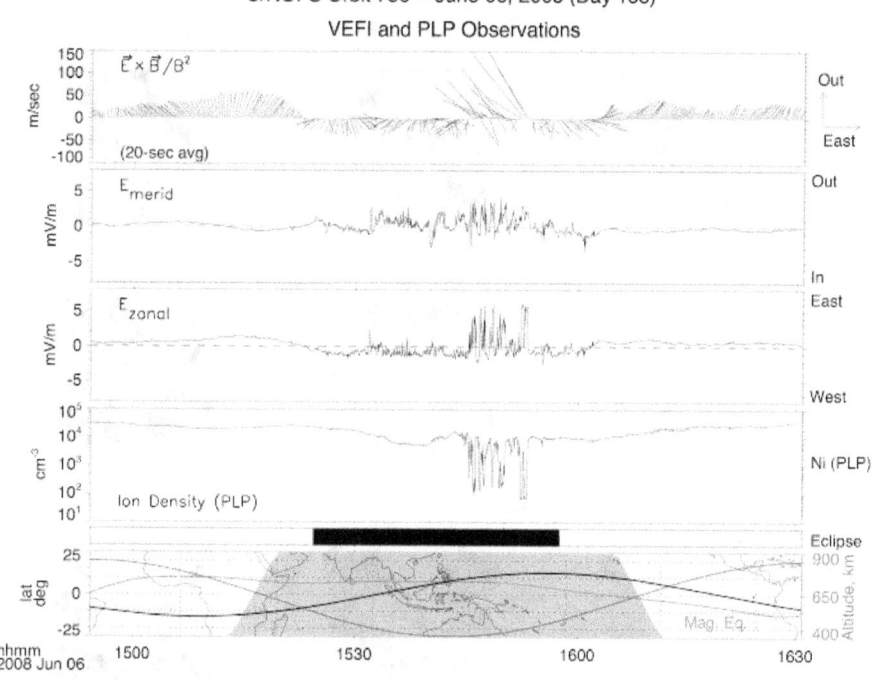

C/NOFS electric field and density data showing spread-F perturbations

Antti Pulkkinen

Dr Pulkkinen is currently Associate Research Scientist at the University of Maryland, Baltimore County, Goddard Earth Sciences and Technology Center operated at GSFC. Dr Pulkkinen received his PhD in theoretical physics from the University of Helsinki, Finland in 2003. Subsequently he joined the nonlinear dynamics group at NASA GSFC to carry out his postdoctoral research in 2004-2006. Since 2006 Dr Pulkkinen has been working in the Community Coordinated Modeling Center (CCMC) operated at NASA/GFSC. Recently Dr Pulkkinen has been working on utilizing the established modeling capabilities in semi-operational nowcasting and forecasting of space weather. Dr Pulkkinen is the main or co-author of about 50 peer-reviewed scientific articles.

In his role as part of CCMC staff, Dr Pulkkinen supports various community functions that pertain to CCMC's activities. Most significant of these was the modeling and model validation support for the most recent Geospace Environment Modeling Challenge. Dr Pulkkinen also reviewed numerous manuscripts submitted to journals such as *Journal of Geophysical Research, Geophysical Research Letters,* and *Philosophical Transactions A*.

The main objective of Dr Pulkkinen's current scientific activities is improved modeling and forecasting of space weather. Over the past year he and his colleagues developed new theoretical models and techniques applicable to modeling of geomagnetic field fluctuations at mid- and low latitudes. They also developed a new technique for automatic 3D characterization of CMEs from white-light coronagraph images. Dr Pulkkinen's more general (public) physics work included theoretical investigation on the usage of space weather-related magnetic field fluctuations as a potential source of electric power.

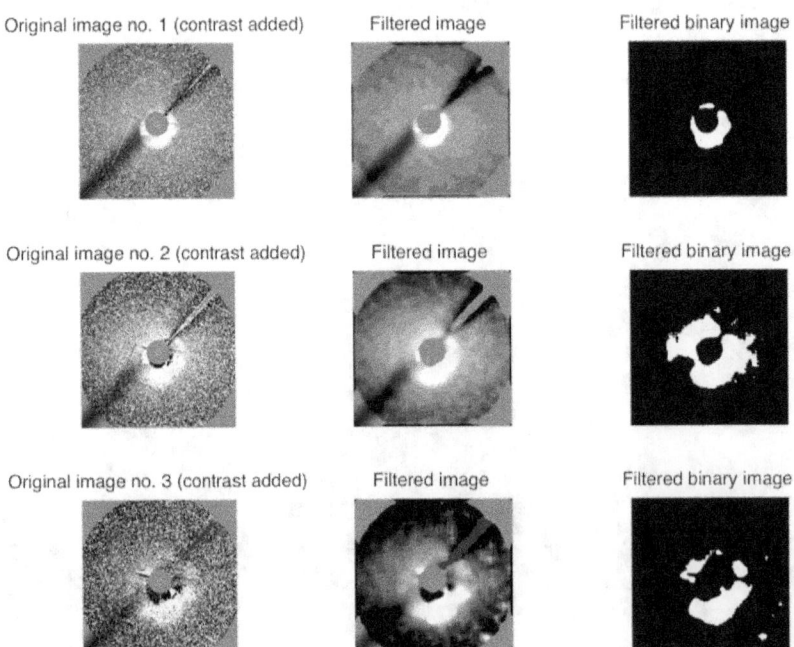

Image processing steps to extract "CME mass" from sequence of white-light coronagraph images. Adopted from Pulkkinen et al.

GSFC Heliophysics Science Division 2009 Science Highlights

Douglas Rabin

Dr Rabin has been at GSFC for nine years; he is the chief of the Solar Physics Laboratory. He received his AB from Harvard College and PhD from the California Institute of Technology. His research is focused on the structure and dynamics of the solar corona.

Dr Rabin is the Deputy Project Scientist of the Solar Radiation and Climate Experiment (SORCE) and Instrument Scientist of the Reflected Solar Instrument on the Climate Absolute Radiance and Refractivity Observatory (CLARREO) which is in development. He serves on the Users Committee of the National Solar Observatory.

Dr Rabin is PI of the Extreme Ultraviolet Normal Incidence Spectrograph (EUNIS) sounding rocket experiment, which is scheduled to fly for the third time in late 2010, following two successful flights in 2006 and 2007. The unprecedented sensitivity of EUNIS allows it to probe intensity and velocity variations in solar plasmas on timescales as short as 10 s. Transient and small-scale events are increasingly recognized as an important source of outer atmospheric heating. The figure shows the detection in EUNIS-06 data of a 26-s velocity oscillation in He II, interpreted as a fast MHD wave.

The EUNIS Team received a 2009 NASA Group Achievement Award; Dr Rabin was awarded the NASA Exceptional Service Medal.

Dr Rabin is involved with instrument and technology development in several areas, including heliospheric imaging and integral field spectroscopy.

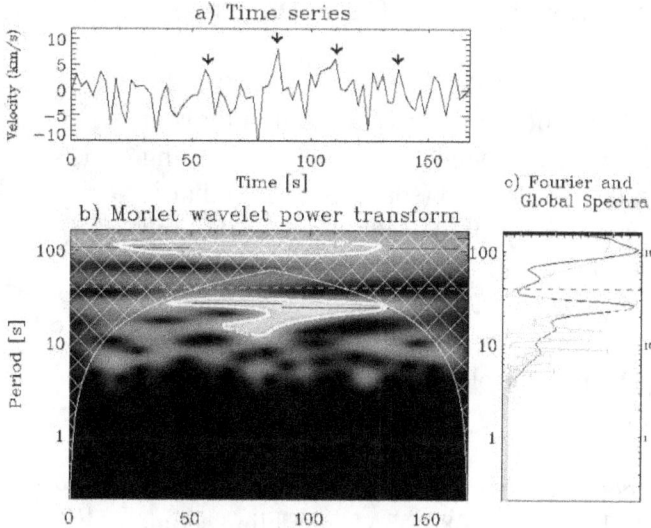

Wavelet diagram showing the detection of a 26±4 s periodic velocity oscillation in He II. The original He II velocity time series is plotted in (a) with arrows indicating the peaks of the oscillation. The wavelet power transform, along with locations where detected power is at, or above, the 99% confidence level, are contained within the contours in (b). Plot (c) shows the summation of the wavelet power transform over time (solid line) and the fast Fourier power spectrum (plus signs) over time, plotted as a function of period. Both methods identify a significant 26-s oscillation. The cone of influence (COI), the cross-hatched area in the plot, defines an area in the wavelet diagram where edge effects become important, and as such any frequencies inside the COI are disregarded.

Lutz Rastaetter

Dr Lutz Rastätter is a space physicist working at GSFC since 1997. He graduated with a PhD in theoretical astrophysics from the Ruhr University of Bochum, Germany, in 1997 and joined Michael Hesse as a post-doctoral research associate working on magnetotail plasma sheet thinning and reconnection. Since the beginning of the Community-Coordinated Modeling Center (CCMC) in 2000 he has been applying his experience in magnetohydrodynamic modeling and magnetospheric physics to implement, test, and run various models and implement the visualization that forms the backbone of online services at CCMC. Since 2004 he has been employed by CUA. Working with the CCMC his primary research areas are the validation of numerical research models that cover the solar corona, heliosphere, magnetosphere, ionosphere/thermosphere, scientific visualization of model results and new features in the online access to CCMC simulation runs.

In FY09, Dr Rastätter continued the development of online visualization tools at the CCMC, which now include plotting in the 4D domain (space and time) that is provided by model simulation runs and the ability to interactively generate and visualize modeled data at any satellite trajectory. The time series utility enables 1D timeline plots and 2D Keogram-type plots.

Recent model validation study work includes the 2008 GEM modeling challenge, in which multiple magnetospheric models were run for four specific events and compared to each other and observation data, for example magnetic field strength at GOES-10 in geosynchronous orbit. Improvements between runs of a particular model or differences between models can be analyzed.

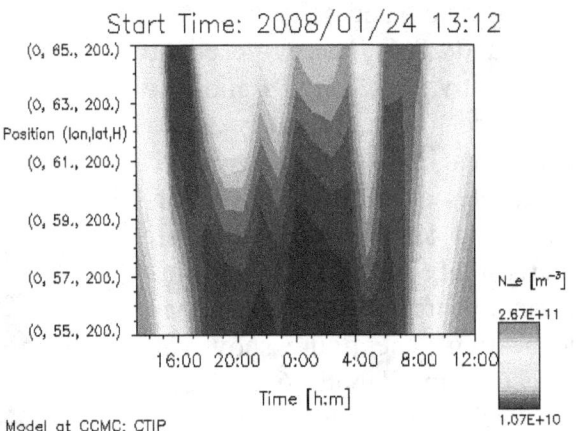

Keogram-style plot obtained from model outputs. Here, electron density from the CTIP ionosphere/thermosphere model on a line at constant altitude is followed over time.

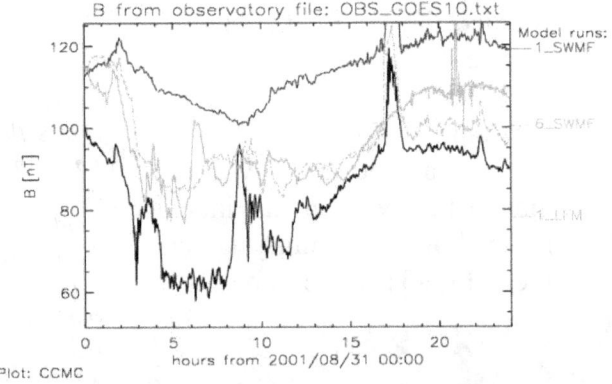

Model run evaluation: Two runs by the SWMF model and one by the LFM model are shown in comparison with the magnetic field magnitude from GOES-10.

Nelson Leslie Reginald

Dr Reginald is an astrophysicist employed at GSFC through CUA. At GSFC, Dr Reginald is engaged in designing experimental techniques to map the thermal electron temperature and its bulk flow speed in the low solar corona. So far these experiments were conducted in conjunction with total solar eclipses. In addition, his work involves in theoretical modeling to interpret the data. Dr Reginald has been at GSFC since January 1998 with the first three years as a graduate student conducting dissertation research work with Dr Joseph Davila. Dr Reginald earned his PhD and MS in Physics from the University of Delaware in 2001 and 1997, respectively, and his BSc (Honors) in physics and pure mathematics from University of Peradeniya, Sri Lanka in 1993. In addition Dr Reginald earned a degree in management accountancy from the Chartered Institute of Management Accountants, UK, in 1990.

Dr Reginald conducts lectures and demonstrations to students enrolled in public schools in Washington DC who visit GSFC for a week of activities under the *Students Enthusiastic About Science and Math* (SUNBEAMS) program.

Dr Reginald's main research focus is on developing experimental techniques to produce simultaneous and global maps of thermal-electron temperature and its bulk flow speed in the solar corona. He compares the electron based physics with similar parameters that have been measured using ion-based physics, to check whether they complement each other. This work culminated in a map of the thermal-electron temperature in the low corona and was accepted for publication in *Solar Physics* on 2009 September 17. His research work also involves theoretical modeling of Thomson scattering off the Earth's magnetosphere electrons as well as a study of the feasibility of extracting this information from the background signals comprising the Thomson scattered component from the solar wind electrons and zodiacal light that would contaminate such measurements.

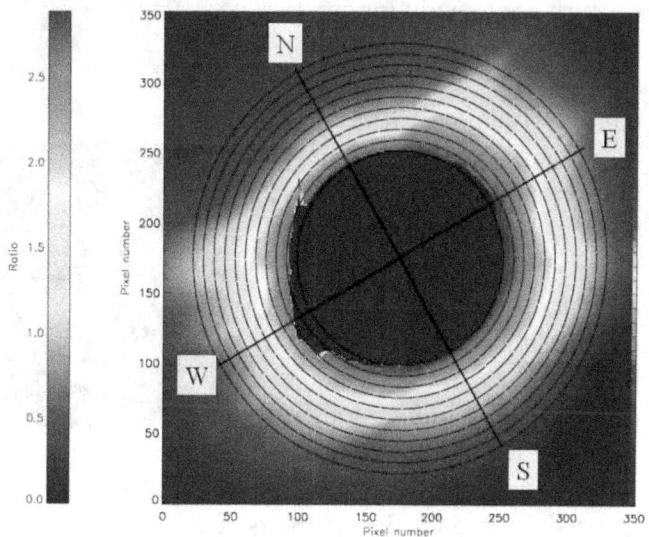

The ratio between two coronal images, taken during the total solar eclipse of 2006 March 29 in Libya, through two filters centered at 4100 and 3850 Å with a bandpass of approximately 40 Å. Through theoretical modeling the ratios can be related to the thermal electron temperature in the corona on a pixel by pixel basis.

D. Aaron Roberts

Dr Roberts has been a physicist with the Heliospheric Physics Laboratory since 1989. He is actively involved in research on interplanetary turbulence and the modeling of nonlinear processes. His observational and MHD simulation efforts in these areas have shown that such processes are necessary to explain the observations and trace the evolution of the interplanetary turbulence from 0.3 to 20 AU. He is the author of over 100 publications in this and other areas. He received both SB and PhD degrees from MIT, the latter in 1983.

As the Project Scientist for the Heliophysics Data and Model Consortium (HDMC), Dr Roberts works with other NASA HP Data Centers to make data easily accessible and usable with Virtual Observatories and other means. He continues his work with NASA HQ on issues such as the Heliophysics Science Data Management Policy. He led the effort on a revision of this policy that was well received by the Heliophysics Subcommittee to the NASA Advisory Council. The HDMC work involves other agencies, such as NSF, as well as foreign partners, to provide worldwide integration of HP data systems. Dr Roberts wrote a NASA Senior Review Proposal for HDMC in July 2009. In addition, he has served on NASA proposal review panels and as referee for various journals (*JGR, ApJ*, etc.).

Dr Roberts continues to work on turbulent evolution in the solar wind, using observations and MHD simulations, including issues such as how the fluctuations respond to the turning of the interplanetary field. Recently submitted work has given very strong arguments, including the lack of bending of rays on open field lines in images (such as that at the right [modified from M. Druckmüller], that waves and turbulence do not accelerate the high-speed solar wind. This and other work will be important for the design of near-Sun missions such as Solar Orbiter and SP+. Roberts will also be the science lead for a proposal for a plasma instrument suite on SP+. Work on the Heliophysics Data Environment and the Data Policy resulted in a NASA Special Act or Service Award.

Douglas Rowland

Douglas Rowland has worked as a space physicist in HSD since November 2003, first as a National Research Council postdoc and as a civil servant since 2005. Dr Rowland works in the electric fields group, developing instrumentation to measure the electric fields responsible for charged-particle transport and energization, as well as leading projects dedicated to studying energetic particle acceleration in regimes ranging from auroral zone ion outflow to thunderstorm-driven electron acceleration.

Dr Rowland's research continues to focus on topics relating to magnetosphere-ionosphere coupling, and energetic particle acceleration processes. This past year he delivered instrumentation for the ACES sounding rocket and has worked to produce electron density altitude profiles from the pair of ACES sounding rockets, to better determine the ionospheric conductivity and current closure in the neighborhood of an auroral arc. In addition, he led a joint Code 600 and Code 500 effort to deliver the Plasma Impedance Spectrum Analyzer (PISA) instrument to the MSFC FASTSat spacecraft. PISA will measure ionospheric electron density and temperature when it is launched on FASTSat in May 2010. The fabrication and test of the instrument took place in <9 months.

Dr Rowland is the PI for the upcoming VISIONS sounding rocket (launch in 2012). This experiment uses GSFC expertise in electric fields and ENA imaging to study the mechanisms by which thermal ions gain more than two orders of magnitude in energy and achieve escape velocity, allowing them to reach high altitudes and populate the magnetosphere.

The PISA instrument, delivered to MSFC Aug 2009.

Dr Rowland continues to develop the NSF Firefly CubeSat, for which he is the PI. Firefly promises to provide critical measurements about the processes by which MeV electrons are accelerated by lightning discharges. This year, the hardware design has matured rapidly following the Mission Requirements Review in January and Design Review in June. The subsystems are undergoing final design and engineering unit testing, with delivery of all flight subsystems to the spacecraft integrator in February 2010. Launch date is TBD (2010 expected).

During FY09 Dr Rowland served a 6-month part-time detail at NASA HQ, working with the Living With a Star Targeted Research and Technology program. Dr Rowland is active in mentoring activities, supervising one NASA postdoctoral fellow in 2009 and mentoring eight undergraduate students at GSFC as they learned about spacecraft design and engineering. In 2009 Dr Rowland served on a NASA review panel, as well as the Sounding Rocket Working Group. He was recently named to the Geospace Management Operations Working Group (G/MOWG) for 2010.

Julia Saba

Dr Saba, a senior staff physicist of the Lockheed Martin Advanced Technology Center Solar & Astrophysics Laboratory in Palo Alto, California, has been a long-term member of the GSFC solar group. Originally an English major, she brings strong editing skills to the group and was a scientific editor for the 2008 HSD Scientific Highlights. Her physics training was initially in non-solar astrophysics. She received three degrees in physics from the University of Maryland in College Park, obtaining her PhD in 1983 based on research performed in the GSFC Laboratory for High-Enegy Astrophysics, analyzing spectroscopic measurements of coronae of accretion disks around the compact member of binary X-ray sources. Between degrees, she worked in the lab at GSFC on prototypes of the satellite detectors that aquired the data she analyzed for her thesis. Beginning her solar career with science planning and operations for the X-ray Polychromator shortly after the on-orbit repair of the Solar Maximum Mission in 1984, Dr Saba has helped with science planning and operations for the Michelson Doppler Imager on SOHO since 1995.

On the research side, her interests include active region and flare dynamics and coronal composition. Most recently she has been working with Dr Strong to uncover observational clues to understanding and potentially predicting the solar activity cycle. These clues include a recurring pattern in the evolution of the spatial distribution of magnetic flux at the photosphere, the sudden onset to both cycles 22 and 23 (most noticeable as an order-of-magnitude jump in the soft X-ray emission baseline and a widespread outburst of magnetic flux, but also accompanied by a sea change in other activity properties), and the possibility that the solar cycle is essentially composed of a series of such bursts, and the observed asymmetries in the magnetic flux in the two hemispheres. Recent work with Dr Strong suggests that the long-overdue outburst of solar cycle 24 may reflect in part the persistence of old-cycle strong-field (B > 50 G) magnetic flux in the southern hemisphere. Dr Saba has presented various aspects of this research at three meetings this year.

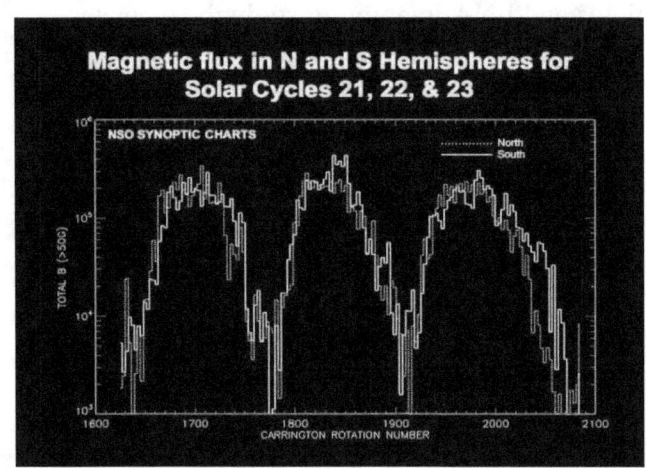

Total strong-field (B>50 G) magnetic flux in latitude bands 60S to 60N from NSO synoptic magnetic charts. The plots show a distinct persistence of flux in the South relative to the North during the cycle 23 decay, not seen for cycles 21 and 22. (Note semi-logarithmic ordinate scale; multiply by $3 \times 10^{22} cm^2$ to convert to Maxwells.)

Fouad Sahraoui

Dr Sahraoui is a space plasma physicist, who joined Code 673 in October 2007 as visitor from the Centre National de Recherche Scientifique (CNRS/LPP, France).

He has been doing research at GSFC to understand plasma turbulence in the magnetosphere and in the solar wind, both theoretically and observationally. Working with Dr M. Goldstein on small-scale solar wind turbulence using the Cluster data, they discovered for the first time the process by which the solar wind turbulence is dissipated at electron scale (~1 km). This process may explain other observations of heating and particle acceleration in astrophysics (e.g., solar corona).

Dr Sahraoui gave several presentations and seminars (e.g., UMd) and three invited talks at the EGU meeting in Vienna (Austria, April 2009), at the conference on "Modern Challenges in Nonlinear Plasma Physics" in Greece (June 2009), and at the PNST meeting (Paris, France, 2009). He is also invited to the 2009 AGU fall meeting in San Francisco.

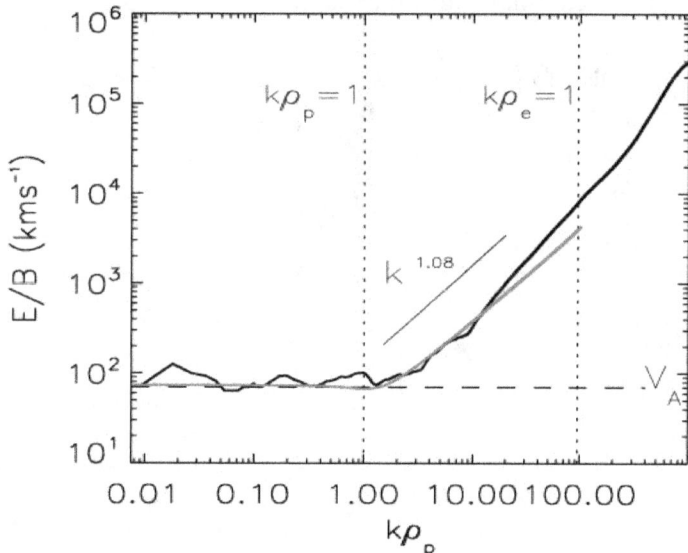

Identification on the nature of the energy cascade and dissipation at electron scale in the solar wind using the Cluster data (black is observations, red is the Kinetic Alfvén Wave theory).

Menelaos Sarantos

Dr Sarantos, previously a NASA Postdoctoral Program Fellow, joined UMBC/GEST this year as an Assistant Research Scientist. His research focuses on the interaction of the solar wind with Mercury and the Moon. He is analyzing plasma, magnetometer and UV spectrometry data obtained by MESSENGER during its recent Mercury flybys, and investigating the detectability limits of several lunar exospheric constituents that are to be targets of the proposed LADEE mission to the Moon.

This past year, he was the primary author of three papers that investigated the following questions:

- How is the solar-wind-magnetosphere interaction at Mercury modified by low-mach-number flows which are common during CME events?
- How do models of Mercury's exosphere-magnetosphere-surface system and of the associated planetary ion transport compare to the data obtained during the MESSENGER flybys?
- What are the relative roles of the solar wind, ultraviolet photons, and micrometeorites in producing the observed lunar sodium exosphere?

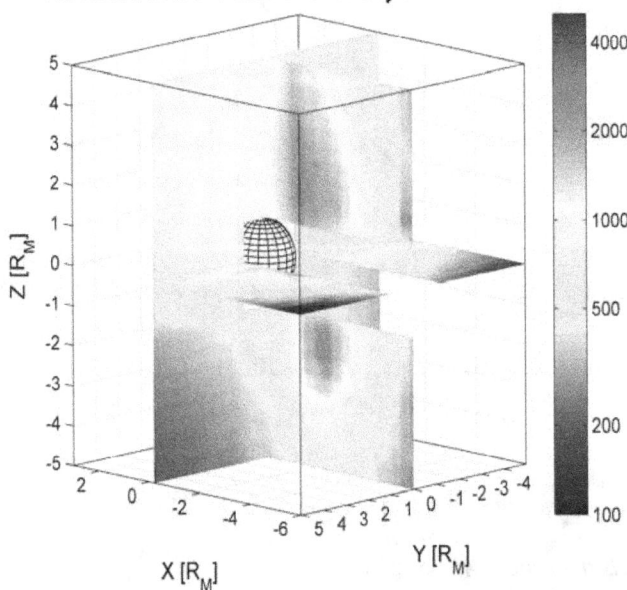

Tomographic reconstruction of Mercury's sodium exosphere [in Rayleighs] from line-of-sight spectroscopic measurements at the time of MESSENGER's second flyby. [View from the tail] Polar enhancements indicate the impact of the solar wind onto Mercury's surface.

Dr Sarantos is a Co-I of the DREAM (Dynamic Response of the Environment At the Moon) Team, a node of the newly established NASA Lunar Science Institute. This assignment entails a multi-institutional collaboration of GSFC, UC Berkeley, JHU/APL and others. He is also a Co-I of the ESA-JAXA-NASA Bepi-Colombo mission to Mercury. Since February 2009, he has been serving as the HSD Seminar Organizer.

GSFC Heliophysics Science Division 2009 Science Highlights

Chris St. Cyr

Dr St. Cyr has been a solar and space weather researcher, an operations scientist, and a Project Scientist at GSFC since 1984. Over the years he has worked as a .edu and .com contractor, and he has been a civil servant since 2002. He has a BS in astrophysics from the University of Oklahoma and a PhD in astronomy from the University of Florida.

Presently Dr St. Cyr serves as Senior Project Scientist for heliophysics missions in development, and as US Project Scientist for Solar Orbiter, a joint ESA/NASA mission to explore the Sun and its connection to the environment of the inner solar system. Over the past year he has worked with NASA Headquarters Office of the Chief Engineer to inventory space weather requirements (especially those pertaining to radiation) across the Agency. He was Co-Chair for the recently completed the triennial Community Roadmap strategic planning activity for the Heliophysics Division. He frequently serves on review panels and briefs advisory groups both internal and external to NASA.

Dr St. Cyr's research interests include the initiation and propagation of solar CMEs, testing new instrumental techniques at total solar eclipses, solar cycle studies of the Sun's white-light corona, and the quantification of economic impacts of space weather in electric power grids. He has recently published a manuscript describing the use of the STEREO spacecraft payload as a detector of micrometer interplanetary dust particles.

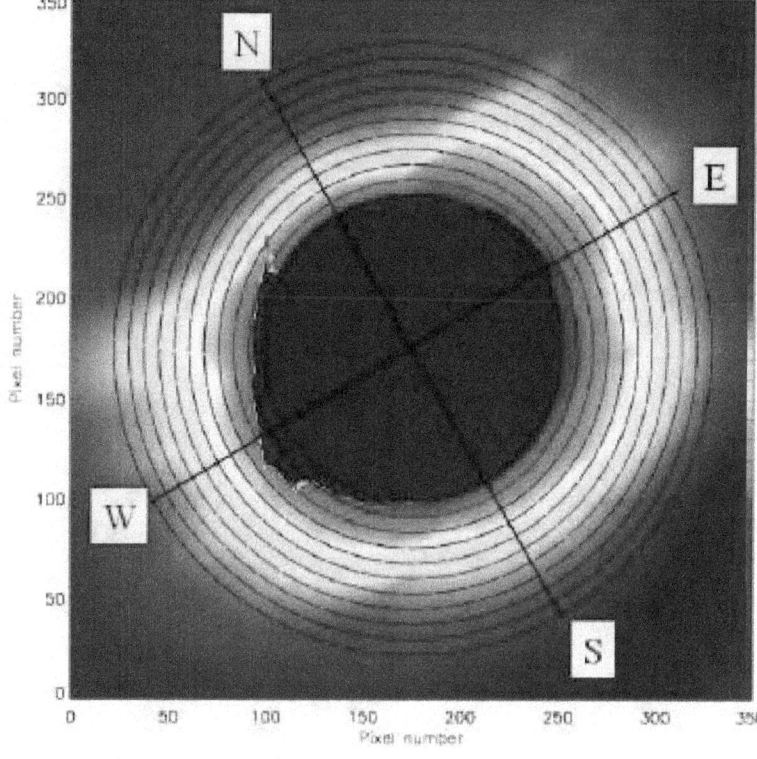

Ratio of two filtergrams acquired at the 2006 total solar eclipse in Libya (from Reginald et al., 2009). These data were used to derive electron temperatures globally in the corona.

Richard Schwartz

Dr Schwartz has a PhD in physics from UCB where he flew the first high-resolution germanium detectors to observe solar flare hard X-rays in 1980.

At GSFC he works on RHESSI data analysis, as well as the analysis of every data set from as many solar X-ray observing missions as possible and to facilitate that analysis for everyone in the international solar community.

Dr Schwartz has been at GSFC from 1986 after a post-doc in Bud Jacobson's HEAO-3 group at JPL. As an undergrad he attended CMU.

For RHESSI his work centers around making certain that the software models the real instrument as closely as possible and that the scientists using the data can work in an efficient and intuitive fashion. While much of his work appears to be coding, he is still intensely interested in the detector physics, the details of the instrument/telemetry interface, and the nature of the energetic particles we observe from the Sun

Some of the other instruments that he has helped bring to the community from the past, present and future: SMM/HXRBS, GRS, Yohkoh/HXT, NEAR PIN, MESSENGER PIN, CGRO/BATSE, FERMI/LAT, GBM, KORONAS/SPHINX and RHESSI.

He continues to be interested in future missions and is hoping to assist at some level with the focusing optics X-ray systems and the advanced γ-ray imaging spectrometers.

The most visible work this past year was the capture of a 3-year FERMI award to bring the solar data to the community. The RHESSI team has worked diligently to bring the UV-Smooth imaging method to the software suite.

This year he served on a review panel downtown.

GSFC Heliophysics Science Division 2009 Science Highlights

Ilgin Seker

Dr Seker is a NASA Postdoctoral Program Fellow in the Geospace Physics Laboratory (673), working with Dr Fung. He started the NASA Postdoctoral Program in August 2009. He completed his major in electrical engineering and minor in physics at the Middle East Technical University in Ankara, Turkey, in 2003. He obtained his MS and PhD degrees in electrical engineering from PSU in 2006 and 2009, respectively.

Dr Seker is responsible for the PSU All-sky Imager at Arecibo Observatory in Puerto Rico. Currently, he is collaborating with researchers at the Pennsylvania State University and Cornell University.

Dr Seker is working with Dr Fung on the irregularities in the night-time mid-latitude F-region of the ionosphere. His current focus is on the relationship between the geomagnetic state parameters and the occurrence of medium-scale traveling ionospheric disturbances (MSTIDS) and spread-F plumes which are found to occur during low and high geomagnetic activity, respectively. For this study, he uses the PSU All-sky Imager data and the geomagnetic state parameters obtained from the Magnetospheric State Query System (MSQS) created by Dr Fung. The results will be useful for determining the conditions under which these irregularities, which adversely affect satellite communications, are triggered, which is currently unknown and is important as it might help forecast these events.

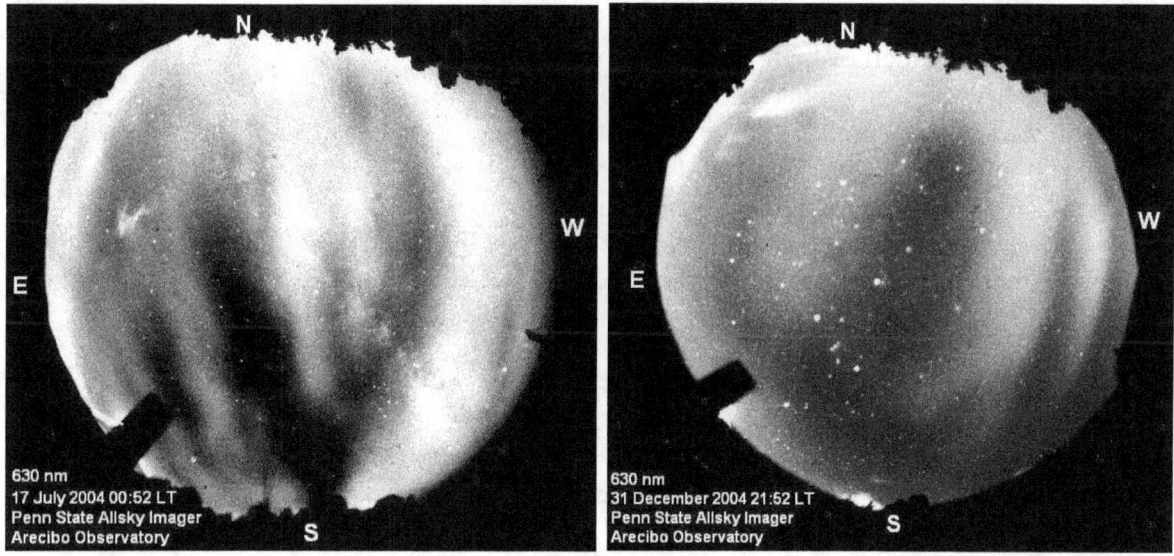

Two all-sky images from Arecibo Observatory showing an equatorial spread-F plume (left panel) reaching mid-latitudes during a geomagnetic storm and MSTIDS bands (right panel) that usually occur when the geomagnetic activity is low.

Albert Shih

Dr Shih is a NASA Postdoctoral Program fellow who started work at GSFC in September 2009. He has a BS in physics and mathematics from the California Institute of Technology and a PhD in physics from the University of California, Berkeley.

He studies the acceleration of ions by solar flares using observable signatures, primarily γ rays. Large amounts of stored magnetic energy can be transiently released in flares, and a significant fraction of this energy (tens of percent) goes into energetic electrons and ions. Understanding how solar flares efficiently accelerate particles to high energies improves our understanding of related energy-release processes occurring throughout the universe.

The main observable signature of accelerated ions is γ-ray line emission, and Dr Shih analyzes data collected by RHESSI. RHESSI observations show that ions and electrons are roughly proportionally accelerated in flares above certain energies, which suggests a common acceleration mechanism. Also, the temporal variability of γ-ray lines can be used to probe changes in ambient abundances as a flare progresses.

Dr Shih is a Co-I on the NASA Low-Cost-Access-to-Space balloon mission called the Gamma-Ray Imager/Polarimeter for Solar flares (GRIPS). GRIPS is being developed and built primarily at the University of California, Berkeley, and will use γ-ray observations to address unanswered questions about particle acceleration in flares. GRIPS combines a new germanium detector technology with a novel γ-ray imaging design to provide a near-optimal combination of high-resolution imaging, spectroscopy, and polarimetry of photons from ~20 keV to >~10 MeV, with an unparalleled γ-ray angular resolution of 12.5 arcseconds. In addition to the direct science return from balloon flights, the technologies on GRIPS will be proven for future space-based instruments.

He is also actively participates in designing and developing future solar instruments to study ion acceleration. Advancements can be made in observing γ–rays, and accelerated ions can also produce ENAs which can provide a new window into ion acceleration.

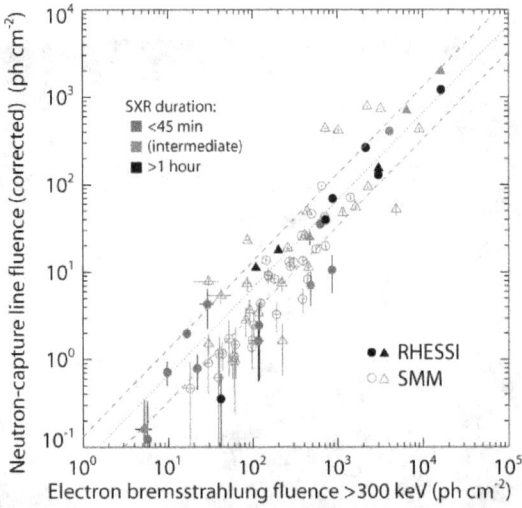

The correlation between ion-associated and electron-associated emissions indicates that >0.3 MeV electrons and >30 MeV protons are proportionally accelerated in solar flares over >3 orders of magnitude in fluence.

GSFC Heliophysics Science Division 2009 Science Highlights

Dave Sibeck

Dr Sibeck is a magnetospheric physicist in the Space Weather Laboratory (Code 674) at GSFC. With the exception of a year when he was detailed to NASA/HQ, he has worked at GSFC since 2002. He was employed at nearby JHU/APL from 1985 to 2002 after receiving a PhD from UCLA in 1984.

As project scientist, Dr Sibeck leads efforts to publicize science discoveries from the THEMIS mission. As mission scientist, he advises HQ, GSFC, and JHU/APL on science requirements for RBSP. He became president-elect of the steering committee for the NSF's Geospace Environment Modeling program in 2009. He works closely with researchers at Solana Scientific, Sciberquest, the University of Maryland, Technical Universities in Braunschweig and Istanbul, University of St. Petersburg, and JHU/APL.

Dr Sibeck serves on the advisory committee for ESA's Cluster Active Archive. He is the heliospherics editor for *Advances in Space Physics* and the corresponding editor for heliophysics at *EOS*. He frequently fields calls to GSFC from the press, radio, and television on heliophysics. Each summer he lectures visiting high school teachers on the aurora; and each fall AGU he presents a lively lecture on the aurora at the NASA booth. He runs the dayside science study group at GSFC, which meets on Mondays at noon.

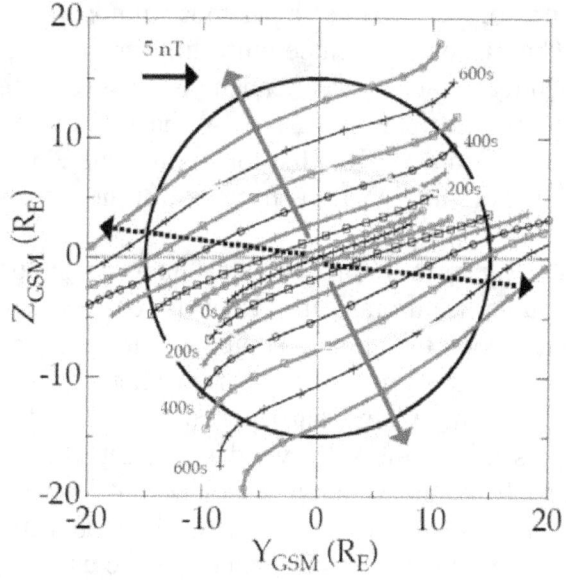

Modeling results: FTEs move northward and dawnward or southward and duskward away from a tilted subsolar component reconnection line

Dr Sibeck is presently studying the structure of the magnetosheath and the characteristics of flux transfer events, two phenomena which may control the nature of the solar wind-magnetosphere interaction. He works closely with Mike Collier on instrument proposals to image the magnetosheath.

He submitted 12 proposals in 2009 and had 13 papers published.

John B. Sigwarth

Dr Sigwarth serves currently as the chief technologist for science in HSD where he leads the efforts for technology development in the division. He serves also as the project scientist for the recently retired Polar spacecraft mission and the deputy project scientist for the TIMED and AIM spacecraft missions. He has been employed at GSFC for over 5 years. Dr Sigwarth received his BS, MS, and PhD degrees from The University of Iowa in 1983, 1988, and 1989, respectively.

This year Dr Sigwarth completed a fourth and final year as member on the Geospace MOWG for NASA Headquarters Heliophysics Science Division. He is a member of the American Geophysical Union.

Dr Sigwarth is the PI leading the effort for a new camera to remotely sense the temperature of the Earth's upper-most atmosphere or "thermosphere". This new camera, the Thermospheric Temperature Imager (TTI), is being developed in collaboration with the United States Naval Academy. Because of this collaboration, the TTI has been included as one of the experiments on the Department of Defense (DoD) Space Experiment Review Board (SERB) list for fiscal years 2007 – 2009. As a result, the TTI will fly in May 2010 on the FASTSat spacecraft of the DoD's Space Test Program.

Dr Sigwarth's other research interests include the conjugate nature of the northern and southern auroras; the impacts of geomagnetic storms on the atomic oxygen and molecular nitrogen composition of the thermosphere; the driving of the aurora by sharp increases in the dynamic pressure in the solar wind impacting the magnetosphere; the coupling efficiency of energy from the solar wind to the magnetosphere as a function of the state of the solar wind; and the use of global auroral images to retrieve the energy Deposition by auroral particles precipitating into the ionosphere.

The Thermospheric Temperature Imager to be launched in May, 2010 on the FAST-Sat spacecraft of the Department of Defense Space Test Program.

Dr Sigwarth has authored or co-authored 136 presentations at scientific conferences and 74 papers for publication. In honor of the 50th anniversary of Alaska Statehood, Dr Sigwarth presented "Determining from Earth Orbit the Causes of the Aurora" at the National Archives in Washington, DC on January 29, 2009.

GSFC Heliophysics Science Division 2009 Science Highlights

Edward C. Sittler, Jr.

Dr Sittler is an astrophysicist in Code 673. He continues a broad spectrum of interests including heliospheric physics, planetary magnetospheres, and astrobiology. He received a BS degree in Physics (magna cum laude) at Hofstra University in 1972 and PhD in Physics at MIT in 1978. He served on several NASA review panels, and the TSSM science definition panel. He was a member of the SP+ panel but that has been disbanded in preparation for its AO.

He is (1) a Co-I on the Cassini Plasma Spectrometer (CAPS) team; (2) PI on two Cassini Data Analysis Program (CDAP)-funded efforts to analyze CAPS plasma data within Saturn's inner magnetosphere and its moon Enceladus and to analyze CAPS plasma data at Titan within the context of a hybrid model of Titan's interaction with Saturn's magnetosphere; (3) PI of an LWS TR&T funded effort to develop a semi-empirical MHD model of the solar corona and solar wind; (4) PI on a funded instrument development effort under the astrobiology instrument development (ASTID) program to build a 3D ion mass spectrometer (IMS) and 3D Ion Neutral Mass Spectrometer (INMS) for a mission to Europa, in collaboration with Code 699, headed by Paul Mahaffy; (5) PI on mid-term IRAD to develop a high precision electric gate (HPEG) for the ASTID developed IMS & INMS; (6) PI on mid-term IRAD to build for the in situ suite of plasma instruments for SP+ with a nadir viewing capability (in collaboration with Lockheed Martin); (7) PI of two IRADs that have been selected for FY10 with the first to develop a radiation hardened version of the ASTID IMS and INMS for Europa and the other to further develop the HPEG and build miniature IMS. He has also submitted numerous invention disclosures to GSFC's patent office this past year for which GSFC is in the process of patenting his HPEG design. The HPEG has the potential of providing mass resolution $M/\Delta M > 1000$ and with time-of-flight reflectometer $M/\Delta M > 10,000$. In addition to the above he has written a chapter in the Titan Cassini-Huygens books on energy Deposition into Titan's upper atmosphere and Titan's induced magnetosphere. He has written two papers on Titan to *Planetary and Space Science* selected for publication with one of the two presenting evidence for exobiology at Titan. He has been at GSFC since 1978.

His main research questions are: 1) what mechanisms drive the solar wind, 2) what are the physics of Saturn's magnetosphere and its interaction with Titan and Enceladus, and 3) what is the astrobiological potential of Titan, Enceladus, and Europa. The nature of the mechanisms driving the solar wind has been a long unsolved problem since its prediction by Parker. Saturn is a fast-rotating planet which has turned out to be more dynamic than originally thought, with its interaction with Titan and Enceladus of intense scientific interest. Titan, Enceladus, and Europa are observationally thought to have subsurface oceans, with Titan being of special interest due to observed organic chemistry in its ionosphere and the ethane/methane lakes on its surface.

GSFC Heliophysics Science Division 2009 Science Highlights

James Slavin

Dr Slavin is the Director of HSD. Space science research and flight missions have been a central focus for him since he joined his first instrument team, the Pioneer Venus magnetic field investigation, in 1978. He thoroughly enjoys not only the opportunity to make new discoveries and advance scientific understanding, but also the sense of community and spirit of team work that pervades space science. However, he also treasures his time with family and friends, enjoy playing tennis and jogging, and spend more time than he should reading science fiction and mysteries. He was hired by GSFC in 1987 to take over the Dynamics Explorer magnetic fields investigation. All told he has been a Project Scientist and/or Science Investigator on 20 space flight missions, a few of which were cancelled or failed, but most of which were highly successful and wonderful experiences.

The vast majority of his time is spent supporting HSD and its highly accomplished and always engaging staff who serve as secretaries, managers, resource analysts, scientists, technicians, engineers, system administrators, and many other important functions. Our Division leads ~20 flight missions and instrument activities with a broad range of science goals. He mentors two post-doctoral scientists, two NASA Graduate Student Research Program and one Co-operative Program graduate students. He also serves on the NASA Advisory Committee's Planetary Science Subcommittee.

His research time is limited, but spans a range of topics all relating to magnetic reconnection and the solar wind interaction with the planets. The most significant scientific result for the previous year was the first direct measurements of reconnection at Mercury's magnetopause.

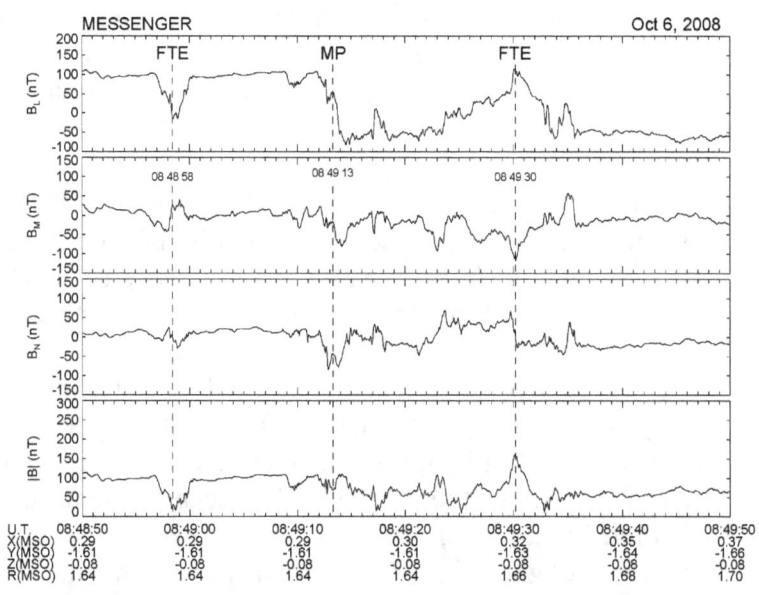

MESSENGER magnetic field observations from MESSENGER's second flyby of the dawn flank magnetopause and a giant flux transfer event (FTE) are shown. Analysis of these measurements has led to the conclusion that reconnection at Mercury proceeds at a rate 10 times that at Earth.

GSFC Heliophysics Science Division 2009 Science Highlights

Keith Strong

Dr Strong is solar physicist who has worked in the areas of X-ray spectroscopy of active regions and flares, coronal abundance variability, coronal dynamics, and, most recently, solar cycle characterization and prediction. He received his BSc from University College London in 1971 and his PhD from the Mullard Space Science Laboratory, University of London, in 1979. He worked at the Lockheed Martin Advanced Technology Center for 34 years initially as a research assistant and eventually as the senior manager of the Space Science departments before retiring in 2007. He joined the GSFC solar group as a part-time scientist to get back to analyzing solar data, which is his first love.

He is particularly interested in E/PO activities. He has given two public lectures this year: one to a standing-room only crowd on the solar cycle, organized by a local astronomy club and the other on the potential link between solar variability and climate change, at the Sir Patrick Moore Planetarium.

Drs Saba and Strong have been working on the analysis and interpretation of solar magnetic data (mainly from SOHO MDI and NSO) and comparing it to coronal activity (data from Yohkoh, SOHO, and GOES). They are currently looking at asymmetries between the timing and amplitude of flux emergence in the northern and southern hemispheres as well as trying to identify active longitudes. They have given several presentations at various meetings on this subject in FY09.

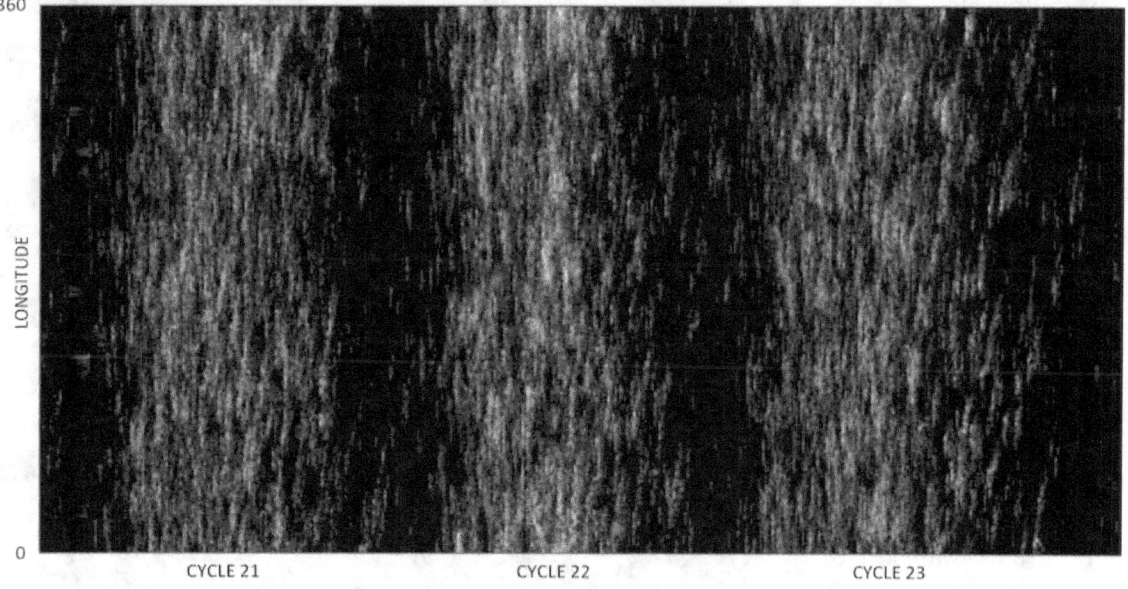

We have dubbed this the Tree Trunk plot. It shows the longitudinal distribution of magnetic flux for each Carrington rotation of the Sun over the last 30 years (so is the longitudinal equivalent of the butterfly diagram). Time goes from right to left on the abscissa, and solar longitude is plotted on the ordinate. Active longitudes should show up as a steeply positive or negative gradient in the plot and change systematically throughout the cycle as the mean latitude of the emerging active regions reduces as the cycle progresses. We find no strong evidence of any active longitudes for the last three cycles. (NSO/SOHO MDI data)

Tim Stubbs

Dr Stubbs has been an assistant research scientist with UMBC since 2005, and is currently pursuing research at GSFC on the lunar environment. He came to GSFC at the end of 2002 as an NRC research associate. Prior to this, he had been at Imperial College, London, where he completed both his PhD in space physics and his undergraduate MSc in physics. He has recently joined the Solar System Exploration Division (Code 695).

Dr Stubbs is a participating scientist with the Lunar Reconnaissance Orbiter (LRO) mission and affiliated with the Cosmic Ray Telescope for the Effects of Radiation (CRaTER) instrument team. He is also affiliated with the NASA Lunar Science Institute (NLSI) as a Co-I with the Dynamic Response of the Environment At the Moon (DREAM) team. He has recently served on the following advisory panels: Lunar Airborne Dust Toxicity Advisory Group, LADTAG (2005–present); NAC Heliophysics Lunar Subpanel (2006–2008); and Deputy Director's Council on Science, DDCS (2007–present). Tim regularly reviews journal articles and proposals for NASA, NSF, and international research programs.

Dr Stubbs' main research interest is the dynamic lunar environment and how its various components interact with each other (see figure). In particular, he has studied lunar surface charging and the electrostatic transport of charged lunar dust. He is using LRO data to model lunar surface electric fields and search for evidence of exospheric dust. Understanding the fundamental processes of the interaction of airless bodies, such as the Moon, with the surrounding space environment is vital for understanding both their surface evolution and the potential hazards that may be faced by future explorers. He has previously studied the injection, energization, and transport of plasma in the Earth's magnetosphere and auroral dynamics.

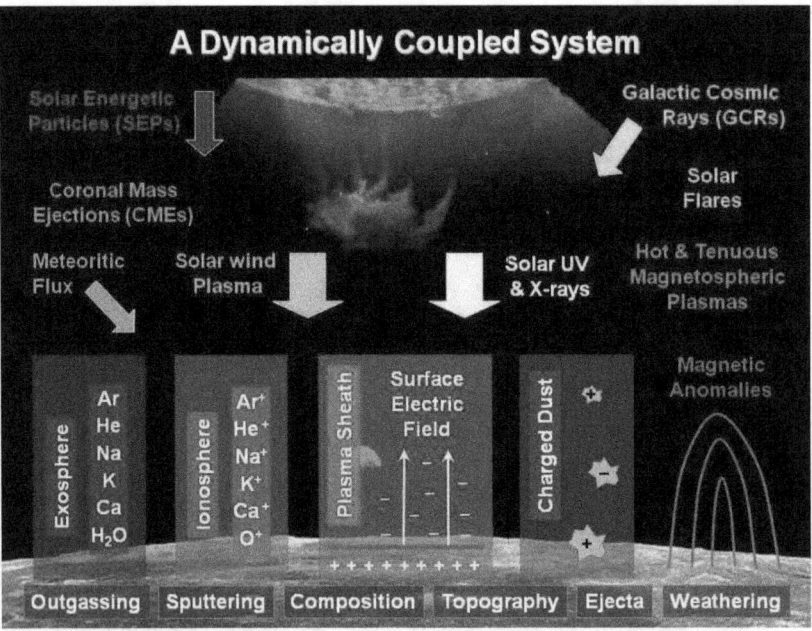

An overview of the dynamic lunar environment, showing how the well-known planetary processes (examples listed at the bottom) couple with external drivers (listed at the top) to produce the various interdependent components of the lunar environment [Credit NLSI/DREAM].

Errol Summerlin

Dr Summerlin began his secondary education at the Georgia Institute of Technology, where he received a BS in physics in 2002. After that he attended Rice University, where he worked under Dr Matthew Baring. His thesis was comprised of results from a Monte-Carlo simulation of first-order Fermi shock acceleration in both heliospheric and astrophysical environments. He wrote this simulation during his tenure at Rice starting from a blank '.c' file. He received his MS there in 2006 and completed all the requirements for his PhD shortly before coming to GSFC. He began work at GSFC in July 2009. His first task is to be, essentially, the omega user for VEPO. Since he is an expert in energetic particles, but has no experience in data analysis or instrumentation, he is exactly the type of scientist that will typically be using VEPO. In using this data for comparison with his simulation, he will be able to provide feedback on how the VEPO documentation can be improved. In the process of understanding these data sets, he also essentially acts as a quality assurance officer for the data set that he is working with (Ulysses HI-SCALE) ensuring that the documentation on VEPO reflects the caveats and nuances of the data set as accurately as possible. In this capacity, he discovered a flaw in the Compton-Getting transformation that was performed 10 years ago and is currently working with the data provider to correct the affected data set.

His primary research interests are in the field of shock acceleration physics. Shocks are expected to be the dominant source of energetic particles in the universe. The process is ubiquitous and powerful. Understanding of this process is a critical but often overlooked part of space weather's predictive capabilities. Many of the largest energetic-particle events observed at Earth are the result of shock acceleration as the shock travels through the interplanetary medium. Understanding this process outside the heliosphere provides new meaning to observed spectral distributions. With some understanding of the emission process in these distant plasmas, one can infer a particle distribution in these objects. Then, with understanding of the acceleration process, one can infer the physical parameters necessary to produce the inferred particle distribution, providing new information about the object in question that cannot be discovered observationally with current technology. Dr Summerlin has been involved in several publications in both contexts.

GSFC Heliophysics Science Division 2009 Science Highlights

Adam Szabo

Dr Szabo is the Chief of the Heliospheric Physics Laboratory, Code 672. He has been working at GSFC since 1994. He received his BA in physics from the University of Chicago and his PhD in physics from MIT.

Dr Szabo serves as the Wind spacecraft project scientist and as deputy project scientist for the ESA/NASA Solar orbiter mission. He is also the PI on the Wind and IMP 8 magnetometers along with the Virtual Heliospheric Observatory (VHO).

Dr Szabo specializes in heliospheric structures such as shocks, discontinuities and ICMEs/magnetic clouds. In a recent study using Wind magnetic field and energetic electron observations within magnetic clouds that are well fitted by a force-free, constant-alpha flux rope model, he has shown that the 1-AU arrival time of the various energy electrons, that closely follow field lines, yield similar 2–4-AU field line lengths from the Sun to those one can compute using the fitted magnetic cloud flux rope model at 1 AU and assuming magnetic flux and current conservation. The significance of this result is that it implies that ICMEs close to the Sun are not nearly as twisted as previously thought.

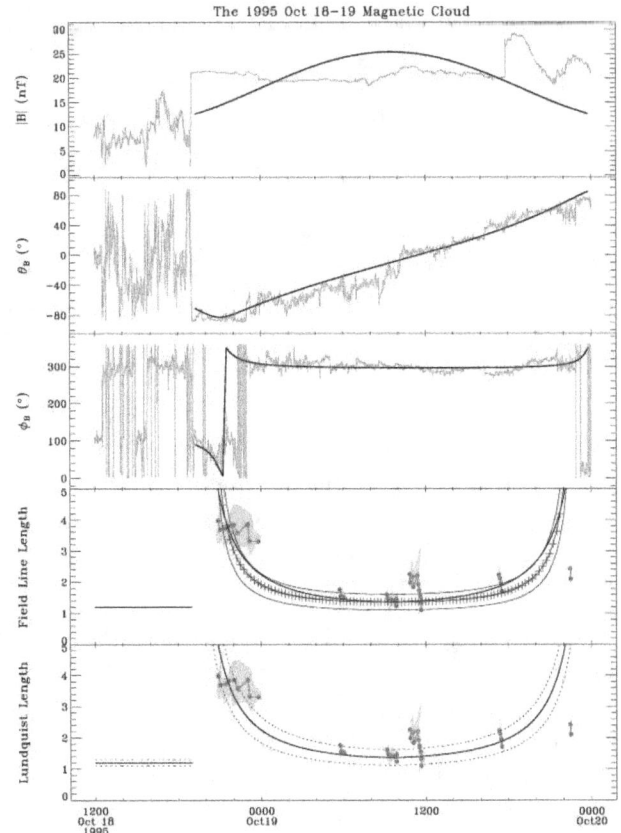

Computation of interplanetary magnetic field line lengths, using the arrival times of energetic electrons and fitted model magnetic flux rope field line lengths in the 1995 Oct 18-19 magnetic cloud. The top three panels show the Wind magnetic field observations along with a force-free flux rope fit. The bottom panels show the comparison of the computed field line length.

Roger Thomas

Dr Thomas is an internationally recognized expert on the design and scientific use of imaging EUV spectrographs. He created the optical designs for CDS/NIS on SOHO and EIS on Hinode, as well as for numerous sounding rocket instruments including SERTS, EUNIS, MOSES, SUMI, RAISE, VERIS, and UVSC, on each of which he serves as Co-I. He has been with GSFC ever since obtaining his PhD in astrophysics in 1970 from the University of Michigan.

Dr Thomas has been the Orbiting Solar Observatory Project Scientist (1976–1983), the Study Scientist for the Solar Cycle and Dynamics mission (1978–1981), the Deputy Chief of NASA's Solar and Heliospheric Physics Office (1983–1984), and the Deputy Project Scientist for the Orbiting Solar Laboratory mission (1990–1992). He has been the science advisor for several NASA red teams monitoring the progress of solar-satellite missions.

Dr Thomas is presently pursuing studies of spatially imaged EUV high-resolution spectra of coronal structures. Specific topics of interest include determination of elemental abundances and their possible variations, investigation of proposed coronal heating mechanisms, and quantitative characterizations of physical plasma conditions in different solar features. He is actively involved in the design, ray-trace optimization, and fabrication of space-flight optical systems, especially XUV imaging spectrographs using varied line-space gratings ruled on aspheric surfaces. He also leads the effort to obtain an absolute radiometric calibration for the EUNIS sounding rocket experiment, a key aspect of its scientific value. (The

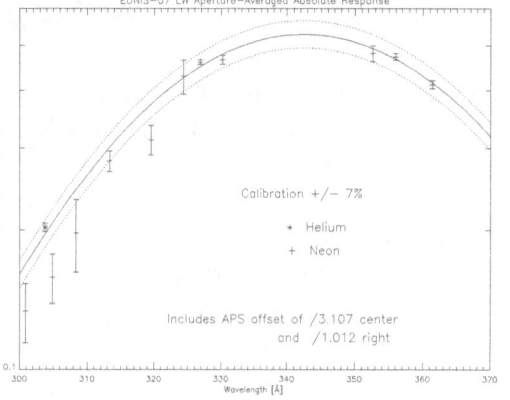

Absolute Radiometric Calibration of the EUNIS-07 EUV Rocket Spectrograph

figure shows the recent EUNIS-07 longwave-channel calibration.) He has authored or co-authored more than 80 scientific publications in refereed journals, and at least 174 other scientific or technical papers.

In 2009, Dr Thomas received the prestigious NASA Exceptional Service Medal, and was a member of the EUNIS team that won a NASA Group Achievement Award.

Barbara Thompson

Dr Thompson, solar physicist, joined Code 671 in August 1998. Prior to that, she served as a mission support scientist on the SOHO mission. Dr Thompson graduated from the University of Pennsylvania in 1991, majoring in physics and mathematics and minoring in geology. She received her PhD in physics from the University of Minnesota in 1996. She divides her time between scientific mission support, science community activities, and research.

Dr Thompson's scientific leadership has emphasized data coordination initiatives. She has served as an organizer of the International Solar-Terrestrial Physics Workshops (1997 and 1999), SOHO-Yohkoh Workshops (1997 and 1999), Coordinated Data Analysis Workshops (1999 and 2002), "Living with a Star" workshops (2000 and 2002), the "Whole Sun Month" Workshops (1997 and 1998), and the "Whole Heliosphere Interval" Workshops (August 2008 and September 2009).

She was co-chair of IAGA's "Sun and Heliosphere Division" from 2005 to 2007 and Director of Operations for the International Heliophysical Year from 2003 to 2009; she serves as the Executive Secretary for the International Living With a Star program.

Dr Thompson has devoted the majority of her solar research to the study of CMEs and associated phenomena, particularly EIT waves. Her collaborations involve analyzing data from multiple sources and the development of model interpretations of observations. She has authored or co-authored more than 50 papers using data from more than one instrument, and many papers combining observations with model interpretations. The majority of her recent published work has involved either the STEREO observations or the data from the Whole Heliosphere Interval (see below).

Dr Thompson wrote the 100-page IHY final report that was published and distributed by the United Nations Committee on the Peaceful Use of Space. She was lead editor of the 370-page "Putting the I in IHY" book, a full-color summary of IHY activities in 60 nations published by Springer Press. The book is available on amazon.com.

She also coordinated analysis efforts for the Whole Heliosphere Interval (WHI) Campaign, and co-organized the WHI Science Workshop that occurred in November 2009. WHI is an internationally coordinated month-long observing and modeling campaign covering a whole solar rotation from the solar interior to Earth's mesosphere and the outer heliosphere. Currently, 250 scientists are registered participants of WHI.

William Thompson

Dr Thompson has been employed as a contractor at GSFC, supporting the solar physics division, since 1984. He currently works for Adnet Systems, Inc. in the position of Senior Scientist. He received a BA in physics and mathematics from New College of the University of South Florida in 1974, and a PhD in astronomy from the University of Massachusetts in 1982. Between 1982 and 1984, he served as a Lecturer in the department of Physics and Astronomy at San Francisco State University.

Dr Thompson serves as the Chief Observer for the NASA Solar Terrestrial Relations Observatory (STEREO) mission. As such, he oversees the STEREO Science Center, which serves as the primary archive for the mission, and is the processing point for the STEREO space weather beacon data. As Chief Observer, he is also responsible for coordinating scientific activities between the STEREO instrument teams. Dr Thompson is also a member of the team operating the COR1 telescope aboard STEREO, and is responsible for characterizing the instrumental calibration. Additionally he serves on the IAU working group on the FITS standard.

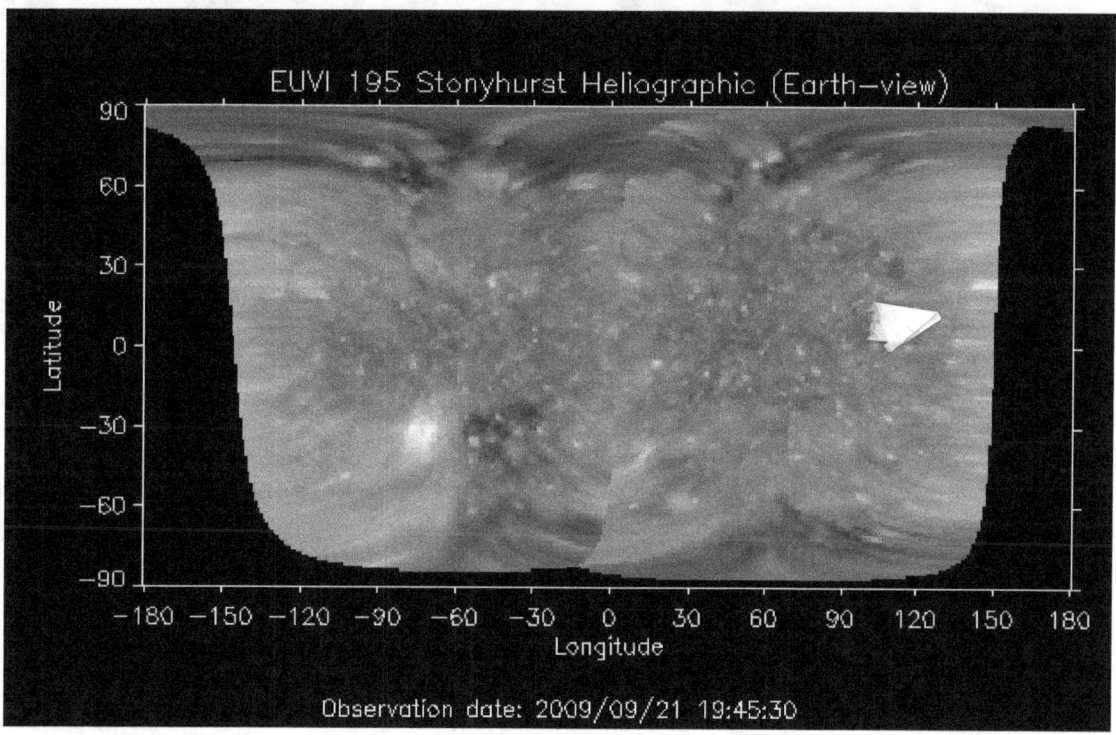

Heliographic map made from real-time STEREO beacon images.

Dr Thompson's research areas include EUV spectroscopy, coronagraphy, stereo triangulation, and space weather monitoring. His work covers all phases of a mission, from instrument development, calibration, software development, through to mission operations and data analysis. Along with his continuing work on the STEREO mission, he has become involved with the SPICE instrument development program for the upcoming Solar Orbiter mission.

GSFC Heliophysics Science Division 2009 Science Highlights

Arcadi Usmanov

Dr Usmanov has been working at GSFC as a Senior NRC Associate from 2001 to 2004 and since 2005 as a research scientist under a grant to the University of Delaware. He received his MS and PhD degrees from the St.-Petersburg State University in St. Petersburg, Russia.

The main topic of Dr Usmanov's research is understanding and quantitatively modeling the evolution of the solar wind on a global scale, from the coronal base to the boundary of the heliosphere. The focus of Dr Usmanov's work over the last year was the development of a global 3D solar wind model that is based on numerical solutions of large-scale Reynolds-averaged magnetohydrodynamic equations coupled with a set of subgrid-scale equations for turbulence transport and dissipation.

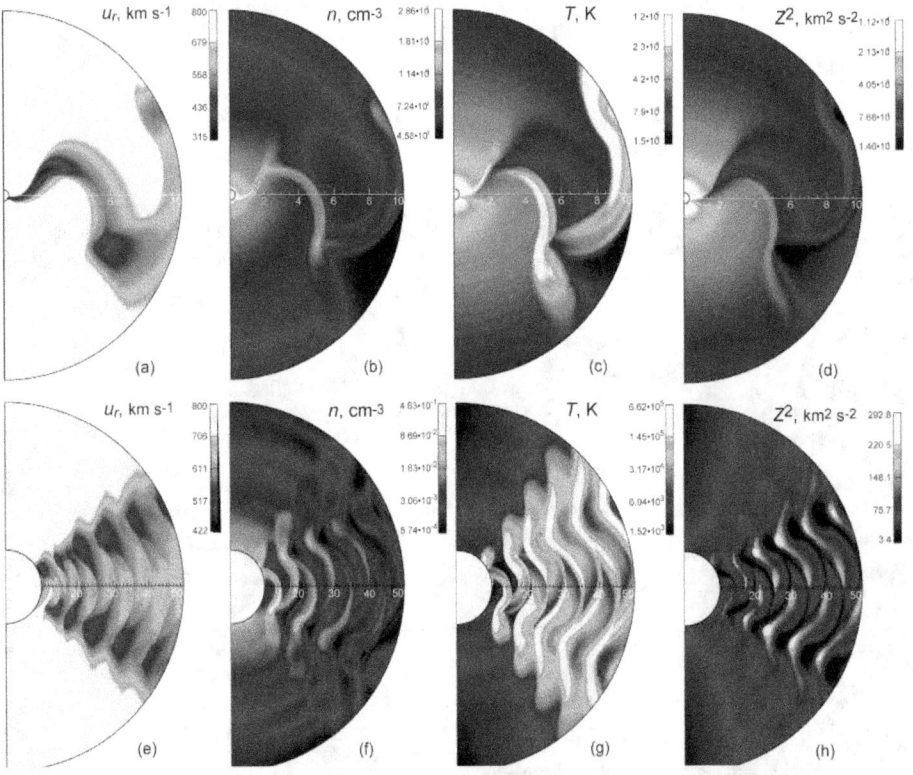

Contour plots in the meridional plane of the solar wind parameters: radial velocity, density, temperature, and turbulent energy from 0.3 to 10 AU (a-d) and from 10 to 100 AU (e-h). A dipole source field on the Sun is tilted by 30 degrees.

The model was used to study (a) properties of the large-scale solar wind, interplanetary magnetic field, and turbulence throughout the heliosphere; (b) the effects of pickup protons in the physical processes of the outer heliosphere; (c) the interaction between the large-scale solar wind and smaller-scale turbulence; and (d) the role of the turbulence in the large scale heliospheric structure and temperature distribution in the solar wind. Dr Usmanov has given scientific presentations at the Solar Wind 12 Conference in St. Malo and at the Mini Symposium on Solar Wind, Brussels.

Adolfo Viñas

Dr Viñas is a space plasma physicist with HSD. He has also served as a research scientist at other GSFC's divisions since 1980. He is a Co-I on the Wind/SWE and Cluster/PEACE plasma electron experiments, and currently is a Co-I with the Fast-Plasma Investigation of the Magnetospheric Multiscale Satellites (MMS) mission. Dr Viñas' research interests include the study of kinetic and MHD processes, plasma instabilities, kinetic turbulence, and shocks and discontinuities in the solar corona, solar wind, and magnetosphere. He is currently involved in the simulation and testing of new analysis techniques for the Fast Plasma Investigation of the MMS mission.

Dr Viñas developed a method to describe the geometrical and physical properties of shocks and disturbances based on the conservation of physical conditions. It has been implemented in a visualization and analysis tool named SDAT. Recently, he has developed a new model independent spherical harmonic spectral method for the calculation and modeling of particle velocity distribution functions (VDFs) and the estimation of its moments and anisotropies for fast and high-angular-energy-resolution plasma spectrometers.

He is currently supervising a doctoral student/fellow from Brazil.

His most recent work uses Cluster/PEACE electron measurements to examine the properties and characteristics of non-thermal three-dimensional velocity distributions of the *strahl* electrons in the solar wind. An example of such non-thermal distributions, in a frame relative to the local magnetic field, in the upstream solar wind region.

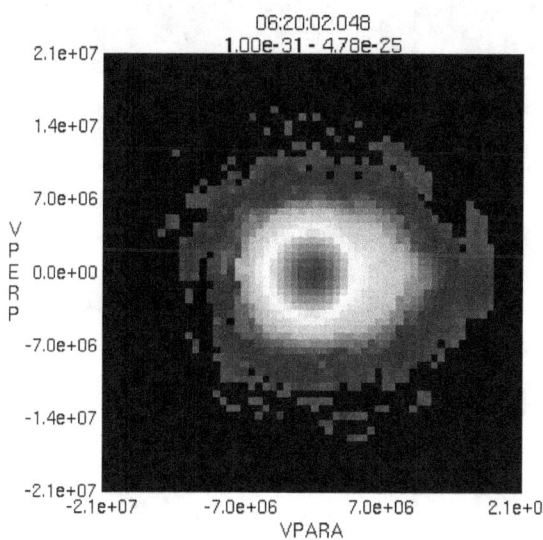
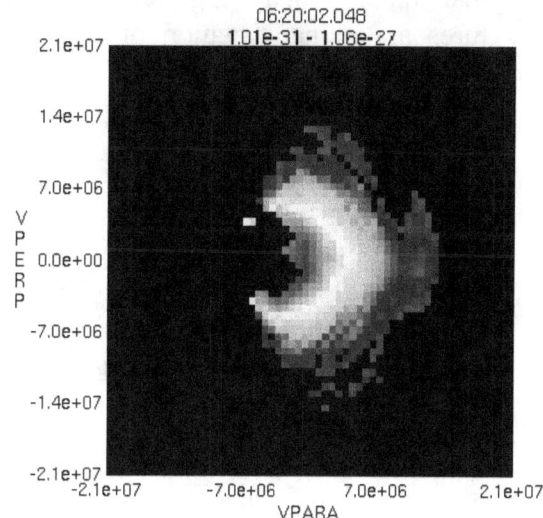

Tongjiang Wang

Dr Wang, solar physicist, has been a research associate at CUA and working at GSFC, Code 671 since 2007.

The objective of his research is to understand dynamics and heating of the solar corona based on observations of various wave phenomena in coronal loops. These waves carry the signature of the emitting source and can diagnose the physical parameters of highly structured coronal loops. He has been also working on radiometric calibration of EUNIS.

In 2009, Dr Wang and his coauthors detected two long-period (12 and 25 min) harmonic waves propagating along fanlike coronal loops with observations from the EUV Imaging Spectrometer (EIS) onboard Hinode. It is the first time that such kinds of waves have been simultaneously detected in both intensity and Doppler shift in a coronal emission line, Fe XII. Their amplitude relationship provides convincing evidence that these propagating features are a manifestation of slow magnetoacoustic waves. Dr Wang proposed a new application of coronal seismology based on this observation, which derived the true sound speed and temperature near the loop's footpoint. This work was presented as a talk at the SPD meeting in Boulder, Colorado. In addition, Dr Wang has been working on underflight radiometric calibration of Hinode/EIS using EUNIS-07.

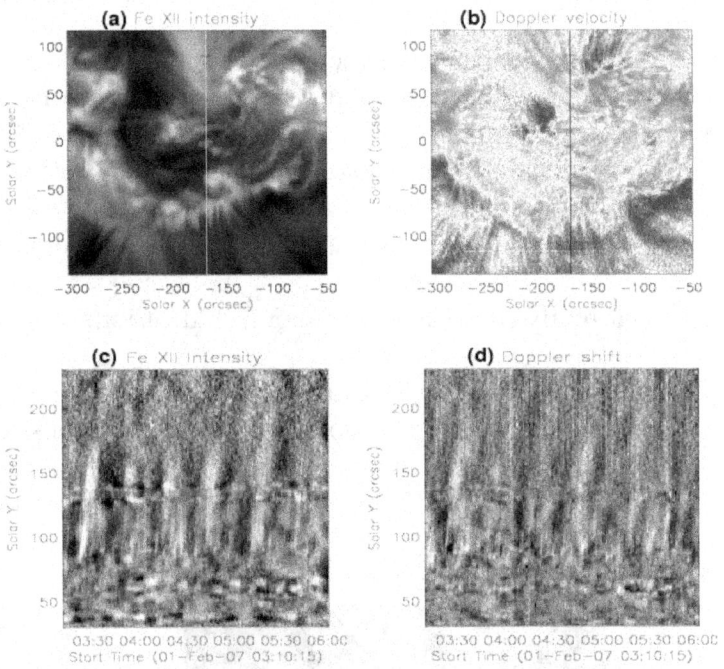

The 12 and 25 min upwardly-propagating slow magnetoacoustic waves in coronal loops observed by the EUV Imaging Spectrometer (EIS) onboard Hinode..

Yongli Wang

Dr Wang is a research faculty of UMBC/GEST and is stationed at GSFC. He received a PhD in Geophysics and Space Physics from University of California Los Angeles in 2003. He joined GSFC in January 2006, working on satellite data calibration, support (i.e., Space Technology 5), as well as NASA and NSF projects in various topics covering the ionosphere, the inner and exterior magnetosphere, the magnetopause and magnetosheath, and the Moon.

Currently Dr Wang is the PI of two NASA grants:
- Automating electron density determinations from magnetospheric dynamic spectra
- Generalized 3D magnetosheath specification model.

He is also a Co-Investigator of four NASA and NSF grants:
- Characterizing the composition of large mid-latitude topside-ionospheric-plasmaspheric gradients
- Characteristics of flux transfer events
- Establishing Links between Solar Wind and Topside Ionospheric Parameters
- Magnetospheric electron density

Meanwhile, Dr Wang is working closely with Dr Timothy J. Stubbs, LRO/LOLA team, and Dynamic Response of the Environment at the Moon (DREAM) team on Moon sunlight and plasma shadow and surface charging modeling. He has been a regular reviewer for NASA and NSF proposals, as well as scientific journals including *J. Geophys. Res., Geophys. Res. Lett., Adv. Space Res., Annales Geophysicae*, and *J. of Atmos. and Solar-Terr. Phys.*

Currently he is working on modeling of 3D magnetopause/bow shock/magnetosheath and their dependence on solar wind and geophysical conditions; building the largest flux transfer event database for its generation and dynamics; studying multiple spacecraft dynamic spectrum and ionosonde observations for magnetosphere- ionosphere density; and building the lunar plasma shadow and surface charging model to understand the lunar environment. All this work uses the best observations/modeling techniques available to help better understand the Earth's ionosphere-magnetosphere system and the Moon.

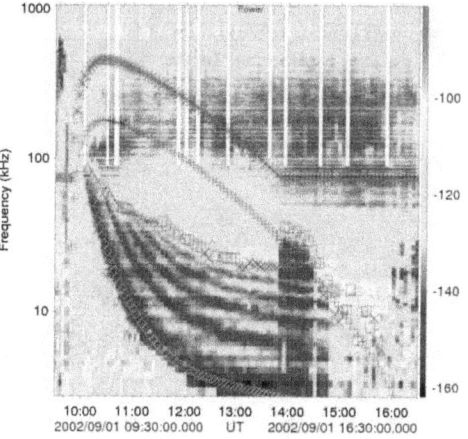

An example of auto fitting of 5 years of IMAGE power spectrum with Z-mode correction to obtain electron density.

Peter Williams

Dr Williams is a second-year NPP and has continued his work with Dr Pesnell analyzing Doppler velocity images from the Michelson Doppler Imager (MDI) aboard SOHO. These images can be reduced to extract surface manifestations of internal convection mechanisms, seen as either granule or supergranule cells. Supergranules are significant in such studies as there is clear evidence of magnetic field interactions within, and at the boundaries of, such cells. The past year has involved studying various characteristics of these features, such as their sizes, lifetimes and advection properties; and then comparing our results between the past two minima, using MDI data from 1996 and 2008. Although the supergranule lifetimes tend to be similar for both years, we find that supergranules are smaller and their flows stronger in 2008 than in 1996.

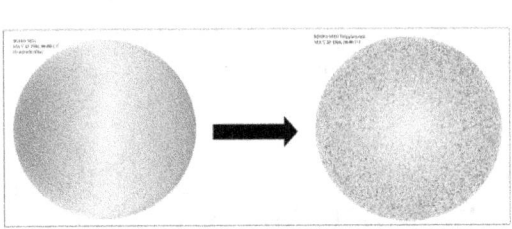
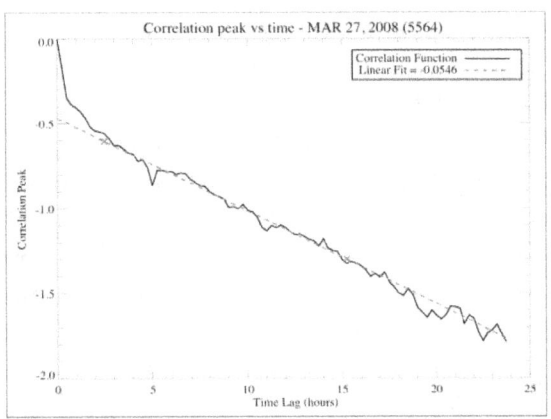
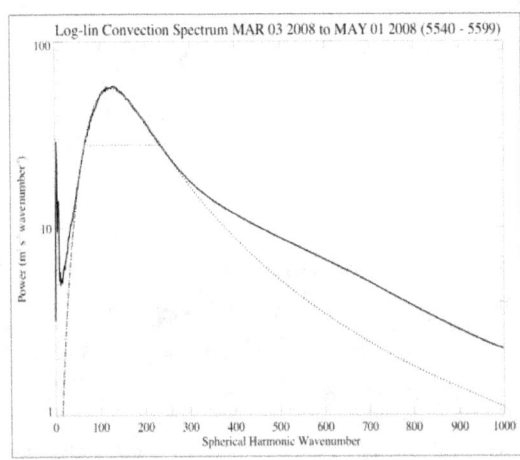
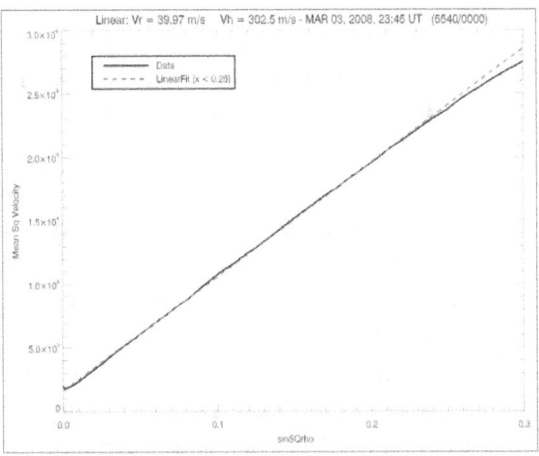

Doppler velocity images (upper left), that illustrate surface flows toward (blue) and away (red) the observer at the surface of the Sun, are reduced to display only those flows due to surface convection features. A spectrum of convection cell sizes can be derived (lower left), showing a peak corresponding to supergranules. The results show typical sizes of around 35 Mm. Correlation techniques can be applied to pairs of images to estimate the typical lifetime of supergranules (upper right). The derived 1/e lifetime is around 18 hours. The velocity patterns may be decomposed into radial and horizontal components (lower right). Horizontal flows are normally a whole magnitude stronger than radial flows.

Charles Wolff

Dr Wolff is a member of the Solar Physics Branch. His research aims to explain why the Sun's neutrino flux and luminous output at all wavelengths is not constant. He seeks answers deep inside the Sun, which supplements the many other efforts that study detailed behavior at the surface. He received his PhD in physics from the University of Illinois. He has been at GSFC for decades and, in recent years, served on several NSF proposal review panels.

He predicted that g-modes (with periods of hours) should couple to form rigidly rotating, oscillating structures like the one in the accompanying figure. Each structure rotates slowly at a different rate, following well-established asymptotic theory. When the high-power regions of several structures overlap (periodically on timescales of months or years), they cause extranuclear burning in the Sun's core causing more neutrino flux. At about the same time the overlap region feeds extra energy into the Sun's convective envelope, which can excite a large-scale upflow that will soon cause more solar activity at the surface. This would explain previously puzzling reports that solar activity levels correlate somewhat with changes in neutrino flux.

Wolff has found observational evidence for the g-modes or their structures by two independent methods. The most recent detection is in a paper. It also contains observational and theoretical reasons why conventional solar structure models are deficient and need to add a rapidly mixed shell about one sixth of the way from center to surface.

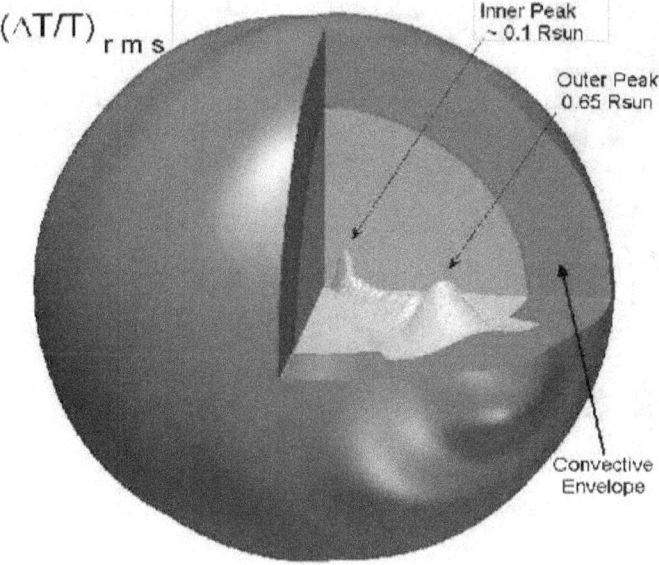

Temperature fluctuations caused inside the Sun by a set of coupled g-mode oscillations are concentrated in angle (red surface) and at two points along the prime radius (yellow). This pattern of oscillatory power is sustained by nuclear burning at the inner peak and it dE/POsits energy mainly at the outer peak, stimulating convection in the Sun's envelope.

Hong Xie

Dr Xie, a solar physicist with CUA, joined Code 671 in April 2003. She has been working on data analysis of CMEs, CME-driven shocks, and associated geomagnetic storms. She is currently maintaining the STEREO COR1 preliminary CME catalog. Her research interests include the origin, 3D structure, and evolution of CMEs. She has developed new analytical CME cone model fitting procedure, and incorporated a geometrical flux-rope model fitting for STEREO/COR coronagraph images.

In 2009, working with STEREO team, Dr St. Cyr and Dr Gopalswamy, she has studied the origin, 3D structure, and kinematic evolution of CMEs near solar minimum using the STEREO COR1 and COR2 data, flux-rope models, and 3D triangulations. She also improved the shock prediction technique based on the empirical shock arrival (ESA) model, where CME deceleration is included in the corrected ESA and synthesized with kilometric Type II-derived shock velocities.

In addition, working with Dr Holly Gilbert, she investigated the relationship between the nature of a prominence eruption including total, partial, failed eruption, the amount of mass loss (draining) occurring prior to and/or during an eruption, and the associated CME properties such as speed, mass, and energy.

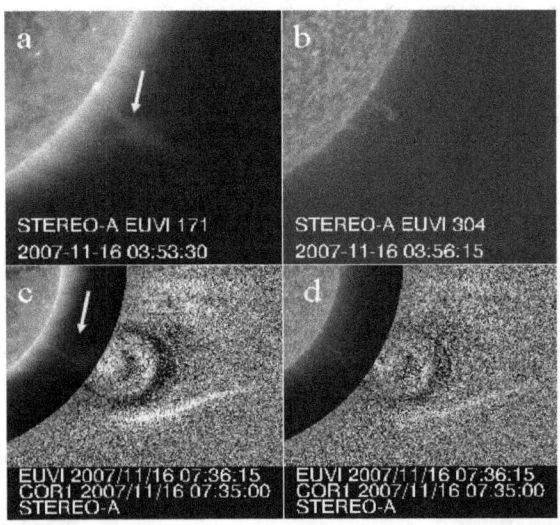

First direct observations of the X-line structures (a well-known location of current sheets and reconnections) in EUVI 171 images, with associated PEPs in 304 and flux rope CME in COR1. (a) and (b) An X-type magnetic structure (yellow arrow) and PEP associated with the 11-16-2007 CME in EUVI-A 171 and 304. (c) and (d) Superposition images of STEREO-A COR1 and EUVI 171 and 304.

GSFC Heliophysics Science Division 2009 Science Highlights

Seiji Yashiro

Dr Yashiro's research in solar physics started in 1995 as an undergraduate student of Prof Uchida. In March 2000, he received his PhD from the University of Tokyo, Japan, on the investigation of the evolution of coronal active regions using Yohkoh data. His thesis supervisor was Prof. Shibata. Since April 2000, He has worked at GSFC with Dr Gopalswamy as a contractor of CUA. His main field has been changed to the CMEs.

He has maintained the CDAW Data Center which was built to organize CDAW workshops (http://cdaw.gsfc.nasa.gov). One product of the CDAW Data Center is an event list of CMEs which is called the SOHO/LASCO CME Catalog. He has compiled the catalog as a core member of the Catalog Team.

Dr Yashiro's main research interest is to understand the relationship between solar flares and CMEs. The two phenomena are thought to be different manifestations of the same process.

In FY09, he examined the statistical relationship between solar flares and CMEs and found a good correlation between flare fluence and CME kinetic energy. Their energetic connection is tighter for larger events. The kinetic energy of associated CMEs ranges over three orders of magnitude for a given C-class flare, while it is under two orders of magnitude for a given X-class flare. The result indicates that the magnetic configuration of the C-class flare-CME events could have large variations, while that of the X-class events might be uniform.

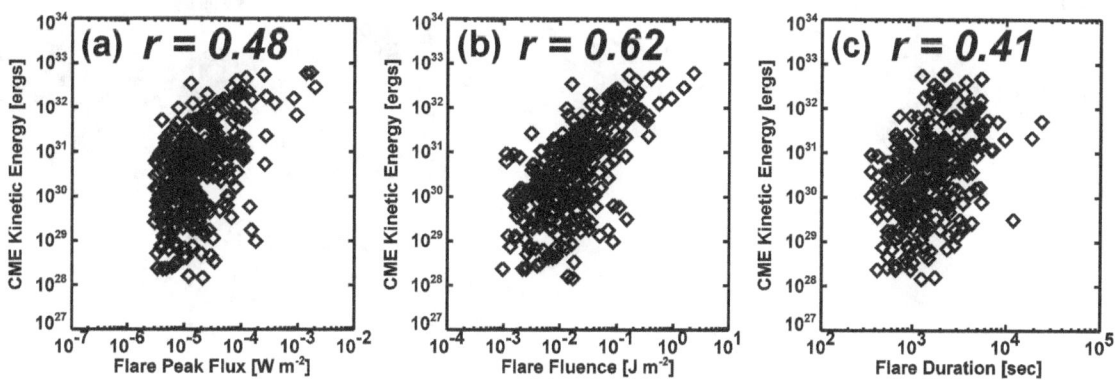

Scatter plots of CME kinetic energy versus flare peak flux (a), fluence (b), and duration (c).

C. Alex Young

Dr Young, ADNET Systems Inc., works to develop signal and image processing methods and software to facilitate a more complete extraction of scientific information from solar physics data. This is to aid both the community as a whole and his own research into the prediction and understanding of dynamic phenomena in the solar corona such as solar flares and CMEs. In 2009, Dr Young released a set of multiscale IDL tools in the Solar Soft library, which he presented at the Fall AGU meeting in San Francisco. Dr Young was a guest editor of *Solar Physics* special issue in April 2008 that contained papers from some of the material presented at the 3rd Solar Image Processing Workshop. A hard cover edition of this journal volume, "Solar Image Analysis and Visualization," was published in 2009. He is currently planning another special volume of *Solar Physics* covering work from the 4th Solar Image Processing Workshop. A 5th workshop is planned for September 2010 in Switzerland. His work has included education both in the form of mentoring several undergraduate and graduate students and giving interviews for Discovery Channel and Tokyo TV.

Dominic Zarro

Dr Zarro obtained a BSc (honors in applied mathematics) from Sydney University and a PhD (astronomy) from the Australian National University, Mt Stromlo Observatory. He worked as a post-doc at Caltech's Big Bear Solar Observatory before joining the Solar Maximum Mission project at GSFC. He has since supported the Yohkoh, SOHO, RHESSI, and Hinode solar missions in operations, scientific research, and developing data analysis software tools.

He is currently employed by ADNET Systems, Inc. as a Group manager on the SESDA-II contract at GSFC. He is a Co-I on a NASA proposal entitled "VxO for heliophysics Data – Extending the VSO to Incorporate Data Analysis Capabilities."

Seiji Zenitani

Dr Zenitani joined Code 674 in November 2006, after completing his PhD work in space physics at University of Tokyo. His dissertation work is on particle acceleration processes and kinetic instabilities of the current sheet structure in relativistic electron-positron pair plasmas in astrophysical settings. He is currently an NPP fellow with Dr Michael Hesse. He has been collaborating with Dr Alex Klimas and Dr Masha Kuznetsova as well.

His ultimate goal is to seamlessly connect heliophysics and relativistic astrophysics. As a first step, he produced a detailed discussion of magnetic reconnection in relativistic astrophysics by means of kinetic PIC (particle-in-cell) simulations in his PhD and postdoc studies. He has given several invited and review talks on these topics. In FY09, he recently developed and carried out the first-ever fluid simulations of relativistic magnetic reconnection, which were presented in two refereed journal papers. Currently he is working on large-scale kinetic modeling of nonrelativistic collisionless magnetic reconnection, trying to import some insights from astrophysics. Other projects on velocity-shear problems of relativistic jets are also in progress.

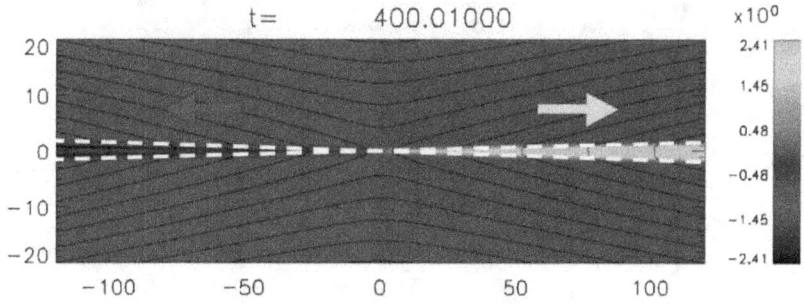

Two-fluid simulations of relativistic magnetic reconnection, in a quasi-steady Petschek-type structure

Hui Zhang

Dr Zhang joined code 674 as an NPP fellow in August 2008 after completing her PhD in astronomy at Boston University. She has been working with Dr David Sibeck on THEMIS data analysis.

Dr Zhang serves as a guest a guest editor for the *Journal of Atmospheric and Solar–Terrestrial Physics* special issue on Magnetospheric Response to Solar Wind Discontinuities. She is a convener and chair of a session for the coming 2009 Fall AGU meeting. She also serves as a THEMIS Tohban, which is a THEMIS scientist who helps the spacecraft operators in the Mission Operation Center (MOC) to optimize the science quality of the observations, and maintain the working order of the scientific instruments on board the five THEMIS spacecraft.

Dr Zhang's current research interest focuses on interaction of solar wind discontinuities (including interplanetary shocks) with the Earth's bow shock and magnetosphere, which is one of the fundamental modes of interaction. By analyzing data obtained by five THEMIS spacecraft, she demonstrated that no fast shock has been observed after the interaction of a weak interplanetary shock with the Earth's bow shock. Instead, the transmitted interplanetary shock took the form of a discontinuity, where the total magnetic field and density increase and the temperature decreases, and propagated earthward with a speed of 90 km/s.

In 2009, Dr Zhang served as PI or Co-I on seven research proposals (three as PI and four as Co-PI or Co-I) submitted to NASA or NSF and was either the lead author or co-author on three papers.

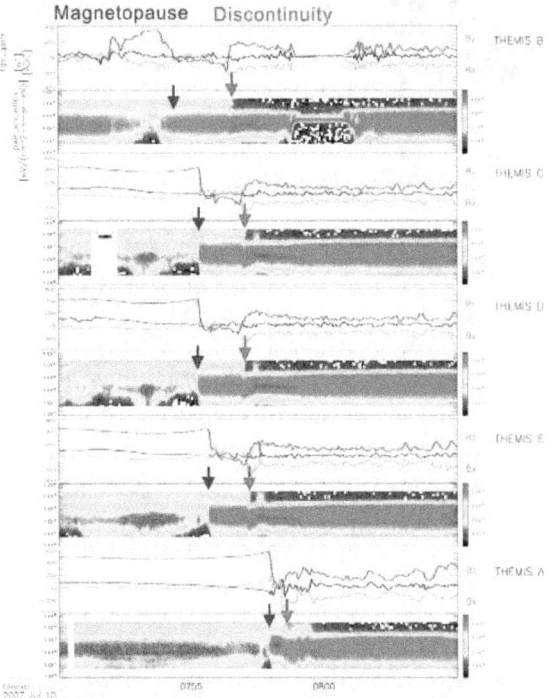

The magnetopause and discontinuity observed by all five THEMIS probes. The top panels show the magnetic field components and the ion spectrum from THEMIS B. The following panels show the same parameters from THEMIS C, D, E and A, respectively. The blue and red arrows mark the magnetopause and discontinuity crossings, respectively.

Qiuhua Zheng

Dr Zhang is currently a research scientist at GSFC studying dynamics and modeling of radiation belts particles of the Earth. He received his PhD in physics from University of New Hampshire in 2004 and worked at Michigan Aerospace Corporation as a research scientist until joining CRESST UMD/GSFC in January 2009.

He works in Dr Fok's group here at GSFC. In addition to research, he helps to maintain the GSFC radiation belts flux nowcast Web page and run some necessary codes.

His current research is focusing on the modeling of Earth radiation belt particle dynamics. The behavior of these particles has significant impacts on the space weather, which is important for space travels, protection of integrated circuits in instruments, etc.

APPENDIX 2: HSD PUBLICATIONS AND PRESENTATIONS

In FY09, HSD has published a total of 298 papers in a wide variety of scientific journals and proceedings, including two papers in both Nature and Science. Of these, 39% have been with HSD scientists as first author. HSD has submitted or has in press a further 98 articles. A comprehensive list of these papers is listed in the next two sections of this appendix.

The HSD group gave a total of 230 presentations (talks and posters) at a total of 74 different science meetings spread all across the US and around the World. The bulk of the HSD presentations were given at the AGU Fall, AGU/SPD Spring, EGU, and Space Weather Workshop meetings.

These figures may include some publications from before October 2008 but probably overall represent a considerable underestimate of the body of work published and presented by the HSD group at GSFC.

Journal Articles

Abers, E.S., **Bhatia**, A.K., Dicus, D.A., Repko, W.W., Rosenbaum, D.C., Teplitz, V.L., Charges on strange quark nuggets in space, *Physical Review D*, **79**(2), 023513, doi:10.1103/PhysRevD.79.023513, 2009.

Akturk, A., Goldsman, N., Aslam, S., **Sigwarth**, J., Herrero, F., Comparison of 4H-SiC impact ionization models using experiments and self-consistent simulations, *Journal Of Applied Physics*, **104**(2), 026101, doi:10.1063/1.2958320, 2008.

Alexeev, I.I., Belenkaya, E.S., Bobrovnikov, S.Y., **Slavin**, J.A., **Sarantos**, M., Paraboloid model of Mercury's magnetosphere, *Journal Of Geophysical Research-Space Physics*, **113**(A12), A12210, doi:10.1029/2008JA013368, 2008.

Anderson, B.J., Acuna, M.H., Korth, H., Purucker, M.E., Johnson, C.L., **Slavin**, J.A., Solomon, S.C., McNutt, R.L., The structure of Mercury's magnetic field from MESSENGER's first flyby, *Science*, **321**(5885), 82-85, doi:10.1126/science.1159081, 2008.

Anderson, B. J., M. H. **Acuña**, H. Korth, J. A. **Slavin**, H. Uno, C. L. Johnson, M. E. **Purucker**, S. C. Solomon, J.M. Raines, T. H. Zurbuchen, G. Gloeckler, and R. L. McNutt, Jr. (2009), The magnetic field of Mercury, *Space Sci. Rev.*, doi:10.1007/s11214-009-9544-3.

Andre, N., Blanc, M., Maurice, S., Schippers, P., Pallier, E., Gombosi, T.I., Hansen, K.C., Young, D.T., Crary, F.J., Bolton, S., **Sittler**, E.C., Smith, H.T., Johnson, R.E., Baragiola, R.A., Coates, A.J., Rymer, A.M., Dougherty, M.K., Achilleos, N., Arridge, C.S., Krimigis, S.M., Mitchell, D.G., Krupp, N., Hamilton, D.C., Dandouras, I., Gurnett, D.A., Kurth, W.S., Louarn, P., Srama, R., Kempf, S., Waite, H.J., Esposito, L.W., Clarke, J.T., Identification of Saturn's magnetospheric regions and associated plasma processes: Synopsis of CASSINI observations during orbit insertion, *Reviews Of Geophysics*, **46**(4), RG4008, doi:10.1029/2007RG000238, 2008.

Antonille, S., Content, D., **Rabin**, D., Wake, S., Wallace, T., Figure verification of a precision ultra-lightweight mirror: Techniques and results from the SHARPI/PICTURE mirror at NASA/GSFC - art. no. 70110Z, *Space Telescopes And Instrumentation 2008: Ultraviolet To Gamma Ray, Pts 1 And 2*, 7011, Z110-Z110, 2008.

Arridge, C.S., Gilbert, L.K., Lewis, G.R., **Sittler**, E.C., Jones, G.H., Kataria, D.O., Coates, A.J., Young, D.T., The effect of spacecraft radiation sources on electron moments from the Cassini CAPS electron spectrometer, *Planetary And Space Science*, **57**(7), 854-869, doi:10.1016/j.pss.2009.02.011, 2009.

Asai, A., Shibata, K., Ishii, T.T., Oka, M., Kataoka, R., Fujiki, K., **Gopalswamy**, N., Evolution of the anemone AR NOAA 10798 and the related geo-effective flares and CMEs, *Journal Of Geophysical Research-Space Physics*, **114**, A00A21, doi:10.1029/2008JA013291, 2009.

Aschwanden, M.J., **Burlaga**, L.F., **Kaiser**, M.L., Ng, C.K., Reames, D.V., Reiner, M.J., Gombosi, T.I., Lugaz, N., Manchester, W., Roussev, I.I., Zurbuchen, T.H., Farrugia, C.J., Galvin, A.B., Lee, M.A., Linker, J.A., Mikic, Z., Riley, P., Alexander, D., Sandman, A.W., Cook, J.W., Howard, R.A., Odstrcil, D., Pizzo, V.J., Kota, J., Liewer, P.C., Luhmann, J.G., Inhester, B., Schwenn, R.W., Solanki, S.K., Vasyliunas, V.M., Wiegelmann, T., Blush, L., Bochsler, P., Cairns, I.H., Robinson, P.A., Bothmer, V., Kecskemety, K., Llebaria, A., Maksimovic, M., Scholer, M., Wimmer-Schweingruber, R.F., Theoretical modeling for the STEREO mission, *Space Science Reviews*, **136**(1-4), 565-604, doi:10.1007/s11214-006-9027-8, 2008.

Baker, D. N., D. Odstrcil, B. J. Anderson, C. N. Arge, M. Benna, G. Gloeckler, J. M. Raines, D. Schriver, J. A. **Slavin**, S. C. Solomon, R. M. Killen, and T. H. Zurbuchen (2009), Space environment of Mercury at the time of the first MESSENGER flyby: Solar wind and interplanetary magnetic field modeling of upstream conditions, *J. Geophys. Res.*, **114**, A10101, doi:10.1029/2009JA014287.

Balikhin, M.A., Sagdeev, R.Z., Walker, S.N., Pokhotelov, O.A., **Sibeck**, D.G., Beloff, N., Dudnikova, G., THEMIS observations of mirror structures: Magnetic holes and instability threshold, *Geophysical Research Letters*, **36**, L03105, doi:10.1029/2008GL036923, 2009.

Benna, M., Acuna, M.H., Anderson, B.J., Barabash, S., **Boardsen**, S.A., Gloeckler, G., Gold, R.E., Ho, G.C., Korth, H., Krimigis, S.M., McNutt, R.L., Raines, J.M., **Sarantos**, M., **Slavin**, J.A., Solomon, S.C., Zhang, T.L.L., Zurbuchen, T.H., Modeling the response of the induced magnetosphere of Venus to changing IMF direction using MESSENGER and Venus Express observations, *Geophysical Research Letters*, **36**, L04109, doi:10.1029/2008GL036718, 2009.

Bilitza, D., Evaluation of the IRI-2007 model options for the topside electron density, *J. Adv. Space Res.*, **44**, #6, 701-706, doi:10.1016/j.asr.2009.04.036 2009.

Billingham, L., Schwartz, S.J., **Sibeck**, D.G., The statistics of foreshock cavities: results of a Cluster survey, *Annales Geophysicae*, **26**(12), 3653-3667, 2008.

Birn, J., **Hesse**, M., Schindler, K., Zaharia, S., Role of entropy in magnetotail dynamics, *Journal Of Geophysical Research-Space Physics*, **114**, A00D03, doi:10.1029/2008JA014015, 2009.

Birn, J., Fletcher, L., **Hesse**, M., Neukirch, T., Energy release and transfer in solar flares: Simulations of three-dimensional reconnection, *Astrophysical Journal*, **695**(2), 1151-1162, doi:10.1088/0004-637X/695/2/1151, 2009.

Birn, J., **Hesse**, M., Reconnection in substorms and solar flares: analogies and differences, *Annales Geophysicae*, **27**(3), 1067-1078, 2009.

Boardsen, S.A., Anderson, B.J., Acuna, M.H., **Slavin**, J.A., Korth, H., Solomon, S.C., Narrow-band ultra-low-frequency wave observations by MESSENGER during its January 2008 flyby through Mercury's magnetosphere, *Geophysical Research Letters*, **36**(1), L01104, doi:10.1029/2008GL036034, 2009.

Boardsen, S. A., J. A. **Slavin**, B. J. Anderson, H. Korth, and S. C. Solomon (2009), Comparison of ultra-low-frequency waves at Mercury under northward and southward IMF, *Geophys. Res. Lett.*, **36**, L18106, doi:10.1029/2009GL039525.

Borovsky, J.E., Lavraud, B., **Kuznetsova**, M.M., Polar cap potential saturation, dayside reconnection, and changes to the magnetosphere, *Journal Of Geophysical Research-Space Physics*, **114**, A03224, doi:10.1029/2009JA014058, 2009.

Borovsky, J.E., **Hesse**, M., Birn, J., **Kuznetsova**, M.M., What determines the reconnection rate at the dayside magnetosphere?, *Journal Of Geophysical Research-Space Physics*, **113**(A7), A07210, doi:10.1029/2007JA012645, 2008.

Bougeret, J.L., Goetz, K., **Kaiser**, M.L., Bale, S.D., Kellogg, P.J., Maksimovic, M., Monge, N., Monson, S.J., Astier, P.L., Davy, S., Dekkali, M., Hinze, J.J., Manning, R.E., Aguilar-Rodriguez, E., Bonnin, X., Briand, C., Cairns, I.H., Cattell, C.A., Cecconi, B., Eastwood, J., Ergun, R.E., **Fainberg**, J., Hoang, S., Huttunen, K.E.J., Krucker, S., Lecacheux, A., MacDowall, R.J., Macher, W., Mangeney, A., Meetre, C.A., Moussas, X., Nguyen, Q.N., Oswald, T.H., Pulupa, M., Reiner, M.J., Robinson, P.A., Rucker, H., Salem, C., Santolik, O., Silvis, J.M., Ullrich, R., Zarka, P., Zouganelis, I., S/WAVES: The radio and plasma wave investigation on the STEREO Mission, *Space Science Reviews*, **136**(1-4), 487-528, doi:10.1007/s11214-007-9298-8, 2008.

Bougeret, J.-L., von Steiger, R., Webb, D. F., Ananthakrishnan, S., Cane, H. V., **Gopalswamy**, N., Kahler, S. W., Lallement, R., Sanahuja, B., Shibata, K., Vandas, M., and

Verheest, F., Commission 49: Interplanetary Plasma and Heliosphere, *IAUTA*, **27**, 124, 2008.

Bradshaw, S.J., A numerical tool for the calculation of non-equilibrium ionisation states in the solar corona and other astrophysical plasma environments, *Astronomy & Astrophysics*, **502**(1), 409-418, doi:10.1051/0004-6361/200810735, 2009.

Bradshaw, S. J., A reinterpretation of the energy balance in active region loops following new results from Hinode EIS, *Astron. & Astrophys. Lett.*, **406**, 5, 2008.

Breech, BA, Matthaeus, W.H., Cranmer, S.R., Kasper, J.C., Oughton, S., Electron and proton heating by solar wind turbulence, *Journal of Geophysical Research*, **114**, A09103, 2009.

Brosius, J. W., and G. D. **Holman**, Observations of the thermal and dynamic evolution of a solar microflare, *Ap. J.*, **692**, 492, 2009.

Brosius, J. W., Conversion from explosive to gentle chromospheric evaporation during a solar flare, *ApJ*, **701**, 1209, 2009.

Burke, W. J., O. de La Beaujardiere, L. C. Gentile, D. E. Hunton, R. F. **Pfaff**, P. A. Roddy, Y.-J. Su, and G. R. Wilson, "C/NOFS observations of plasma density and electric field irregularities at post-midnight local times," *Geophys. Res. Lett.*, **36**, 2009.

Burlaga, L.F., Ness, N.F., Compressible "turbulence" observed in the heliosheath by Voyager 2, *Astrophysical Journal*, **703**(1), 311-324, doi:10.1088/0004-637X/703/1/311, 2009.

Burlaga, L.F., Ness, N.F., Acuna, M.H., Wang, Y.M., Sheeley, N.R., Radial and solar cycle variations of the magnetic fields in the heliosheath: Voyager 1 observations from 2005 to 2008, *Journal Of Geophysical Research-Space Physics*, **114**, A06106, doi:10.1029/2009JA014071, 2009.

Burlaga, L.F., Tsallis statistics for models and observations of the heliospheric magnetic field, *Numerical Modeling Of Space Plasma Flows: Astronum-2008*, **406**, 181-188, 2009.

Burlaga, L.F., Ness, N.F., Acuna, M.H., Richardson, J.D., Stone, E., McDonald, F.B., Observations of the heliosheath and solar wind near the termination shock by Voyager 2, *Astrophysical Journal*, **692**(2), 1125-1130, doi:10.1088/0004-637X/692/2/1125, 2009.

Burlaga, L.F., Ness, N.F., Acuna, M.H., Magnetic field strength fluctuations and temperature in the heliosheath, *Astrophysical Journal Letters*, **691**(2), L82-L86, doi:10.1088/0004-637X/691/2/L82, 2009.

Burlaga, L.F., Ness, N.F., Acuna, M.H., Magnetic fields in the termination shock, heliosheath and solar wind, *Particle Acceleration And Transport In The Heliosphere And Beyond*, **1039**, 329-334, 2008.

Burlaga, L.F., Ness, N.F., Acuna, M.H., Wang, Y.M., Sheeley, N.R., Wang, C., Richardson, J.D., Global structure and dynamics of large-scale fluctuations in the solar wind: Voyager 2 observations during 2005 and 2006, *Journal Of Geophysical Research-Space Physics*, **113**(A2), A02104, doi:10.1029/2007JA012796, 2008.

Carpenter, K.G., **Airapetian**, V., The atmospheric dynamics of alpha Tau (K5 III) - Clues to understanding the magnetic dynamo in late-type giant stars, cool stars, *Stellar Systems And The Sun*, **1094**, 712-715, 2009

Chamberlin, P. C., T. N. Woods, D. A. Crotser, F. G. Eparvier, R. A. Hock, and D. L. Wodraska, Solar cycle minimum measurements of the solar extreme ultraviolet spectral irradiance on 14 April 2008, *Geophys. Res. Lett.*, **36**, L05102, doi:10.1029/2008GL037145, 2009.

Chandler, M.O., Avanov, L.A., Craven, P.D., Mozer, F.S., **Moore**, T.E., Observations of the ion signatures of double merging and the formation of newly closed field lines, *Geophysical Research Letters*, **35**(10), L10107, doi:10.1029/2008GL033910, 2008.

Chappell, C.R., Huddleston, M.M., **Moore**, T.E., Giles, B.L., Delcourt, D.C., Observations of the warm plasma cloak and an explanation of its formation in the magnetosphere, *Journal Of Geophysical Research-Space Physics*, **113**(A9), A09206, doi:10.1029/2007JA012945, 2008.

Chaston, C.C., Johnson, J.R., Wilber, M., Acuna, M., **Goldstein**, M.L., Reme, H., Kinetic Alfvén wave turbulence and transport through a reconnection diffusion region, *Physical Review Letters*, **102**(1), 015001, doi:10.1103/PhysRevLett.102.015001, 2009.

Chen, P.C., Lyon, R.G., Van Steenberg, M.E., Optical design and in situ fabrication of large telescopes on the Moon - art. no. 70104D, *Space Telescopes And Instrumentation 2008: Optical, Infrared, And Millimeter, Pts 1 And 2*, 7010, D104-D104, 2008.

Christian, E.R., **Kaiser**, M.L., **Kucera**, T.A., **StCyr**, O.C., van Driel-Gesztelyi, L., Mandrini, C.H., STEREO science results at solar minimum - Preface, *Solar Physics*, **256**(1-2), 1-2, doi:10.1007/s11207-009-9365-2, 2009.

Collier, M.R., **Stubbs**, T.J., Neutral solar wind generated by lunar exospheric dust at the terminator, *Journal Of Geophysical Research-Space Physics*, **114**, A01104, doi:10.1029/2008JA013716, 2009.

Collier, M. R., T. F. Abbey, N. P. Bannister, J. A. Carter, M. **Choi**, T. Cravens, M. Evans, G. W. Fraser, H. K. **Hills**, K. **Kuntz**, J. **Lyons**, N. Omidi, F. S. **Porter**, A. M. Read, I. Robertson, P. **Rozmarynowski**, S. Sembay, D. G. **Sibeck**, S. L. **Snowden**, T.

Stubbs, and P. Travnicek, The Lunar X-ray Observatory (LXO)/Magnetosheath Explorer in X-rays (MagEX), The Local Bubble and Beyond II: *Proceedings of the International Conf.*, **1156**(1), 105–111, doi:10.1063/1.3211802, 2009.

Cooper, J. F., P. D. **Cooper**, E. C. **Sittler**, S. J. **Sturner**, and A. M. Rymer, Old Faithful model for radiolytic gas-driven cryovolcanism at Enceladus, *Planetary and Space Science*, doi:10.1016/j.pss.2009.08.002, 2009.

Cooper, J. F., R. E. **Hartle**, E. C. **Sittler Jr.**, R. M. **Killen**, S. J. **Sturner**, C. Paranicas, M. E. Hill, A. M. Rymer, P. D. Cooper, D. Pascu, R. E. Johnson, T. A. Cassidy, T. M. Orlando, K. D. Retherford, N. A. Schwadron, R. I. Kaiser, F. Leblanc, L. J. Lanzerotti, C. J. Alexander, H. B. Garrett, A. R. Hendrix, and W. H. Ip, Space weathering impact on solar system surfaces and mission science, Community White Paper submitted to Planetary Science Decadal Survey, 2013—2022, National Research Council, Washington, D.C., Sept. 15, 2009.

Coustenis, A., Atreya, S.K., Balint, T., Brown, R.H., Dougherty, M.K., Ferri, F., Fulchignoni, M., Gautier, D., Gowen, R.A., Griffith, C.A., Gurvits, L.I., Jaumann, R., Langevin, Y., Leese, M.R., Lunine, J.I., Mckay, C.P., Moussas, X., Muller-Wodarg, I., Neubauer, F., Owen, T.C., Raulin, F., **Sittler**, E.C., Sohl, F., Sotin, C., Tobie, G., Tokano, T., Turtle, E.P., Wahlund, J.E., Waite, J.H., Baines, K.H., Blamont, J., Coates, A.J., Dandouras, I., Krimigis, T., Lellouch, E., Lorenz, R.D., Morse, A., Porco, C.C., Hirtzig, M., Saur, J., Spilker, T., Zarnecki, J.C., Choi, E., Achilleos, N., Amils, R., Annan, P., Atkinson, D.H., Benilan, Y., Bertucci, C., Bezard, B., Bjoraker, G.L., Blanc, M., Boireau, L., Bouman, J., Cabane, M., Capria, M.T., Chassefiere, E., Coll, P., Combes, M., **Cooper**, J.F., Coradini, A., Crary, F., Cravens, T., Daglis, I.A., de Angelis, E., de Bergh, C., de Pater, I., Dunford, C., Durry, G., Dutuit, O., Fairbrother, D., Flasar, F.M., Fortes, A.D., Frampton, R., Fujimoto, M., Galand, M., Grasset, O., Grott, M., Haltigin, T., Herique, A., Hersant, F., Hussmann, H., Ip, W., Johnson, R., Kallio, E., Kempf, S., Knapmeyer, M., Kofman, W., Koop, R., Kostiuk, T., Krupp, N., Kuppers, M., Lammer, H., Lara, L.M., Lavvas, P., Le Mouelic, S., Lebonnois, S., Ledvina, S., Li, J., Livengood, T.A., Lopes, R.M., Lopez-Moreno, J.J., Luz, D., Mahaffy, P.R., Mall, U., Martinez-Frias, J., Marty, B., McCord, T., Salvan, C., Milillo, A., Mitchell, D.G., Modolo, R., Mousis, O., Nakamura, M., Neish, C.D., Nixon, C.A., Mvondo, D., Orton, G., Paetzold, M., Pitman, J., Pogrebenko, S., Pollard, W., Prieto-Ballesteros, O., Rannou, P., Reh, K., Richter, L., Robb, F.T., Rodrigo, R., Rodriguez, S., Romani, P., Bermejo, M., Sarris, E.T., Schenk, P., Schmitt, B., Schmitz, N., Schulze-Makuch, D., Schwingenschuh, K., Selig, A., Sicardy, B., Soderblom, L., Spilker, L.J., Stam, D., Steele, A., Stephan, K., Strobel, D.F., Szego, K., Szopa, C., Thissen, R., Tomasko, M.G., Toublanc, D., Vali, H., Vardavas, I., Vuitton, V., West, R.A., Yelle, R., Young, E.F., TandEM: Titan and Enceladus mission, *Experimental Astronomy*, **23**(3), 893-946, doi:10.1007/s10686-008-9103-z, 2009.

Cranmer, S.R., Matthaeus, W.H., **Breech**, BA, Kasper, J.C., Empirical constraints on proton and electron heating in the fast solar wind, *Astrophysical Journal*, **702**(2), 1604-1614, doi:10.1088/0004-637X/702/2/1604, 2009.

Crooker, N.U., Kahler, S.W., Gosling, J.T., **Lepping**, R.P., Evidence in magnetic clouds for systematic open flux transport on the Sun, *Journal Of Geophysical Research-Space Physics*, **113**(A12), A12107, doi:10.1029/2008JA013628, 2008.

Dahlburg, R. B., Liu, J.-H., **Klimchuk**, J. A., & Nigro, G., Explosive instability and coronal heating, *Astrophys. J.*, **704**, 1059, 2009.

Daoudi, H., Blush, L. M., Bochsler, P., Galvin, A. B., Giammanco, C., Karrer, R., Opitz, A., Wurz, P., Farrugia, C., Kistler, L. A., Popecki, M. A., Möbius, E., Singer, K., Klecker, B., Wimmer-Schweingruber, R. F., **Thompson, B.**, The STEREO/PLASTIC response to solar wind ions (Flight measurements and models), *Astrophys. and Spa. Sci. Trans.*, **5**, 1, 2009

Davila, J., **Gopalswamy**, N., **Thompson**, B., Haubold, H.J., International Heliophysical Year 2007: A report from the UN/NASA Workshop Bangalore, India, 27 November-1 December 2006, *Earth Moon And Planets*, **103**(1-2), 9-24, doi:10.1007/s11038-008-9231-5, 2008

Davila, J.M., **Gopalswamy**, N., **Thompson**, B.J., Universal processes in heliophysics, *Universal Heliophysical Processes*, **257**, 11-16, 2009.

Davila, J. M., **Gopalswamy**, N., **Thompson**, B. J., Universal processes in heliophysics, *Proc. of IAU*, v. 257, pp. 11 – 16, 2009.

De La Beaujardiere, O., Retterer, J.M., **Pfaff**, R.F., Roddy, P.A., Roth, C., Burke, W.J., **Su**, Y.J., Kelley, M.C., Ilma, R.R., Wilson, G.R., Gentile, L.C., Hunton, D.E., Cooke, D.L., C/NOFS observations of deep plasma depletions at dawn, *Geophysical Research letters*, **36**, L00C06, doi:10.1029/2009GL038884, 2009.

De Moortel, I., **Bradshaw**, S. J., Forward modeling of corona intensity perturbations, *Sol. Phys.*, **252**, 101, 2008.

Demoulin, P., **Pariat**, E., Modelling and observations of photospheric magnetic helicity, *Advances In Space Research*, **43**(7), 1013-1031, doi:10.1016/j.asr.2008.12.004, 2009.

Dennis, B.R., **Pernak**, R.L., Hard x-ray flare source sizes measured with the Ramaty High Energy Solar Spectroscopic Imager, *Astrophysical Journal*, **698**(2), 2131-2143, doi:10.1088/0004-637X/698/2/2131, 2009.

Eastwood, J.P., Bale, S.D., Maksimovic, M., Zouganelis, I., Goetz, K., **Kaiser**, M.L., Bougeret, J.L., Measurements of stray antenna capacitance in the STEREO/WAVES instrument: Comparison of the radio frequency voltage spectrum with models of the galactic nonthermal continuum spectrum, *Radio Science*, **44**, RS4012, doi:10.1029/2009RS004146, 2009.

Ebihara, Y., Nishitani, N., Kikuchi, T., Ogawa, T., Hosokawa, K., **Fok**, M.C., Thomsen, M.F., Dynamical property of storm time subauroral rapid flows as a manifestation of complex structures of the plasma pressure in the inner magnetosphere, *Journal Of Geophysical Research-Space Physics*, **114**, A01306, doi:10.1029/2008JA013614, 2009.

Ebihara, Y., Nishitani, N., Kikuchi, T., Ogawa, T., Hosokawa, K., **Fok**, M.C., Two-dimensional observations of overshielding during a magnetic storm by the Super Dual Auroral Radar Network (SuperDARN) Hokkaido radar, *Journal Of Geophysical Research-Space Physics*, **113**(A1), A01213, doi:10.1029/2007JA012641, 2008

Engebretson, M.J., Posch, J.L., Westerman, A.M., Otto, N.J., **Slavin**, J.A., **Le**, G., Strangeway, R.J., Lessard, M.R., Temporal and spatial characteristics of Pc1 waves observed by ST5, *Journal Of Geophysical Research-Space Physics*, **113**(A7), A07206, doi:10.1029/2008JA013145, 2008.

Ergun, R.E., Malaspina, D.M., Cairns, I.H., Goldman, M.V., Newman, D.L., Robinson, P.A., Eriksson, S., Bougeret, J.L., Briand, C., Bale, S.D., Cattell, C.A., Kellogg, P.J., **Kaiser**, M.L., Eigenmode structure in solar-wind Langmuir waves, *Physical Review Letters*, **101**(5), 051101, doi:10.1103/PhysRevLett.101.051101, 2008.

Farrell, W.M., Kurth, W.S., Gurnett, D.A., Johnson, R.E., **Kaiser**, M.L., Wahlund, J.E., Waite, J.H., Electron density dropout near Enceladus in the context of water-vapor and water-ice, *Geophysical Research Letters*, **36**, L10203, doi:10.1029/2008GL037108, 2009.

Farrell, W.M., **Kaiser**, M.L., Gurnett, D.A., Kurth, W.S., Persoon, A.M., Wahlund, J.E., Canu, P., Mass unloading along the inner edge of the Enceladus plasma torus, *Geophysical Research Letters*, **35**(2), L02203, doi:10.1029/2007GL032306, 2008.

Farrugia, C.J., Gratton, F.T., Jordanova, V.K., Matsui, H., Muhlbachler, S., Torbert, R.B., **Ogilvie**, K.W., Singer, H.J., Tenuous solar winds: Insights on solar wind-magnetosphere interactions, *Journal Of Atmospheric And Solar-Terrestrial Physics*, **70**(2-4), 371-376, doi:10.1016/j.jastp.2007.08.032, 2008.

Feofilov, A. G., **A. A. Kutepov**, W. D. **Pesnell**, R. A. **Goldberg**, B. T. Marshall, L. L. Gordley, M. García-Comas, M. López-Puertas, R. O. Manuilova, V. A. Yankovsky, S. V. Petelina, and J. M. Russell III, Daytime SABER/TIMED observations of water vapor in the mesosphere: Retrieval approach and first results, *Atmos. Chem. Phys. Discuss.*, **9**, 13,943-13,997, 2009.

Fischer, G., Gurnett, D.A., Kurth, W.S., Akalin, F., Zarka, P., Dyudina, U.A., Farrell, W.M., **Kaiser**, M.L., Atmospheric electricity at Saturn, *Space Science Reviews*, **137**(1-4), 271-285, doi:10.1007/s11214-008-9370-z, 2008.

Fujii, H.A., Watanabe, T., Kojima, H., Oyama, K-I, Kusagaya, T., Yamagiwa, Y., Ohtsu, H., Cho, M., Sasaki, S., Tanaka, K., Williams, J., Rubin, B., Johnson, C. L., **Khazanov**, G., Sanmartin, J.R., Lebreton, J-P., van der Heidek, E., Kruijff, M., De Pascale, F., Trivailo, P.M., Sounding rocket experiment of bare electrodynamic tether system, *Acta Astronautica*, Elsevier Ltd., **64**, pp. 313–324, 2009.

Fuselier, S.A., Claflin, E.S., **Moore**, T.E., Magnetic local time extent of ion outflow during substorm recovery, *Journal Of Geophysical Research-Space Physics*, **113**(A6), A06204, doi:10.1029/2007JA012811, 2008.

Fuselier, G. Gloeckler, M. Gruntman, J. Heerikhuisen, V. Izmodenov, P. Janzen, P. Knappenberger, S. Krimigis, H. Kucharek, M. Lee, G. Livadiotis, S. Livi, R. J. **MacDowall**, D. Mitchell, E. Möbius, T. **Moore**, N. V. Pogorelov, D. Reisenfeld, E. Roelof, L. Saul, N. A. Schwadron, P. W. Valek, R. Vanderspek, P. Wurz, G. P. Zank, Global observations of the interstellar interaction from the Interstellar Boundary Explorer (IBEX), *www.sciencexpress.org*, 15 October 2009, Page 1, 10.1126, science.1180906.

Fuselier, S. A., F. Allegrini, H. O. Funsten, A. G. Ghielmetti, D. Heirtzler, H. Kucharek, O. W. Lennartsson, D. J. McComas, E. Möbius, T. E. **Moore**, S. M. Petrinec, L. A. Saul, J. A. Scheer, N. Schwadron, P. Wurz, Width and variation of the ena flux ribbon observed by the Interstellar Boundary Explorer, *www.sciencexpress.org*, 15 October 2009, Page 1, 10.1126, science.1180981.

Galli, A., Wurz, P., Kallio, E., Ekenback, A., Holmstrom, M., Barabash, S., Grigoriev, A., Futaana, Y., **Fok**, M.C., Gunell, H., Tailward flow of energetic neutral atoms observed at Mars, *Journal Of Geophysical Research-Planets*, **113**(E12), E12012, doi:10.1029/2008JE003139, 2008.

Galli, A., **Fok**, M.C., Wurz, P., Barabash, S., Grigoriev, A., Futaana, Y., Holmstrom, M., Ekenback, A., Kallio, E., Gunell, H., Tailward flow of energetic neutral atoms observed at Venus, *Journal Of Geophysical Research-Planets*, **113**, E00B15, doi:10.1029/2008JE003096, 2008.

Gamayunov, K.V., **Khazanov**, G.V., Liemohn, M.W., **Fok**, M.C., Ridley, A.J., Self-consistent model of magnetospheric electric field, ring current, plasmasphere, and electromagnetic ion cyclotron waves: Initial results, *Journal Of Geophysical Research-Space Physics*, **114**, A03221, doi:10.1029/2008JA013597, 2009.

Gamayunov, K.V., **Khazanov**, G.V., Crucial role of ring current H+ in electromagnetic ion cyclotron wave dispersion relation: Results from global simulations, *Journal Of Geophysical Research-Space Physics*, **113**(A11), A11220, doi:10.1029/2008JA013494, 2008.

Gibson, S. E., Kozyra, J. U., de Toma, G., Emery, B. A., Onsager, T., **Thompson**, B. J., If the Sun is so quiet, why is the Earth ringing? A comparison of two solar minimum intervals, *Jour. Geophys. Res.*, **114**, A9, 2009.

Gilbert, H. R., K. **Strong**, J. **Saba**, GSFC Heliophysics Science Division 2008 Science Highlights, NASA/TM-2009-214178, 2009.

Gizon, L., Schunker, H., Baldner, C.S., Basu, S., Birch, A.C., Bogart, R.S., Braun, D.C., Cameron, R., **Duvall**, T.L., Hanasoge, S.M., Jackiewicz, J., Roth, M., Stahn, T., Thompson, M.J., Zharkov, S., Helioseismology of sunspots: A Case Study of NOAA Region 9787, *Space Science Reviews*, **144**(1-4), 249-273, doi:10.1007/s11214-008-9466-5, 2009.

Gjerloev, J.W., Hoffman, R.A., **Sigwarth**, J.B., Frank, L.A., Baker, J.B.H., Typical auroral substorm: A bifurcated oval, *Journal Of Geophysical Research-Space Physics*, **113**(A3), A03211, doi:10.1029/2007JA012431, 2008.

Glocer, A., Toth, G., Gombosi, T., Welling, D., Polar Wind Ouflow Model (PWOM): Modeling ionospheric outflows and their impact on the magnetosphere, initial results, *Journal Of Geophysical Research-Space Physics*, **114**, A05216, doi:10.1029/2009JA014053, 2009.

Glocer, A., G. Toth, M. **Fok**, T. Gombosi, M. Liemohn, Integration of the radiation belt environment model into the space weather modeling framework, *Journal of Atmospheric and Solar-Terrestrial Physics*, doi:10.1016/j.jastp.2009.01.003, 2009.

Goldstein, M.L., Observations and Modeling of Turbulence in the Solar Wind, *Turbulence, Dynamos, Accretion Disks, Pulsars And Collective Plasma Processes*, 21-33, 2009.

Goldstein, M. L. (2008), Observations and modeling of turbulence in the solar wind, First Kodai-Trieste Workshop on Plasma Astrophysics, Kodaikanal, India, August 27–September 7, 2007, *Astrophysics and Space Science Proceedings*, ISSN 1570-6591 (Print) 1570-6605 (Online), Springer Netherlands, DOI 10.1007/978-1-4020-8868-1.

Gopalswamy, N., Halo coronal mass ejections and geomagnetic storms, *Earth Planets And Space*, **61**(5), 595-597, 2009.

Gopalswamy, N., **Akiyama**, S., **Yashiro**, S., Major solar flares without coronal mass ejections, *Universal Heliophysical Processes*, **257**, 283-286, 2009.

Gopalswamy, N., **Makela**, P., Xie, H., **Akiyama**, S., **Yashiro**, S., CME interactions with coronal holes and their interplanetary consequences, *Journal Of Geophysical Research-Space Physics*, **114**, A00A22, doi:10.1029/2008JA013686, 2009.

Gopalswamy, N., Introduction to special section on large geomagnetic storms, *Journal Of Geophysical Research-Space Physics*, **114**, A00A00, doi:10.1029/2008JA014026, 2009.

Gopalswamy, N., Eichhorn, G., Sakurai, T., Haubold, H., Preface to the Proceedings of the European General Assembly and the United Nations Workshop, *Earth Moon And Planets*, **104**(1-4), 139-140, doi:10.1007/s11038-008-9278-3, 2009.

Gopalswamy, N., **Yashiro**, S., Michalek, G., **Stenborg**, G., Vourlidas, A., Freeland, S., Howard, R., The SOHO/LASCO CME Catalog, *Earth Moon And Planets*, **104**(1-4), 295-313, doi:10.1007/s11038-008-9282-7, 2009.

Gopalswamy, N., **Yashiro**, S., Temmer, M., **Davila**, J., **Thompson**, W.T., Jones, S., McAteer, R.T.J., Wuelser, J.P., Freeland, S., Howard, R.A., EUV wave reflection from a coronal hole, *Astrophysical Journal Letters*, **691**(2), L123-L127, doi:10.1088/0004-637X/691/2/L123, 2009.

Gopalswamy, N., Solar connections of geoeffective magnetic structures, *Journal Of Atmospheric And Solar-Terrestrial Physics*, **70**(17), 2078-2100, doi:10.1016/j.jastp.2008.06.010, 2008.

Gopalswamy, N., Xie, H., Comment on "Prediction of the 1-AU arrival times of CME-associated interplanetary shocks: Evaluation of an empirical interplanetary shock propagation model" by K.-H. Kim et al., *Journal Of Geophysical Research-Space Physics*, **113**(A10), A10105, doi:10.1029/2008JA013030, 2008.

Gopalswamy, N., **Yashiro**, S., **Akiyama**, S., **Makela**, P., Xie, H., **Kaiser**, M.L., Howard, R.A., Bougeret, J.L., Coronal mass ejections, type II radio bursts, and solar energetic particle events in the SOHO era, *Annales Geophysicae*, **26**(10), 3033-3047, 2008.

Gopalswamy, N., Type II radio emission and solar energetic particle events, *Particle Acceleration And Transport In The Heliosphere And Beyond*, **1039**, 196-202, 2008.

Gopalswamy, N., **Akiyama**, S., **Yashiro**, S., Michalek, G., **Lepping**, R.P., Solar sources and geospace consequences of interplanetary magnetic clouds observed during solar cycle 23, *Journal Of Atmospheric And Solar-Terrestrial Physics*, **70**(2-4), 245-253, doi:10.1016/j.jastp.2007.08.070, 2008.

Gopalswamy, N., Dal Lago, Yashiro, S., and Akiyama, S., The expansion and radial speeds of coronal mass ejections, *Cent. Eur. Astrophys. Bull.*, **33**, 115, 2009.

Gopalswamy, N. and Webb, D. F., Universal Heliophysical Processes, IAUS, 257, 2009.

Gopalswamy, N., Coronal mass ejections and space weather, in Climate and Weather of the Sun-Earth System (CAWSES) edited by Ed. T. Tsuda, R. Fujii, K. Shibata, and M. A. Geller, *TERRAPUB*, Tokyo, pp. 77-120, 2009.

Gopalswamy, N., W. T. **Thompson**, J. M. **Davila**, M. L. **Kaiser**, S. **Yashiro**, P. **Mäkelä**, G. Michalek, J.-L. Bougeret, and R. A. Howard, Relation between type II bursts and CMEs Inferred from STEREO observations, *Sol.Phys.*, **259**, 227, 2009. (Online).

Gopalswamy, N., S. **Akiyama**, S. **Yashiro**, and P. **Makela**, Coronal Mass Ejections from sunspot and non-sunspot regions, To appear in "Magnetic Coupling between the Interior and the Atmosphere of the Sun", eds. S. S. Hasan and R. J. Rutten, *Astrophysics and Space Science Proceedings*, Springer-Verlag, Heidelberg, Berlin, 2009.

Gosling, J.T., McComas, D.J., **Roberts**, D.A., Skoug, R.M., A one-sided aspect of Alfvénic fluctuations in the solar wind, *Astrophysical Journal LetterS*, **695**(2), L213-L216, doi:10.1088/0004-637X/695/2/L213, 2009.

Grebowsky J., R. **Benson**, P. Webb, V. Truhlik, and D. **Bilitza**, Altitude variation of the plasmapause signature in the main ionospheric trough, *J. Atmos. Sol.-Terr. Phys.*, **71**, 1669–1676 doi:10.1016/j.jastp.2009.05.016 2009.

Grebowsky, J. M., R. F. **Benson**, P. A. **Webb**, V. Truhlik, and D. **Bilitza**, Altitude variation of the plasmapause signature in the main ionospheric trough, *J. Atmos. Solar-Terr. Phys.*, doi:10.1016/j.jastp.2009.1005.1016, 2009.

Halekas, J.S., Delory, G.T., Lin, R.P., **Stubbs**, T.J., Farrell, W.M., Lunar surface charging during solar energetic particle events: Measurement and prediction, *Journal Of Geophysical Research-Space Physics*, **114**, A05110, doi:10.1029/2009JA014113, 2009.

Halekas, J.S., Delory, G.T., Lin, R.P., **Stubbs**, T.J., Farrell, W.M., Lunar Prospector measurements of secondary electron emission from lunar regolith, *Planetary And Space Science*, **57**(1), 78-82, doi:10.1016/j.pss.2008.11.009, 2009.

Hamilton, K., Smith, C.W., Vasquez, B.J., **Leamon**, R.J., Anisotropies and helicities in the solar wind inertial and dissipation ranges at 1 AU, *Journal Of Geophysical Research-Space Physics*, **113**(A1), A01106, doi:10.1029/2007JA012559, 2008.

Hanasoge, S.M., **Duvall**, T.L., Sub-wavelength resolution imaging of the solar deep interior, *Astrophysical Journal*, **693**(2), 1678-1685, doi:10.1088/0004-637X/693/2/1678, 2009.

Hesse, M., **Zenitani**, S., **Klimas**, A., The structure of the electron outflow jet in collisionless magnetic reconnection, *Physics Of Plasmas*, **15**(11), 112102, doi:10.1063/1.3006341, 2008.

Hill, F., Martens, P., Yoshimura, K., **Gurman**, J., **Hourcle**, J., **Dimitoglou**, G., Suarez-Sola, I., Wampler, S., Reardon, K., Davey, A., Bogart, R., Tian, K., The Virtual Solar Observatory-A resource for International Heliophysics Research, *Earth Moon And Planets*, **104**(1-4), 315-330, doi:10.1007/s11038-008-9274-7, 2009.

Holmstrom, M., **Collier**, M.R., Barabash, S., Brinkfeldt, K., **Moore**, T.E., **Simpson**, D., Mars Express/ASPERA-3/NPI and IMAGE/LENA observations of energetic neutral atoms in Earth and Mars orbit, *Advances In Space Research*, **41**(2), 343-350, doi:10.1016/j.asr.2007.09.011, 2008.

Hosokawa, K., Taguchi, S., Suzuki, S., **Collier**, M.R., **Moore**, T.E., Thomsen, M.F., Estimation of magnetopause motion from low-energy neutral atom emission, *Journal Of*

Geophysical Research-Space Physics, **113**(A10), A10205, doi:10.1029/2008JA013124, 2008

Huang, T.S., E. Romashets, G. **Le**, Y. Wang, J.A. **Slavin**, A new time-dependent ionosphere–magnetosphere coupling model: Comparison of field-aligned currents against ST5 observations, *Journal of Atmospheric and Solar-Terrestrial Physics*, doi:10.1016/j.jastp.2009.03.020, 2009.

Hughes, P.P., Coplan, M.A., DeFazio, J.N., **Chornay**, D.J., **Collier**, M.R., **Ogilvie**, K.W., **Shappirio**, M.D., Scattering of neutral hydrogen at energies less than 1 keV from tungsten and diamondlike carbon surfaces, *JOurnal Of Vacuum Science & Technology A*, **27**(5), 1188-1195, doi:10.1116/1.3196788, 2009.

Huttunen, K.E.J., Kilpua, S.P., **Pulkkinen**, A., Viljanen, A., Tanskanen, E., Solar wind drivers of large geomagnetically induced currents during the solar cycle 23, *Space Weather-The International Journal Of Research And Applications*, **6**(10), S10002, doi:10.1029/2007SW000374, 2008.

Hysell, D.L., Hedden, R.B., Chau, J.L., Galindo, F.R., Roddy, P.A., **Pfaff**, R.F., Comparing F region ionospheric irregularity observations from C/NOFS and Jicamarca, *Geophysical Research Letters*, **36**, L00C01, doi:10.1029/2009GL038983, 2009.

Ieda, A., **Fairfield**, D.H., **Slavin**, J.A., Liou, K., Meng, C.I., Machida, S., Miyashita, Y., Mukai, T., Saito, Y., Nose, M., Shue, J.H., Parks, G.K., Fillingim, M.O., Longitudinal association between magnetotail reconnection and auroral breakup based on Geotail and Polar observations, *Journal Of Geophysical Research-Space Physics*, **113**(A8), A08207, doi:10.1029/2008JA013127, 2008.

Jacobsen, K.S., Phan, T.D., Eastwood, J.P., **Sibeck**, D.G., Moen, J.I., Angelopoulos, V., McFadden, J.P., Engebretson, M.J., Provan, G., Larson, D., Fornacon, K.H., THEMIS observations of extreme magnetopause motion caused by a hot flow anomaly, *Journal Of Geophysical Research-Space Physics*, **114**, A08210, doi:10.1029/2008JA013873, 2009.

Jennings, D.E., Flasar, F.M., Kunde, V.G., Samuelson, R.E., Pearl, J.C., Nixon, C.A., Carlson, R.C., Mamoutkine, A.A., Brasunas, J.C., Guandique, E., Achterberg, R.K., Bjoraker, G.L., Romani, P.N., Segura, M.E., Albright, S.A., **Elliott**, M.H., Tingley, J.S., Calcutt, S., Coustenis, A., Courtin, R., TITAN's surface brightness temperatures, *Astrophysical Journal letters*, **691**(2), L103-L105, doi:10.1088/0004-637X/691/2/L103, 2009.

Jeong, H., J. Chae, and Y-J. Moon, Magnetic helicity injection during the formation of an intermediate filament, *Journal of the Korean Astronomical Society*, **42**, 9, 2009.

Johnson, R.E., Fama, M., Liu, M., Baragiola, R.A., **Sittler**, E.C., Smith, H.T., Sputtering of ice grains and icy satellites in Saturn's inner magnetosphere, *Planetary And Space Science*, **56**(9), 1238-1243, doi:10.1016/j.pss.2008.04.003, 2008.

Jones, H.P., Chapman, G.A., Harvey, K.L., **Pap**, J.M., Preminger, D.G., Turmon, M.J., Walton, S.R., A comparison of feature classification methods for modeling solar irradiance variation, *Solar Physics*, **248**(2), 323-337, doi:10.1007/s11207-007-9069-4, 2008.

Jones, S.I., **Davila**, J.M., Localized plasma density enhancements observed in STEREO COR1, *Astrophysical Journal*, **701**(2), 1906-1910, doi:10.1088/0004-637X/701/2/1906, 2009.

Juckett, D.A., **Wolff**, C.L., Correspondence between solar variability (0.6-aEuro parts per thousand 7.0 years) and the theoretical positions of rotating sets of coupled g modes, *Solar Physics*, **257**(1), 13-36, doi:10.1007/s11207-009-9340-y, 2009.

Juckett, D.A., **Wolff**, C.L., Evidence for long-term retrograde motions of sunspot patterns and indications of coupled g-mode rotation rates, *Solar Physics*, **252**(2), 247-266, doi:10.1007/s11207-008-9265-x, 2008.

Juusola, L., Amm, O., Frey, H.U., Kauristie, K., Nakamura, R., Owen, C.J., Sergeev, V., **Slavin**, J.A., Walsh, A., Ionospheric signatures during a magnetospheric flux rope event, *Annales Geophysicae*, **26**(12), 3967-3977, 2008.

Kaladze, T. D., W. Horton, T. W. Garner, J. W. Van Dam, and M. L. **Mays** (2009), A method for the intensification of atomic oxygen green line emission by internal gravity waves, J. Geophys. Res., 113, A12307, doi:10.1029/2008JA013425.

Karimabadi, H., T. B. Sipes, Y. **Wang**, B. Lavraud, and A. **Roberts**, A new multivariate time series data analysis technique: Automated detection of flux transfer events using Cluster data, *J. Geophys. Res.*, **114**, A06216, doi:10.1029/2009JA014202, 2009.

Kauhanen, J., **Siili**, T., Jarvenoja, S., Savijarvi, H., The Mars limited area model and simulations of atmospheric circulations for the Phoenix landing area and season of operation, *Journal Of Geophysical Research-Planets*, **113**, E00A14, doi:10.1029/2007JE003011, 2008.

Kepko, L., H. E. Spence, D. F. Smart, M. A. Shea, Solar cosmic rays observed in Greenland ice cores:1940-1950, *Adv. Space Res.*, 2009.

Khazanov, G. V., A. A. Tel'nikhin, and T. K. Kronberg, Chaotic motion of relativistic electrons driven by Whistler Waves, Chapter in *Plasma Physics Research Advances*, NOVA Publishers, 2009.

Kilper, G., H. **Gilbert**, and D. Alexander, Mass composition in pre-eruption quiet Sun filaments, *Astrophysical Journal*, **704**, 522, 2009.

Kirk, M. S., W. D. **Pesnell**, C. A. Young, and S. A. Hess Webber, Automated detection of EUV polar coronal holes during solar cycle 23, *Solar Physics*, **257**, 99-112, 2009, 10.1007/s11207-009-9369-y2.

Kiyani, K.H., Chapman, S.C., Khotyaintsev, Y.V., Dunlop, M.W., **Sahraoui**, F., Global scale-invariant dissipation in collisionless plasma turbulence, *Physical Review Letters*, **103**(7), 075006, doi:10.1103/PhysRevLett.103.075006, 2009.

Klimchuk, J. A., van Driel-Gesztelyi, L., Schrijver, C. J., Melrose, D. B., Fletcher, L., **Gopalswamy**, N., Harrison, R. A., Mandrini, C. H., Peter, H., Tsuneta, S., Vrsnak, B., and Wang, J.-X., Commission 10: Solar Activity, *IAUTA*, **27**, 79, 2008.

Klimchuk, J. A., et al., Commission 10: Solar Activity (2009 Triennial report), in *Transactions IAU, reports on Astronomy 2006-2009*, **XXVIIA**, ed. K. A. van der Hucht (Cambridge: Cambridge Univ. Press), 79, 2009.

Kondo, Y., Hudman, R.C., Nakamura, K., Koike, M., Chen, G., Miyazaki, Y., Takegawa, N., Blake, D.R., Simpson, I.J., Ko, M., Kita, K., Shirai, T., Kawakami, S., Mechanisms that influence the formation of high-ozone regions in the boundary layer downwind of the Asian continent in winter and spring, *Journal Of Geophysical Research-Atmospheres*, **113**(D15), D15304, doi:10.1029/2007JD008978, 2008.

Korotova, G.I., **Sibeck**, D.G., Rosenberg, T., Geotail observations of FTE velocities, *Annales Geophysicae*, **27**(1), 83-92, 2009.

Kuznetsova, M.M., **Sibeck**, D.G., **Hesse**, M., **Wang**, Y., **Rastaetter**, L., Toth, G., Ridley, A., Cavities of weak magnetic field strength in the wake of FTEs: Results from global magnetospheric MHD simulations, *Geophysical Research Letters*, **36**, L10104, doi:10.1029/2009GL037489, 2009.

Lakhina, G.S., S.V. Singh, A.P. Kakad, M.L. **Goldstein**, A. F. **Viñas**, J.S. Pickett, (2009),"A mechanism for electrostatic solitary structures in the Earth's magnetosheath", *J. Geophys. Res.*, **114**, A09212, doi:10.1029/2009JA014306.

Landi, E., **Bhatia**, A.K., Atomic data and spectral line intensities for Ni XXV, *Atomic Data And Nuclear Data Tables*, **95**(4), 547-576, doi:10.1016/j.adt.2009.03.001, 2009.

Landi, E., **Bhatia**, A.K., Atomic data and spectral line intensities for Ca XVII, *Atomic Data And Nuclear Data Tables*, **95**(2), 155-183, doi:10.1016/j.adt.2008.10.003, 2009.

Laurenza, M., Cliver, E.W., Hewitt, J., Storini, M., Ling, A.G., Balch, C.C., **Kaiser**, M.L., A technique for short-term warning of solar energetic particle events based on flare location, flare size, and evidence of particle escape, *Space Weather-The InternationaL Journal Of Research And Applications*, **7**, S04008, doi:10.1029/2007SW000379, 2009.

Le Contel, O., Roux, A., Jacquey, C., Robert, P., Berthomier, M., Chust, T., Grison, B., Angelopoulos, V., **Sibeck**, D., Chaston, C.C., Cully, C.M., Ergun, B., Glassmeier, K.H., Auster, U., McFadden, J., Carlson, C., Larson, D., Bonnell, J.W., Mende, S., Russell, C.T., Donovan, E., Mann, I., Singer, H., Quasi-parallel whistler mode waves observed by

THEMIS during near-earth dipolarizations, *Annales Geophysicae*, **27**(6), 2259-2275, 2009.

Le, G., **Wang**, Y., **Slavin**, J.A., Strangeway, R.J., Space Technology 5 multipoint observations of temporal and spatial variability of field-aligned currents, *Journal Of Geophysical Research-Space PHysics*, **114**, A08206, doi:10.1029/2009JA014081, 2009.

Leamon, R.J., McIntosh, S.W., How the solar wind ties to its photospheric origins, *Astrophysical Journal Letters*, **697**(1), L28-L32, doi:10.1088/0004-637X/697/1/L28, 2009.

Leamon, R.J., McIntosh, S.W., Could we have forecast "The day the solar wind died"?, *ASTROPHYSICAL JOURNAL LETTERS*, **679**(2), L147-L150, 2008.

Lee, C.O, Luhmann,J.G., Odstrcil,D., **MacNeice**, P.J., de Pater,I., Riley,P., and Arge,C.N., (2009), The solar wind at 1 AU during the declining phase of solar cycle 23: Comparison of 3D numerical model results with observations, *Solar Phys.*, **254**, 155, DOI 10.1007/s11207-008-9280-y.

Lepping, R.P., **Narock**, T.W., **Wu**, C.C., A scheme for finding the front boundary of an interplanetary magnetic cloud, *Annales Geophysicae*, **27**(3), 1295-1311, 2009.

Lipatov, A.S., Rankin, R., Nonlinear field line resonances. Effect of Hall term on plasma compression: 1D Hall-MHD modeling, *Planetary And Space Science*, **57**(3), 404-414, doi:10.1016/j.pss.2008.12.012, 2009.

Liu, W., **Wang**, T.J., **Dennis**, B.R., **Holman**, G.D., Episodic x-ray emission accompanying the activation of an eruptive prominence: evidence of episodic magnetic reconnection, *Astrophysical Journal*, **698**(1), 632-640, doi:10.1088/0004-637X/698/1/632, 2009.

Liu, W., Petrosian, V., **Dennis**, B.R., **Holman**, G.D., Conjugate hard X-ray footpoints in the 2003 October 29 X10 flare: Unshearing motions, correlations, and asymmetries, *Astrophysical Journal*, **693**(1), 847-867, doi:10.1088/0004-637X/693/1/847, 2009.

Lui, A.T.Y., Angelopoulos, V., LeContel, O., Frey, H., Donovan, E., **Sibeck**, D.G., Liu, W., Auster, H.U., Larson, D., Li, X., Nose, M., Fillingim, M.O., Determination of the substorm initiation region from a major conjunction interval of THEMIS satellites, *Journal Of Geophysical Research-Space Physics*, **113**, A00C04, doi:10.1029/2008JA013424, 2008.

Liu, R., D. Alexander, H. R. **Gilbert**, Asymmetric Eruptive Filaments, *Ap. J.*, **691**, 1079, 2009.

Lyatskaya, S., **Lyatsky**, W., **Khazanov**, G.V., Auroral electrojet AL index and polar magnetic disturbances in two hemispheres, *Journal Of Geophysical Research-Space Physics*, **114**, A06212, doi:10.1029/2009JA014100, 2009.

Lyatskaya, S., **Lyatsky**, W., **Khazanov**, G.V., Relationship between substorm activity and magnetic disturbances in two polar caps, *Geophysical Research Letters*, **35**(20), L20104, doi:10.1029/2008GL035187, 2008.

Lyatsky, W., **Khazanov**, G.V., Effect of geomagnetic disturbances and solar wind density on relativistic electrons at geostationary orbit, *Journal Of Geophysical Research-Space Physics*, **113**(A8), A08224, doi:10.1029/2008JA013048, 2008.

Lynch, B.J., **Antiochos**, S.K., Li, Y., Luhmann, J.G., **DeVore**, C.R., Rotation of Coronal Mass Ejections during eruption, *Astrophysical Journal*, **697**(2), 1918-1927, doi:10.1088/0004-637X/697/2/1918, 2009.

MacNeice, P., Validation of community models: Identifying events in space weather model timelines, *Space Weather-The International Journal Of Research And Applications*, **7**, S06004, doi:10.1029/2009SW000463, 2009.

Mäkelä, P., N., **Gopalswamy**, S.**Yashiro**, S. **Akiyama**, H. **Xie**, E. Valtonen, SEPs and CMEs during cycle 23, Proceedings of the International Astronomical Union, IAU Symposium, 257, p. 475-477, 2009.

Marubashi, K., Sung, S.K., Cho, K.S., **Lepping**, R.P., Impacts of torus model on studies of geometrical relationships between interplanetary magnetic clouds and their solar origins, *Earth Planets And Space*, **61**(5), 589-594, 2009.

Masson, S., **Pariat**, E., Aulanier, G., Schrijver, C.J., the nature of flare ribbons in Coronal null-point topology, *Astrophysical Journal*, **700**(1), 559-578, doi:10.1088/0004-637X/700/1/559, 2009.

Masters, A., Achilleos, N., Dougherty, M.K., **Slavin**, J.A., Hospodarsky, G.B., Arridge, C.S., Coates, A.J., An empirical model of Saturn's bow shock: Cassini observations of shock location and shape, *Journal Of Geophysical Research-Space Physics*, **113**(A10), A10210, doi:10.1029/2008JA013276, 2008.

Mays, M. L., W. Horton, E. Spencer, and J. Kozyra (2009), Real-time predictions of geomagnetic storms and substorms: Use of the Solar Wind Magnetosphere-Ionosphere System model, Space Weather, 7, S07001, doi:10.1029/2008SW000459.

McComas, D.J., F. Allegrini, P. Bochsler, M. Bzowski, **E.R. Christian**, G.B. Crew, R. DeMajistre, H. Fahr, H. Fichtner, P.C. Frisch, H.O. Funsten, S.A. Fuselier, G. Gloeckler, M. Gruntman, J. Heerikhuisen, V. Izmodenov, P. Janzen, P. Knappenberger, S. Krimigis, H. Kucharek, M. Lee, G. Livadiotis, S. Livi, **R.J. MacDowall**, D. Mitchell, E. Moebius, **T. Moore**, N.V. Pogorelov, D. Reisenfeld, E. Roelof, L. Saul, N.A. Schwadron, P.W. Valek, R. Vanderspek, P. Wurz, G.P. Zank, Global observations of the interstellar interaction from the Interstellar Boundary Explorer (IBEX), *Science*, 1180906, 2009.

McIntosh. S.W., Burkepile, J. and R. J. **Leamon**, More of the Inconvenient Truth About

Coronal Dimmings, Proceedings of the Second Hinode Science Meeting, 2009.

McLaughlin, J.A., **Ofman**, L., Three-dimensional magnetohydrodynamic wave behavior in active regions: Individual loop density structure, *Astrophysical Journal*, **682**(2), 1338-1350, 2008.

McPherron, R. L, **L. Kepko**, T. I. Pulkinnen, T. S. Hsu, J. W. Weygand, and L. F. Bargatze, Changes in the response of the AL Index with solar cycle and Epoch within a corotating interaction region, *Ann. Geophys.*, **27**, 8, 2009.

Melrose, D. B., **Klimchuk**, J. A., et al., Commission 10: Solar Activity (2009 Triennial report), in *Transactions IAU, reports on Astronomy 2006-2009*, **XXVIIA**, ed. K. A. van der Hucht (Cambridge: Cambridge Univ. Press), 7, 2009.

Mende, S.B., Frey, H.U., McFadden, J., Carlson, C.W., Angelopoulos, V., Glassmeier, K.H., **Sibeck**, D.G., Weatherwax, A., Coordinated observation of the dayside magnetospheric entry and exit of the THEMIS satellites with ground-based auroral imaging in Antarctica, *Journal Of Geophysical Research-Space Physics*, **114**, A00C23, doi:10.1029/2008JA013496, 2009.

Meyer-Vernet, N., Maksimovic, M., Czechowski, A., Mann, I., Zouganelis, I., Goetz, K., **Kaiser**, M.L., **St Cyr**, O.C., Bougeret, J.L., Bale, S.D., Dust detection by the Wave instrument on STEREO: Nanoparticles picked up by the solar wind?, *Solar Physics*, **256**(1-2), 463-474, doi:10.1007/s11207-009-9349-2, 2009.

Meyer-Vernet, N., Lecacheux, A., **Kaiser**, M.L., Gurnett, D.A., Detecting nanoparticles at radio frequencies: Jovian dust stream impacts on Cassini/RPWS, *Geophysical Research Letters*, **36**, L03103, doi:10.1029/2008GL036752, 2009.

Michalek, G., **Gopalswamy**, N., **Yashiro**, S., Space weather application using projected velocity asymmetry of halo CMEs, *Solar Physics*, **248**(1), 113-123, doi:10.1007/s11207-008-9126-7, 2008.

Milan, S.E., Grocott, A., Forsyth, C., **Imber**, S.M., Boakes, P.D., Hubert, B., A superposed Epoch analysis of auroral evolution during substorm growth, onset and recovery: open magnetic flux control of substorm intensity, *Annales Geophysicae*, **27**(2), 659-668, 2009.

Milligan, R.O., **Dennis**, B.R., Velocity characteristics of evaporated plasma using Hinode/EUV Imaging Spectrometer, *Astrophysical Journal*, **699**(2), 968-975, doi:10.1088/0004-637X/699/2/968, 2009.

Milligan, R.O., A hot microflare observed with RHESSI and Hinode, *Astrophysical Journal Letters*, **680**(2), L157-L160, 2008.

Miyashita, Y., Hosokawa, K., Hori, T., Kamide, Y., Yukimatu, A.S., Fujimoto, M., Mukai, T., Machida, S., Sato, N., Saito, Y., Shinohara, I., **Sigwarth**, J.B., Response of large-scale ionospheric convection to substorm expansion onsets: A case study, *Journal Of Geophysical Research-Space Physics*, **113**(A12), A12309, doi:10.1029/2008JA013586, 2008.

Moore, T. E., (2009), Fifty Years Observing Plasmas in Space, in "Results of Space Exploration in the First 50 Years", *IKI,* 2009.

Mostl, C., Farrugia, C.J., Miklenic, C., Temmer, M., Galvin, A.B., Luhmann, J.G., Kilpua, E.K.J., **Leitner**, M., **Nieves-Chinchilla**, T., Veronig, A., Biernat, H.K., Multispacecraft recovery of a magnetic cloud and its origin from magnetic reconnection on the Sun, *Journal Of Geophysical Research-Space Physics*, **114**, A04102, doi:10.1029/2008JA013657, 2009.

Mueller, D., **Fleck**, B., **Dimitoglou**, G., Caplins, B.W., Amadigwe, D.E., Ortiz, J.P.G., Wamsler, B., Alexanderian, A., **Hughitt**, V.K., **Ireland**, J., JHelioviewer: Visualizing large sets of solar images using JPEG 2000, *Computing In Science & Engineering*, **11**(5), 38-47, 2009.

Nakano, S., Ueno, G., Ebihara, Y., **Fok**, M.C., Ohtani, S., Brandt, P.C., Mitchell, D.G., Keika, K., Higuchi, T., A method for estimating the ring current structure and the electric potential distribution using energetic neutral atom data assimilation, *JOURNAL OF Geophysical Research-SpacE Physics*, **113**(A5), A05208, doi:10.1029/2006JA011853, 2008.

Narita ,Y., K.H. Glassmeier, S.P. Gary, M.L. **Goldstein**, and R.A. Treumann,Wave number spectra in the solar wind, the foreshock, and the magnetosheath, Cluster workshop, Tenerife, Canary Islands, Spain, March, 2008, *Astrophysics and Space Science Proceedings,* Springer.

Nieves-Chinchilla, T., **Vinas**, A.F., Hidalgo, M., Magnetic field profiles within magnetic clouds: A Model-Approach, *Earth Moon And Planets*, **104**(1-4), 109-113, doi:10.1007/s11038-008-9252-0, 2009.

Nieves-Chinchilla, T., Vinas, A.F., Kappa-like distribution functions inside magnetic clouds, *Geofisica Internacional*, **47**(3), 245-249, 2008.

Nose, M., Taguchi, S., Christon, S.P., **Collier**, M.R., **Moore**, T.E., Carlson, C.W., McFadden, J.P., Response of ions of ionospheric origin to storm time substorms: Coordinated observations over the ionosphere and in the plasma sheet, *Journal Of Geophysical Research-Space Physics*, **114**, A05207, doi:10.1029/2009JA014048, 2009.

Nowada, M., Shue, J.H., Lin, C.H., Sakurai, T., **Sibeck**, D.G., Angelopoulos, V., Carlson, C.W., Auster, H.U., Alfvénic plasma velocity variations observed at the inner edge of the low-latitude boundary layer induced by the magnetosheath mirror mode waves: A THE-

MIS observation, *Journal Of Geophysical Research-Space Physics*, **114**, A07208, doi:10.1029/2008JA014033, 2009.

Ofman, L., Three-dimensional magnetohydrodynamic models of twisted multithreaded coronal loop oscillations, *Astrophysical Journal*, **694**(1), 502-511, doi:10.1088/0004-637X/694/1/502, 2009.

Ofman, L., and **Selwa**, M., Three-dimensional MHD modeling of waves in active region loops, Universal Heliophysical Processes Proceedings IAU Symposium No. 257, A. Nindos, etal., eds., DOI: 10.1017/S1743921309029202, pp. 151-154, 2009.

Ofman, L., Progress, challenges, and perspectives of the 3D MHD numerical modeling of oscillations in the solar corona, *Space Science Reviews*, doi: 10.1007/s11214-009-9501-1, 2009.

Ofman, L., Balichin, M., Russel, C.T., and Gedalin, M., Collisionless relaxation of ion distributions downstream of laminar quasi-perpendicular shocks, *Journal of Geophysical Research - Space Physics*, **114**, A09106, doi:10.1029/2009JA014365, 2009.

Omidi, N., **Sibeck**, D.G., Blanco-Cano, X., Foreshock compressional boundary, *Journal Of Geophysical Research-Space Physics*, **114**, A08205, doi:10.1029/2008JA013950, 2009.

Omidi, N., Phan, T., **Sibeck**, D.G., Hybrid simulations of magnetic reconnection initiated in the magnetosheath, *Journal Of Geophysical Research-Space Physics*, **114**, A02222, doi:10.1029/2008JA013647, 2009.

Opitz, A., Karrer, R., Wurz, P., Galvin, A. B., Bochsler, P., Blush, L. M., Daoudi, H., Ellis, L., Giammanco, C., Kistler, L. M., Klecker, B., Kucharek, H., Lee, M. A., Möbius, E., Popecki, M., Sigrist, M., Simunac, K., Singer, K., **Thompson**, B., Wimmer-Schweingruber, R. F., Temporal evolution of the solar wind bulk velocity at solar minimum by correlating the STEREO A and B PLASTIC measurements, *Solar Physics*, Volume 256, Issue 1-2, pp. 365-377, 2009.

Oswald, T.H., Macher, W., Rucker, H.O., Fischer, G., Taubenschuss, U., Bougeret, J.L., Lecacheux, A., **Kaiser**, M.L., Goetz, K., Various methods of calibration of the STEREO/WAVES antennas, *Advances In Space Research*, **43**(3), 355-364, doi:10.1016/j.asr.2008.07.017, 2009.

Orisini, S., et al. (2009), SERENA: A suite of four instruments (ELENA,STROFIO, PICAM and MIPA) onboard BepiColombo-MPO for particle detection in the Hermean environment, *Planet Space Sci.*, doi:10.1016/j.pss.2008.09.012.

Panchenko, M., Khodachenko, M.L., Kislyakov, A.G., Rucker, H.O., Hanasz, J., **Kaiser**, M.L., Bale, S.D., Lamy, L., Cecconi, B., Zarka, P., Goetz, K., Daily variations of auroral kilometric radiation observed by STEREO, *Geophysical Research Letters*, **36**, L06102, doi:10.1029/2008GL037042, 2009.

Pariat, E., Masson, S., Aulanier, G., current buildup in emerging serpentine flux tubes, *Astrophysical JournaL*, **701**(2), 1911-1921, doi:10.1088/0004-637X/701/2/1911, 2009.

Patsourakos, S., **Klimchuk**, J.A., Spectroscopic observations of hot lines constraining coronal heating in solar active regions, *Astrophysical Journal*, **696**(1), 760-765, doi:10.1088/0004-637X/696/1/760, 2009.

Patsourakos, S. and **Klimchuk**, J. A., Static and impulsive models of solar active regions, *Astrophys. J.*, **689**, 1406, 2008.

Pierrard, V., **Khazanov**, G.V., Cabrera, J., Lemaire, J., Influence of the convection electric field models on predicted plasmapause positions during magnetic storms, *Journal Of Geophysical Research-Space Physics*, **113**(A8), A08212, doi:10.1029/2007JA012612, 2008.

Plaschke, F., Glassmeier, K.H., Auster, H.U., Constantinescu, O.D., Magnes, W., Angelopoulos, V., **Sibeck**, D.G., McFadden, J.P., Standing Alfvén waves at the magnetopause, *Geophysical Research Letters*, **36**, L02104, doi:10.1029/2008GL036411, 2009.

Pulkkinen, A., **Taktakishvili**, A., Odstrcil, D., Jacobs, W., Novel approach to geomagnetically induced current forecasts based on remote solar observations, *Space Weather-The International Journal Of Research And Applications*, 7, S08005, doi:10.1029/2008SW000447, 2009.

Pulkkinen, A., Viljanen, A., Pirjola, R., Harnessing celestial batteries, *American Journal Of Physics*, **77**(7), 610-613, doi:10.1119/1.3119172, 2009.

Pulkkinen, A., M. **Hesse**, S. **Habib**, L. Van der Zel, B. Damsky, F. **Policelli**, D. Fugate, and W. Jacobs, Solar Shield: forecasting and mitigating space weather effects on high-voltage power transmission systems, *Natural Hazards*, doi:10.1007/s11069-009-9432-x, 2009.

Pulkkinen, A., A. Viljanen, R. Pirjola, and L Häkkinen, Electromagnetic source equivalence and extension of the complex image method for geophysical applications, *Progress in Electromagnetics Research B*, **16**, 57-84, 2009.

Rabello-Soares, M.C., **Thompson**, B.J., Scherrer, D., Morrow, C., Education and public outreach program for IHY - A global approach, *Advances In Space Research*, **41**(8), 1206-1211, doi:10.1016/j.asr.2007.06.016, 2008.

Rabello-Soares, M. C., Rabiu, A. B., **Gopalswamy**, N., **Thompson**, B. J., **Davila**, J. M., Sobrinho, A. A., Outreach activities during the 2006 total solar eclipse sponsored by the International Heliophysical Year, *Adv. Spac. Res.*, **42**, 11, pp. 1792-1799, 2008.

Raftery, C.L., Gallagher, P.T., **Milligan**, R.O., **Klimchuk**, J.A., Multi-wavelength observations and modelling of a canonical solar flare, *Astronomy & AstrophysicS*, **494**(3), 1127-1136, doi:10.1051/0004-6361:200810437, 2009.

Rauch, B.F., J.T. **Link**, K. Lodders, M.H. Israel, L.M. Barbier, W.R. Binns, E.R. **Christian**, J.R. **Cummings**, G.A. **de Nolfo**, S. Geier, R.A. Mewaldt, J.W. **Mitchell**, S. M. Schindler, L.M. Scott, E.C. Stone, R.E. **Streitmatter**, C.J. Waddington, M.E. Wiedenbeck, Cosmic Ray origin in OB associations and preferential acceleration of refractory elements: Evidence from abundances of elements 26Fe through 34Se, *ApJ*, **697**, 2083, 2009.

Reale, F., Testa, P., **Klimchuk**, J.A., Parenti, S., Evidence of widespread hot plasma in a nonflaring coronal active region from Hinode/X-ray telescope, *Astrophysical Journal*, **698**(1), 756-765, doi:10.1088/0004-637X/698/1/756, 2009.

Robbrecht, E., Hochedez, J.F., **Fleck**, B., **Gurman**, J., Forsyth, R., SOHO 20 - Transient events on the Sun and in the heliosphere - Preface, *Annales Geophysicae*, **26**(10), 2953-2953, 2008.

Rodrigues, F.S., Kelley, M.C., Roddy, P.A., Hunton, D.E., **Pfaff**, R.F., de La Beaujardiere, O., Bust, G.S., C/NOFS observations of intermediate and transitional scale-size equatorial spread F irregularities, *Geophysical Research Letters*, **36**, L00C05, doi:10.1029/2009GL038905, 2009.

Rosenqvist, L., Opgenoorth, H.J., **Rastaetter**, L., Vaivads, A., Dandouras, I., Buchert, S., Comparison of local energy conversion estimates from Cluster with global MHD simulations, *Geophysical Research Letters*, **35**(21), L21104, doi:10.1029/2008GL035854, 2008.

Rouillard, A.P., Savani, N.P., Davies, J.A., Lavraud, B., Forsyth, R.J., Morley, S.K., Opitz, A., Sheeley, N.R., **Burlaga**, L.F., Sauvaud, J.A., Simunac, K.D.C., Luhmann, J.G., Galvin, A.B., Crothers, S.R., Davis, C.J., Harrison, R.A., Lockwood, M., Eyles, C.J., Bewsher, D., Brown, D.S., A multispacecraft analysis of a small-scale transient entrained by solar wind streams, *Solar Physics*, **256**(1-2), 307-326, doi:10.1007/s11207-009-9329-6, 2009.

Sahraoui, F., **Goldstein**, M.L., **Robert**, P., Khotyaintsev, Y.V., Evidence of a cascade and dissipation of solar-wind turbulence at the electron gyroscale, *Physical Review Letters*, **102**(23), 231102, doi:10.1103/PhysRevLett.102.231102, 2009.

Sangalli, L., Knudsen, D.J., Larsen, M.F., Zhan, T., **Pfaff**, R.F., **Rowland**, D., Rocket-based measurements of ion velocity, neutral wind, and electric field in the collisional transition region of the auroral ionosphere, *Journal Of Geophysical Research-Space Physics*, **114**, A04306, doi:10.1029/2008JA013757, 2009.

Sarantos, M., **Slavin**, J.A., **Benna**, M., **Boardsen**, S.A., **Killen**, R.M., Schriver, D., Travnicek, P., Sodium-ion pickup observed above the magnetopause during MESSEN-

GER's first Mercury flyby: Constraints on neutral exospheric models, *Geophysical Research Letters*, **36**, L04106, doi:10.1029/2008GL036207, 2009.

Sarantos, M., **Slavin**, J.A., On the possible formation of Alfvén wings at Mercury during encounters with coronal mass ejections, *Geophysical Research Letters*, **36**, L04107, doi:10.1029/2008GL036747, 2009.

Sarantos, M., Killen, R.M., Sharma, A.S., **Slavin**, J.A., Influence of plasma ions on source rates for the lunar exosphere during passage through the Earth's magnetosphere, *Geophysical Research Letters*, **35**(4), L04105, doi:10.1029/2007GL032310, 2008.

Sarantos, D. Schriver, S. C. Solomon, P. Trávníček, T. H. Zurbuchen (2009), MESSENGER Observations of Magnetic Reconnection in Mercury's Magnetosphere, *Science*, **324**, 606–610, doi:10.1126/science.1172011.

Sarantos, S. C. Solomon, T.-L. Zhang, and T. H. Zurbuchen (2008), MESSENGER and Venus Express observations of the solar wind interaction with Venus, *Geophys. Res. Lett.*, **36**, L09106, doi:10.1029/2009GL037876.

Schmelz, J.T., Saar, S.H., DeLuca, E.E., Golub, L., Kashyap, V.L., Weber, M.A., **Klimchuk**, J.A., HINODE X-ray telescope detection of hot emission from quiescent active regions: A nanoflare signature?, *Astrophysical Journal Letters*, **693**(2), L131-L135, doi:10.1088/0004-637X/693/2/L131, 2009.

Schmelz, J. T., V. L. Kashyap, S. H. Saar, B. R. **Dennis**, P. C. Grigis, L. Lin, E. E. DeLuca, G. D. **Holman**, L. Golub, Weber, M. A., Some like it hot: Coronal heating observations from Hinode X-ray telescope and RHESSI, *Ap. J.*, **704**, 863, 2009.

Schmidt, J.M., **Gopalswamy**, N., Synthetic radio maps of CME-driven shocks below 4 solar radii heliocentric distance, *Journal Of Geophysical Research-Space Physics*, **113**(A8), A08104, doi:10.1029/2007JA013002, 2008.

Schneider, N.M., **Burger**, M.H., Schaller, E.L., Brown, M.E., Johnson, R.E., Kargel, J.S., Dougherty, M.K., Achilleos, N.A., No sodium in the vapour plumes of Enceladus, *NaturE*, **459**(7250), 1102-1104, doi:10.1038/nature08070, 2009.

Schwadron, N. A., M. Bzowski, G. B. Crew, M. Gruntman, H. Fahr, H. Fichtner, P. C. Frisch, H. O. Funsten, S. Fuselier, J. Heerikhuisen, V. Izmodenov, H. Kucharek, M. Lee, G. Livadiotis, D. J. McComas, E. Moebius, T. **Moore**, J. Mukherjee, N.V. Pogorelov, C. Prested, D. Reisenfeld, E. Roelof, G.P. Zank, Comparison of Interstellar Boundary Explorer observations with 3d global heliospheric models, *www.sciencexpress.org*, 15 October 2009, Page 1, 10.1126, science.1180986.

Selwa, M., and **Ofman**, L., 3D numerical simulations of coronal loops oscillations, *Annales Geophysicae*, **27**, 3899, 2009.

Sergeev, V.A., Apatenkov, S.V., Angelopoulos, V., McFadden, J.P., Larson, D., Bonnell, J.W., **Kuznetsova**, M., Partamies, N., Honary, F., Simultaneous THEMIS observations in the near-tail portion of the inner and outer plasma sheet flux tubes at substorm onset, *Journal Of Geophysical Research-Space Physics*, **113**, A00C02, doi:10.1029/2008JA013527, 2008.

Shanmugaraju, A., Moon, Y.J., Cho, K.S., **Gopalswamy**, N., Umapathy, S., Investigation of CME dynamics in the LASCO field of view, *Astronomy & Astrophysics*, **484**(2), 511-516, doi:10.1051/0004-6361:20078978, 2008.

Sharma, A.S., Nakamura, R., Runov, A., Grigorenko, E.E., Hasegawa, H., Hoshino, M., Louarn, P., Owen, C.J., Petrukovich, A., Sauvaud, J.A., Semenov, V.S., Sergeev, V.A., **Slavin**, J.A., Sonnerup, B.U.O., Zelenyi, L.M., Fruit, G., Haaland, S., Malova, H., Snekvik, K., Transient and localized processes in the magnetotail: a review, *Annales Geophysicae*, **26**(4), 955-1006, 2008.

Shi, Q. Q., Z. Y. Pu, J. Soucek, Q.-G. Zong, S. Y. Fu, L. Xie, Y. Chen, H. **Zhang**, L. Li, L. D. Xia, Z. X. Liu, E. Lucek, A. N. Fazakerley and H. Reme, Spatial structures of magnetic depression in the Earth's High-altitude Cusp: Cluster multi-point Observations, *J. Geophys. Res.*, **114**, doi:10.1029/2009JA014283, 2009.

Sibeck, D.G., Concerning the occurrence pattern of flux transfer events on the dayside magnetopause, *Annales Geophysicae*, **27**(2), 895-903, 2009.

Simunac, K. D. C., Kistler, L. M., Galvin, A. B., Lee, M. A., Farrugia, C., Moebius, E., Blush, L. M., Bochsler, P., Wurz, P., Klecker, B., Wimmer-Schweingruber, R. F., **Thompson**, B., Luhmann, J. G., Russell, C. T., Howard, R. A., In Situ observations of solar wind stream interface evolution, Solar *Physics Online First*, 2009.

Siscoe, G.L., **Kuznetsova**, M.M., Raeder, J., Search for an onset mechanism that operates for both CMEs and substorms, *Annales Geophysicae*, **27**(8), 3141-3146, 2009.

Sittler, Jr., E. C., R. E. **Hartle**, J. F. **Cooper**, R. E. Johnson, A. J. Coates, A. Ari, D. G. **Simpson**, and D. T. Young, Heavy ion formation in Titan's ionosphere, magnetospheric introduction of free oxygen and source of Titan's aerosols? *Planetary and Space Science*, doi:10.1016/j.pss.2009.07.017, 2009.

Slavin, J.A., Acuna, M.H., **Anderson**, B.J., Barabash, S., Benna, M., **Boardsen**, S.A., Fraenz, M., Gloeckler, G., Gold, R.E., Ho, G.C., Korth, H., Krimigis, S.M., McNutt, R.L., Raines, J.M., **Sarantos**, M., Solomon, S.C., Zhang, T., Zurbuchen, T.H., MESSENGER and Venus Express observations of the solar wind interaction with Venus, *Geophysical Research Letters*, **36**, L09106, doi:10.1029/2009GL037876, 2009.

Slavin, J.A., Acuna, M.H., Anderson, B.J., Baker, D.N., Benna, M., **Boardsen**, S.A., Gloeckler, G., Gold, R.E., Ho, G.C., Korth, H., Krimigis, S.M., McNutt, R.L., Raines, J.M., **Sarantos**, M., Schriver, D., Solomon, S.C., Travnicek, P., Zurbuchen, T.H., MES-

SENGER observations of magnetic reconnection in Mercury's magnetosphere, *Science*, **324**(5927), 606-610, doi:10.1126/science.1172011, 2009.

Slavin, J.A., **Anderson**, B.J., Zurbuchen, T.H., Baker, D.N., Krimigis, S.M., Acuna, M.H., Benna, M., **Boardsen**, S.A., Gloeckler, G., Gold, R.E., Ho, G.C., Korth, H., McNutt, R.L., Raines, J.M., **Sarantos**, M., Schriver, D., Solomon, S.C., Travnicek, P., MESSENGER observations of Mercury's magnetosphere during northward IMF, *Geophysical Research Letters*, **36**, L02101, doi:10.1029/2008GL036158, 2009.

Slavin, J.A., M. H. **Acuña**, B. J. Anderson, S. Barabash, M. **Benna**, S. A. **Boardsen**, M. Fraenz, G. Gloeckler, R.E. Gold, G. C. Ho, H. Korth, S. M. Krimigis, R. L. McNutt, Jr., J.M. Raines, M. **Sarantos**, S. C. Solomon, T.-L. Zhang, and T. H. Zurbuchen (2008), MESSENGER and Venus Express observations of the solar wind interaction with Venus, *Geophys. Res. Lett.*, **36**, L09106, doi:10.1029/2009GL037876.

Smith, H.T., **Shappirio**, M., Johnson, R.E., Reisenfeld, D., **Sittler**, E.C., Crary, F.J., McComas, D.J., Young, D.T., Enceladus: A potential source of ammonia products and molecular nitrogen for Saturn's magnetosphere, *Journal Of Geophysical Research-Space physics*, **113**(A11), A11206, doi:10.1029/2008JA013352, 2008.

Soares, M.C.R., Rabiu, A.B., **Gopalswamy**, N., **Thompson**, B.J., **Davila**, J.M., Sobrinho, A.A., Outreach activities during the 2006 total solar eclipse sponsored by the International Heliophysical Year, *Advances In Space Research*, **42**(11), 1792-1799, doi:10.1016/j.asr.2007.04.014, 2008.

Sokolov, I.V., Roussev, I.I., Skender, M., Gombosi, T.I., **Usmanov**, A.V., Transport equation for MHD turbulence: Application to particle acceleration at interplanetary shocks, *Astrophysical Journal*, **696**(1), 261-267, doi:10.1088/0004-637X/696/1/261, 2009.

Solomon, S.C., McNutt, R.L., Watters, T.R., Lawrence, D.J., Feldman, W.C., Head, J.W., Krimigis, S.M., Murchie, S.L., Phillips, R.J., **Slavin**, J.A., Zuber, M.T., Return to Mercury: A global perspective on MESSENGER's first mercury flyby, *Science*, **321**(5885), 59-62, doi:10.1126/science.1159706, 2008.

Spencer, E., A. Rao, W. Horton, and M. L. **Mays** (2009), Evaluation of Solar Wind - Magnetosphere Coupling Functions during Geomagnetic Storms with the WINDMI Model, J. Geophys. Res., 114, A02206, doi:10.1029/2008JA013530.

St Cyr, O.C., **Kaiser**, M.L., Meyer-Vernet, N., Howard, R.A., Harrison, R.A., Bale, S.D., **Thompson**, W.T., Goetz, K., Maksimovic, M., Bougeret, J.L., Wang, D., Crothers, S., STEREO SECCHI and S/WAVES observations of spacecraft debris caused by micron-size interplanetary dust impacts, *Solar Physics*, **256**(1-2), 475-488, doi:10.1007/s11207-009-9362-5, 2009.

Stanislavsky, A.A., Konovalenko, A.A., Rucker, H.O., Abranin, E.P., **Kaiser**, M.L., Dorovskyy, V.V., Mel'nik, V.N., Lecacheux, A., Antenna performance analysis for decameter solar radio observations, *Astronomische Nachrichten*, **330**(7), 691-697, doi:10.1002/asna.200911226, 2009.

Stone, E.C., Cummings, A.C., McDonald, F.B., Heikkila, B.C., **Lal**, N., Webber, W.R., An asymmetric solar wind termination shock, *Nature*, **454**(7200), 71-74, doi:10.1038/nature07022, 2008.

Straus, T., **Fleck**, B., Jefferies, S.M., Cauzzi, G., McIntosh, S.W., Reardon, K., Severino, G., Steffen, M., The energy flux of internal gravity waves in the lower solar atmosphere, *Astrophysical Journal Letters*, **681**(2), L125-L128, 2008.

Strong, K.T., **Saba**, J.L.R., A new approach to solar cycle forecasting, *Advances In Space Research*, **43**(5), 756-759, doi:10.1016/j.asr.2008.12.007, 2009.

Su, Y.J., Retterer, J.M., de La Beaujardiere, O., Burke, W.J., Roddy, P.A., **Pfaff**, R.F., Wilson, G.R., Hunton, D.E., Assimilative modeling of equatorial plasma depletions observed by C/NOFS, *Geophysical Research Letters*, **36**, L00C02, doi:10.1029/2009GL038946, 2009.

Suzuki, S., Taguchi, S., Hosokawa, K., Collier, M.R., **Moore**, T.E., Frey, H.U., Mende, S.B., Conjugate observations of ENA signals in the high-altitude cusp and proton auroral spot in the low-altitude cusp with IMAGE spacecraft, *Geophysical Research Letters*, **35**(13), L13103, doi:10.1029/2008GL034543, 2008.

Švanda, M., Klvaňa, M., Sobotka, M., Kosovichev, A. G., **Duvall**, T. L. Jr., Large-scale horizontal flows in the solar photosphere IV. On the vertical structure of large-scale horizontal flows, *New Astronomy*, 2009, **14**, 429-434, doi:10.1016/j.newast.2008.12.003, 2009.

Taguchi, S., Suzuki, S., Hosokawa, K., Ogawa, Y., Yukimatu, A.S., Sato, N., **Collier**, M.R., **Moore**, T.E., Moving mesoscale plasma precipitation in the cusp, *Journal Of Geophysical Research-Space Physics*, **114**, A06211, doi:10.1029/2009JA014128, 2009.

Taguchi, S., Hosokawa, K., Nakao, A., **Collier**, M.R., **Moore**, T.E., Sato, N., Yukimatu, A.S., HF radar polar patch and its relation with the cusp during B-Y-dominated IMF: Simultaneous observations at two altitudes, *Journal Of Geophysical Research-Space Physics*, **114**, A02311, doi:10.1029/2008JA013624, 2009.

Taktakishvili, A., **Kuznetsova**, M., **MacNeice**, P., **Hesse**, M., **Rastatter**, L., **Pulkkinen**, A., Chulaki, A., Odstrcil, D., Validation of the coronal mass ejection predictions at the Earth orbit estimated by ENLIL heliosphere cone model, *Space Weather-The International Journal Of Research And Applications*, **7**, S03004, doi:10.1029/2008SW000448, 2009.

Taktakishvili, A., **Kuznetsova**, M.M., **Hesse**, M., **Fok**, M.C., **Rastaetter**, L., **Maddox**, M., Chulaki, A., Toth, G., Gombosi, T.I., De Zeeuw, D.L., Role of periodic loading-unloading in the magnetotail versus interplanetary magnetic field B-z flipping in the ring current buildup, *Journal Of Geophysical Research-Space Physics*, **113**(A3), A03206, doi:10.1029/2007JA012845, 2008.

Tang, C.L., Lu, L., McKenna-Lawlor, S., Barabash, S., Liu, Z.X., **Fok**, M.C., Li, Z.Y., A comparison of Neutral Atom Detector Unit neutral atom image inversion with a comprehensive ring current model, *Journal Of Geophysical Research-Space Physics*, **113**(A7), A07S32, doi:10.1029/2007JA012680, 2008.

Thomas, R., Frederick, E., Krabill, W., Manizade, S., Martin, C., Recent changes on Greenland outlet glaciers, *Journal Of Glaciology*, **55**(189), 147-162, 2009.

Thomas, R., Davis, C., Frederick, E., Krabill, W., Li, Y.H., Manizade, S., Martin, C., A comparison of Greenland ice-sheet volume changes derived from altimetry measurements, *Journal Of Glaciology*, **54**(185), 203-212, 2008.

Thompson, B.J., Myers, D.C., A catalog of coronal "EIT WAVE" transients, *Astrophysical Journal Supplement Series*, **183**(2), 225-243, doi:10.1088/0004-637X/183/2/225, 2009.

Thompson, B. J., **Gopalswamy**, N., **Davila**, J. M., Haubold, H., Putting the 'I' in IHY: The United Nations report for the International Heliophysical Year 2007, Springer Press, Wein/New York, 2009.

Thompson, W.T., 3D triangulation of a Sun-grazing comet, *Icarus*, **200**(2), 351-357, doi:10.1016/j.icarus.2008.12.011, 2008.

Tokar, R.L., Wilson, R.J., Johnson, R.E., Henderson, M.G., Thomsen, M.F., Cowee, M.M., **Sittler**, E.C., Young, D.T., Crary, F.J., McAndrews, H.J., Smith, H.T., Cassini detection of water-group pick-up ions in the Enceladus torus, *Geophysical Research Letters*, **35**(14), L14202, doi:10.1029/2008GL034749, 2008.

Travnicek, P.M., Hellinger, P., Schriver, D., Hercik, D., **Slavin**, J.A., **Anderson**, B.J., Kinetic instabilities in Mercury's magnetosphere: Three-dimensional simulation results, *Geophysical Research Letters*, **36**, L07104, doi:10.1029/2008GL036630, 2009.

Tripathi, D., S. E. Gibson, J. Qiu, L. Fletcher, R. Liu, H. **Gilbert**, H. E. Mason, On partially erupting prominences, *A&A*, **498**, 295, 2009.

Truhlik, V., D. **Bilitza**, and L. Triskova, Latitudinal variation of the topside electron temperature at different levels of solar activity, *J. Adv. Space Res.*, **44**, #6, 693-700, doi:10.1016/j.asr.2009.04.029, 2009.

Uritsky, V.M., **Davila**, J.M., Jones, S.I., Comment on "Coexistence of self-organized criticality and intermittent turbulence in the solar corona" reply, *Physical Review Letters*, **103**(3), 039502, doi:10.1103/PhysRevLett.103.039502, 2009.

Uritsky, V.M., Donovan, E., **Klimas**, A.J., Spanswick, E., Collective dynamics of bursty particle precipitation initiating in the inner and outer plasma sheet, *Annales Geophysicae*, **27**(2), 745-753, 2009.

Uritsky, V.M., Donovan, E., **Klimas**, A.J., Spanswick, E., Scale-free and scale-dependent modes of energy release dynamics in the nighttime magnetosphere, *Geophysical Research Letters*, **35**(21), L21101, doi:10.1029/2008GL035625, 2008.

Usmanov, A.V., W.H. Matthaeus, B. Breech, M.L. **Goldstein**,(2009), An MHD solar wind model with turbulence transport , Numerical Modeling Of Space Plasma Flows: Astronum-2008, *Astronomical Society of the Pacific Conference Series*, Vol. 406, Nikolai V. Pogorelov, Edouard Audit, Phillip Colella, and Gary P. Zank, eds.

Viall, N. M., L. **Kepko**, and H. E. Spence, Relative occurrence rates and connection of discrete frequency oscillations in the solar wind density and dayside magnetosphere, *J. Geophys. Res.*, **114**, A01201, 2009.

Wang, T.J., **Ofman**, L., **Davila**, J.M., Propagating slow magnetoacoustic waves in coronal loops observed by Hinode/EIS, *Astrophysical JournaL*, **696**(2), 1448-1460, doi:10.1088/0004-637X/696/2/1448, 2009.

Wang, T.J., **Ofman**, L., **Davila**, J.M., Mariska, J., Hinode/EIS observations of propagating low-frequency slow magnetoacoustic waves in fan-like coronal loops, *Astronomy and Astrophysics*, **503**, L25, doi: 10.1051/0004-6361/200912534, 2009.

Wang, Y., **Le**, G., **Slavin**, J.A., **Boardsen**, S.A., Strangeway, R.J., Space Technology 5 measurements of auroral field-aligned current sheet motion, *Geophysical Research Letters*, **36**, L02105, doi:10.1029/2008GL035986, 2009.

Warmuth, A., **Holman**, G.D., **Dennis**, B.R., Mann, G., Aurass, H., **Milligan**, R.O., Rapid changes of electron acceleration characteristics at the end of the impulsive phase of an x-class solar flare, *Astrophysical Journal*, **699**(1), 917-922, doi:10.1088/0004-637X/699/1/917, 2009.

Webb, D. F., Biesecker, D. A., **Gopalswamy**, N., O.C. **St. Cyr**, J.M. **Davila**, B. J. **Thompson** and K.D.C. Simunac, Using STEREO-B as an L5 Space Weather Pathfinder Mission, in press, *Space Weather*, 2009.

Webber, W.R., Cummings, A.C., McDonald, F.B., Stone, E.C., Heikkila, B., **Lal**, N., Transient intensity changes of cosmic rays beyond the heliospheric termination shock as observed at Voyager 1, *Journal Of Geophysical Research-Space Physics*, **114**, A07108, doi:10.1029/2009JA014156, 2009.

West, M. J., **Bradshaw**, S. J., Cargill, P. J., On the lifetime of hot coronal plasmas arising from nanoflares, *Sol. Phys.*, **252**, 89, 2008.

Wik, M., Pirjola, R., Lundstedt, H., Viljanen, A., Wintoft, P., **Pulkkinen**, A., Space weather events in July 1982 and October 2003 and the effects of geomagnetically induced currents on Swedish technical systems, *Annales Geophysicae*, **27**(4), 1775-1787, 2009.

Wincheski, B., **Williams**, P., Simpson, J., Analysis of eddy current capabilities for the detection of outer diameter cracking in small bore metallic structures, *Review Of Progress In Quantitative Nondestructive Evaluation, VOL 27A and 27B*, 975, 384-391, 2008.

Wolff, C.L., Effects of a deep mixed shell on solar g-modes, p-modes, and neutrino flux, *Astrophysical Journal*, **701**(1), 686-697, doi:10.1088/0004-637X/701/1/686, 2009.

Woods, T. N., P. C. **Chamberlin**, Comparison of solar soft X-ray irradiance from broadband photometers to a high spectral resolution rocket observation, *Adv. Space Res.*, **43**, 349-354, 2009.

Woods, T. N., P. C. **Chamberlin**, J. Harder, R. A. Hock, M. Snow, F. G. Eparvier, J. Fontenla, W. E. McClintock, and E. C. Richard, Solar Irradiance Reference Spectra (SIRS) for the Whole Heliosphere Interval (WHI), *Geophys. Res. Lett.*, **36**, L01101, doi:101029/2008GL036373, 2009.

Wright, D.M., Dhillon, R.S., Yeoman, T.K., Robinson, T.R., Thomas, E.C., Baddeley, L.J., **Imber**, S., Excitation thresholds of field-aligned irregularities and associated ionospheric hysteresis at very high latitudes observed using SPEAR-induced HF radar backscatter, *Annales Geophysicae*, **27**(7), 2623-2631, 2009.

Wu, S.T., Wang, A.H., Fry, C.D., Feng, X.S., **Wu**, C.C., Dryer, M., Challenges of modeling solar disturbances' arrival times at the Earth, *Science In China Series E-Technological Sciences*, **51**(10), 1580-1588, doi:10.1007/s11431-008-0266-7, 2008.

Xie, H., N. **Gopalswamy**, O. C. **StCyr**, Modeling and prediction of fast CME/shocks associated with type II bursts, *Proceedings of the International Astronomical Union*, IAU Symposium, **257**, p. 489-491, 2009.

Yashiro, S., Michalek, G., **Gopalswamy**, N., A comparison of coronal mass ejections identified by manual and automatic methods, *Annales Geophysicae*, **26**(10), 3103-3112, 2008.

Yashiro, S., Michalek, G., **Akiyama**, S., **Gopalswamy**, N., Howard, R.A., Spatial relationship between solar flares and coronal mass ejections, *Astrophysical Journal*, **673**(2), 1174-1180, 2008.

Yashiro, S. and **Gopalswamy**, N., Statistical relationship between solar flares and coronal mass ejections, *IAUS*, **257**, 233, 2009.

Zeilhofer, C., Schmidt, M., **Bilitza**, D., Shum, C.K., Regional 4-D modeling of the ionospheric electron density from satellite data and IRI, *Advances In Space Research*, **43**(11), 1669-1675, doi:10.1016/j.asr.2008.09.033, 2009.

Zenitani, S., **Hesse**, M., **Klimas**, A., Two-fluid magnetohydrodynamic simulations of relativistic magnetic reconnection, *Astrophysical Journal*, **696**(2), 1385-1401, doi:10.1088/0004-637X/696/2/1385, 2009.

Zhang, H., Zong, Q.G., **Sibeck**, D.G., Fritz, T.A., McFadden, J.P., Glassmeier, K.H., Larson, D., Dynamic motion of the bow shock and the magnetopause observed by THEMIS spacecraft, *Journal Of Geophysical Research-Space Physics*, **114**, A00C12, doi:10.1029/2008JA013488, 2009.

Zheng, Y.H., Lui, A.T.Y., **Fok**, M.C., Anderson, B.J., Brandt, P.C., Mitchell, D.G., Controlling factors of Region 2 field-aligned current and its relationship to the ring current: Model results, *Advances In Space Research*, **41**(8), 1234-1242, doi:10.1016/j.asr.2007.05.084, 2008.

Zheng, Y.H., Brandt, P.C., Lui, A.T.Y., **Fok**, M.C., On ionospheric trough conductance and subauroral polarization streams: Simulation results, *Journal Of Geophysical Research-Space Physics*, **113**(A4), A04209, doi:10.1029/2007JA012532, 2008.

Zhukov, A.N., Saez, F., Lamy, P., Llebaria, A., **Stenborg**, G., The origin of polar streamers in the solar corona, *Astrophysical Journal*, **680**(2), 1532-1541, 2008.

Zimbardo, G., Greco, A., Veltri, P., Voros, Z., Amata, E., **Taktakishvili**, A., Carbone, V., Sorriso-Valvo, L., Guerra, I., Solar-Terrestrial relations: Magnetic turbulence in the Earth's magnetosphere and geomagnetic activity, *Earth Moon AND Planets*, **104**(1-4), 127-129, doi:10.1007/s11038-008-9251-1, 2009.

Zurbuchen, T.H., Raines, J.M., Gloeckler, G., Krimigis, S.M., **Slavin**, J.A., Koehn, P.L., Killen, R.M., Sprague, A.L., McNutt, R.L., Solomon, S.C., MESSENGER observations of the composition of Mercury's ionized exosphere and plasma environment, *Science*, **321**(5885), 90-92, doi:10.1126/science.1159314, 2008.

Submitted / In Press

Abbo, L., **Ofman**, L., Giordano, S., Streamers study at solar minimum: Combination of observations and numerical modeling, Solar Wind 12, *AIP Conference Proc.*, 2009, submitted.

Airapetian, V., K. Carpenter, L. **Ofman** 2009, Winds from luminous late-type stars: Broadband frequency distribution of Alfvén waves, *ApJ.*, 2009, accepted.

Airapetian, V., W. Liu, Hot and violent mega aurorae in extrasolar giant planets, *ApJ Letter*, 2009, accepted.

Baring, M. G. and **Summerlin**, E. J., Particle acceleration at relativistic shocks in extragalactic systems, to appear in *8th International Astrophysics Conference*, ``Shock Waves in Space and Astrophysical Environments," (2010), eds. X. Ao, R. Burrows & G.~P. Zank (AIP Conf. Proc., New York), held in Kailua-Kona, Hawaii, 1 - 7 May, 2009, submitted.

Bradshaw, S. J., Cargill, P.J., The enthalpy powered transition region, *Astrophys. J. Lett.*, 2009, in press.

Bradshaw, S. J., Cargill, P.J., The cooling of coronal plasmas: III. The importance of enthalpy as a mechanism for energy loss, *Astrophys. J.*, 2009, in press.

Buzulukova, N., M.-C. **Fok**, A. **Pulkkinen**, M. **Kuznetsova**, T. E. **Moore**, A. **Glocer**, P. C. Brandt, G. Toth, and L. **Rastätter**, Dynamics of ring current and electric fields in the inner magnetosphere during disturbed periods: CRCM–BATS-R-US coupled model, Journal of Geophysical Research, 2009, submitted.

Buzulukova, N., M.C. **Fok**, A. **Pulkkinen**, M. **Kuznetsova**, T. E. **Moore**, A. **Glocer**, P. Brandt, G. Toth, L. **Rastaetter,** Dynamics of ring current and electric fields in the inner magnetosphere during disturbed periods: CRCM-BATS-R-US coupled model, *J. Geophys. Res.*, 2009, submitted.

Dennis, B.R. and **Pernak**, R.L., Hard X-ray flare source sizes measured with RHESSI, *ApJ*, **698**, 2131-2143, 2009, submitted.

Dennis, B.R., R.L. **Pernak**, Hard X-ray flare source sizes measured with the Ramaty High Energy Solar Spectroscopic Imager, *ApJ*, 698:2131{2143), June 2009, submitted.

Del Zanna, G., Mitra-Kraev, U., **Bradshaw**, S.J., Mason, H.E., Asahi, A., The 22 May 2007 B-class flare: new insights from Hinode observations, *Astron. & Astrophys.*, 2009, in press.

Del Zanna, G., **Bradshaw**, S.J., Coronal loops: New insights from EIS observations, *Astron. Soc. Pac. Conference Series*, 2009, in press.

Del Zanna, G., V. Andretta, P.C. **Chamberlin**, T.N. Woods, and W.T. **Thompson** (2009), The EUV spectrum of the Sun: SOHO CDS NIS radiometric calibration, *A&A*, Submitted.

Dorelli, J. C., A. Bhattacharjee, On the generation and topology of Flux Transfer Events, *J. Geophys. Res.*, 2009, in press.

Dorelli, J.C., Asymmetric reconnection at Earth's dayside magnetopause, *J. Geophys. Res.*, 2009, in press.

Farrell, W.M., T.J. **Stubbs**, J.S. Halekas, R.M. **Killen**, G.T. Delory, M.R. **Collier**, and R.R. **Vondrak**, The anticipated electrical environment within permanently shadowed lunar craters, *J. Geophys. Res.*, 2009JE003464, 2009, in press.

Feofilov, G., Feofilov, A.G. **A.A. Kutepov**, W.D. **Pesnell**, R.A. **Goldberg**, B.T. Marshall, L.L. Gordley, M. Martcia-Comas, M. López-Puertas, R.O. Manuilova, V.A. Yankovsky, S.V. Petelina, and J.M. Russell III, Daytime Saber/Timed observations of water vapor in the Mesosphere: Retrieval approach and first results, *Atmos. Chem. Phys.*, 2009, in. press.

Fernandez J.R., C.J. Mertens, D. **Bilitza**, X. Xu, J.M. Russell III, and M.G. Mlynczak, Storm/Quiet ratio comparisons between TIMED/SABER no+(v) volume emission rates and incoherent scatter radar electron densities at e-region altitudes, *Adv. Space Res*, 2009, submitted.

Glocer, G., Y. Ma, T. Gombosi, Multi-Fluid BATS-R-US: magnetospheric composition and dynamics during geomagnetic storms, initial results, *J. Geophys. Res.*, doi:10.1029/2009JA014418, in press.

Gopalswamy, N., Large-scale solar eruptions, in "Universal Heliophysical Processes", ed. N. **Gopalswamy**, S. Hasan, and A. Ambastha, *Springer*, in press.

Gopalswamy, N., Akiyama, S., Yashiro, S., and Mäkelä, P., Coronal Mass Ejections from sunspot and non-sunspot regions, To appear in "Magnetic Coupling between the Interior and the Atmosphere of the Sun", eds. S.S. Hasan and R.J. Rutten, *Astrophysics and Space Science Proceedings*, Springer-Verlag, Heidelberg, Berlin, 2009, in press.

Gopalswamy, N., **Xie**, H., Yashiro, S., Akiyama, S., and **Mäkelä**, P., Large geomagnetic storms associated with limb halo coronal mass ejections, *Advances in Geosciences*, 2009, accepted.

Gopalswamy, N., **Xie**, H., Mäkelä, P., Akiyama, S., Yashiro, S., **Kaiser,** M. L., Howard, R. A., and Bougeret, J.-L. , Interplanetary shocks lacking type II radio bursts, *ApJ*, 2009, submitted.

Gopalswamy, N., The Sun and Earth's space environment, *IEEE Proceedings*, 2009, submitted.

Gopalswamy, N., S.**Yashiro**, H. **Xie**, S. **Akiyama**, P. **Makela**, Large geomagnetic storms associated with limb halo coronal mass ejections, *Advances in Geosciences*, 2009, accepted.

Gopalswamy, N., S. **Akiyama**, S. **Yashiro**, and P. **Mäkelä** , Coronal mass ejections from sunspot and non-sunspot regions, in *Magnetic Coupling between the Interior and the Atmosphere of the Sun*, eds. S. S. Hasan and R. J. Rutten, Astrophysics and Space Science Proceedings, *Springer-Verlag*, Heidelberg, Berlin, 2009, in press.

Gopalswamy, N., P. **Mäkelä**, H. **Xie**, S. **Akiyama**, and S. **Yashiro**, Solar sources of "driverless" interplanetary shocks, in *Solar wind 12, AIP Conf.Proc. Ser.*, submitted.
Habbal, S.R., M. Druckm¨uller, H. Morgan, A. **Daw**, J. Johnson, A. Ding, M. Arndt, R. Esser, and V. Ruˇsin, Mapping the distribution of electron temperature and Fe charge states in the corona with total solar eclipse observations, *ApJ*, 2009, submitted.

Hathaway, D.H., P.E. **Williams**, M. Cuntz, "The advection of supergranules by large-scale flows." In: solar-stellar dynamos as revealed by helio- and asteroseismology, *ASP Conf. Proc.*, M. Dikpati et al., (Eds.), 2009, in press.

Hesse, M., S. **Zenitani**, M. Kuznetsova, and A. **Klimas**, Relativistic two-fluid simulations of guide field magnetic reconnection, *Phys. Plasmas*, in press.

Hesse, M., S. Zenitani, M. Kuznetsova, and A. Klimas, A simple, analytical model of collisionless magnetic reconnection in a pair plasma, *Physics of Plasmas*, 2009, in press.

Hock, R. A., P. C. **Chamberlin**, T. N. Woods, D. Crotser, F. G. Eparvier, D. L. Woodraska, and E. C. Woods (2009), Extreme ultraviolet Variability Experiment (EVE) Multiple EUV Grating Spectrographs (MEGS): radiometric calibrations and results, *Sol. Phys.*, submitted.

Holman, G. D., Aschwanden, M. J., Aurass, H., Battaglia, M., Grigis, P. C., Kontar, E. P., **Liu**, W., Saint-Hilaire, P., and Zharkova, V. V., Implications of X-ray observations for electron acceleration and propagation in solar flares, *Space Science Reviews*, 2008, submitted.

Holzworth, R. H., M. P. McCarthy, A. R. Jacobson, R. F. **Pfaff,** and D. E. **Rowland**, "Ionospheric Effects of lightning observed by C/NOFS", *Geophysics Research Letters*, 4/30/2009, submitted.

Huang, T. S., E. Romashets, G. **Le**, Y. **Wang**, and J. A. **Slavin**, Field-aligned currents at low altitude: Dependence on the geomagnetic activity, *J. of Atmos. and Solar-Terr. Phys.*, 2009, in press.

Ireland, J., M. S. Marsh, T. A. **Kucera**, **C.** A. **Young**, Automated detection of oscillating areas in the solar atmosphere, *Sol. Phys.*, 2009, submitted
Israelevich, and **Ofman,** L., Hybrid Simulation of Ion-acoustic waves excitation by a standing alfvén wave, *Annales Geophysicae*, 2009, submitted.

Ji, H., H. Wang, C. Liu, and B. R. **Dennis**, A hard X-ray sigmoidal structure during the initial phase of the 2003 October 29 X10 flare, *ApJ*, 680:734(739), June 2008, submitted.

Kahler, S. W. and **Gopalswamy,** N., CME geometry and the production of shocks and sep events, *Proc. 31st International Cosmic Ray Conference*, 2009, submitted.

Kepko, L., E. Spanswick, V. Angelopoulos, E. Donovan, J. McFadden, K.-H. Glassmeier, J. Raeder, H. J. Singer, Equatorward moving auroral signatures of a flow burst observed prior to auroral onset, *J. Geophys. Res.*, 2009, submitted.

Kilper, G., H. R. **Gilbert**, D. Alexander, Mass composition in pre-eruption quiet Sun filaments, *Ap.J.*, in press.

Klimchuk, J. A., Coronal Loop Models and Those Annoying Observations, Proceedings of the Hinode II Meeting (*ASP Conf. Ser*. Vol. XX), ed. B. Lites, 2009, in press.

Kontar, E. P., Brown, J. C., Emslie, A. G., Hajdas, W., **Holman**, G. D., Hurford, G. J., Kasparova, J., Mallik, P. C. V., Massone, A. M., McConnell, M. L., Piana, M., Prato, M., Schmahl, E. J., and Suarez-Garcia, E., Deducing Electron Properties from Hard X-ray observations, *Space Science Reviews*, 2008, submitted.

Korotova, G. I., D. G. **Sibeck**, V. Kondratovich, and O. D. Constantinescu, THEMIS observations of compressional pulsations in the dawnside magnetosphere: A case study, *Ann. Geophys.*, 2009, in press.

Labrosse, N., Heinzel, P., Vial J.-C., **Kucera,** T., Parenti, S., Gunar, S., Schmieder, B., and **Kilper** G., 2009, Physics of solar prominences: Spectral diagnostics and Non-LTE modelling, *Space Science Rev.,* submitted.

Leamon, R. J., McIntosh. S.W. and B. de Pontieu, The spectroscopic footprint of the fast solar wind, ApJ, submitted (July 2009).

Lin, R. L., X. X. Zhang, S. Q. Liu, Y. **Wang**, and J. C. Gong, A three-dimensional asymmetric magnetopause model, *J. Geophys. Res.,* 2009, in press.

Liu, W., V. Petrosian, B. R. **Dennis**, and G. D. **Holman**, Conjugate hard X-ray footpoints in the 2003 October 29 X10 flare: Unshearing motions, correlations, and asymmetries, *ApJ*, 693:847-867, March 2009, submitted.

Liu, W., V. Petrosian, B. R. **Dennis**, and Y. W. Jiang, Double coronal hard and soft X-ray source observed by RHESSI: Evidence for magnetic reconnection and particle acceleration in solar flares, *ApJ*, 676:704(716), March 2008, submitted.

Liu, W., T.-J. Wang, B. R. **Dennis**, and G. D. **Holman**, Episodic X-ray emission accompanying the activation of an eruptive prominence: Evidence of episodic magnetic reconnection, *ApJ*, 698:632(640), June 2009, submitted.

Mackay, D.H., J.T. **Karpen**, J.L. Ballester, B. Schmieder, & G. Aulanier, Physics of solar prominences: II - Magnetic structure and dynamics, *Space Science Reviews*, 2009, submitted.

MacNeice, P., (2009), Validation of Community Models II: Development of a baseline using the Wang-Sheeley-Arge Model, *Space Weather*, doi:10.1029/2009SW000489, accepted.

Mäkelä, P., N. **Gopalswamy**, H. **Xie**, S. **Akiyama**, and S. **Yashiro**, Radio-quiet shocks and energetic particle enhancements at 1 AU, *Astrophys. J.*, submitted.

Milligan, R. and **Dennis**, B.R., Velocity characteristics of evaporated plasma using Hinode/EIS, *ApJ*, **699**, 968-975, 2009, submitted.

Moore, T.E., plasma escape by centrifugal pick-up, *J. Geophys. Res.*, 16 Sep 2009, submitted.

Narock, T.W., V. Yoon, J. **Merka**, and A. **Szabo**, The semantic web in federated information systems: A space physics case study, *Journal of Information Technology Theory and Application*, 9/8/2009, submitted.

Ofman, L., Acceleration and heating of solar wind ions by nonlinear waves, *Advances in Geosciences*, Solar Terrestrial (ST), Marc Duldig (ed.), 2009, in press.

Omidi, N., J.P. Eastwood, and D.G. **Sibeck**, Foreshock bubbles and their global magnetospheric impacts, *J. Geophys. Res.*, 2009, in press.

Paranicas, C., J.F. **Cooper**, H.B. Garrett, R.E. Johnson, and S.J. **Sturner**. Europa's radiation environment and its effect on the surface, in EUROPA, Editors: R. Pappalardo, W. B. McKinnon, and K. Khurana, *University of Arizona Press Space Science Series*, 2009, in press.

Pei, C, Bieber, J.W., Breech, B., Burger, R.A., Clem, J., Matthaeus, W.H., Cosmic Ray diffusion tensor throughout the heliosphere, 2009, submitted.

Pesnell, W. D., Predicting solar cycle 24 using Ap and F10.7 as a geomagnetic precursor pair, *Solar Physics*, 2010, in press.

Pfaff, R., D. **Rowland**, H. **Freudenreich**, K. **Bromund**, C. **Liebrecht**, G. Le, **M. Acuna**, W. Burke, N. Maynard, G. Wilson, D. Hunton, P. Roddy, Initial observations of vector DC electric fields and associated plasma drifts measured on the C/NOFS satellite, *Geophysical Research Letters*, 2009, submitted.

Plaschke, F., K.H. Glaßmeier, D.G. **Sibeck**, H.U. Auster, O.D. Constantinescu, V. Angelopoulos, W. Magnes, and J.P. McFadden, Magnetopause surface oscillation frequencies at different solar wind conditions, *Ann. Geophys.*, 2009, submitted.

Pogorelov, N.V., S.N. Borovikov, L.F. **Burlaga**, R.W. Ebert, J. Heerikhuisen, Q. Hu, D.J. McComas, S.T. Suess and G.P. Zank. Transient phenomena in the distant solar wind and in the heliosheath, *Astrophys. J. Lett.*, 2009, in press.

Pulkkinen, A., L. **Rastätter, M. Kuznetsova,** M. **Hesse**, A. Ridley, J. Raeder, H.J. Singer, and A. **Chulaki**, Systematic evaluation of ground and geostationary magnetic field predictions generated by global magnetohydrodynamic models, *Journal of Geophysical Research*, 2009, accepted.

Pulkkinen, A., and L. **Rastätter**, Minimum variance analysis-based propagation of the solar wind observations: application to real-time global magnetohydrodynamic simulations, *Space Weather*, 2009, submitted.

Pulkkinen, A., T. Oates, and A. **Taktakishvili**, Automatic determination of the conic coronal mass ejection model parameters, *Solar Physics*, 2009, submitted.

Pulkkinen, A., M. Hesse, S. Habib, L. Van der Zel, B. Damsky, F. Policelli, D. Fugate, and W. Jacobs, Solar Shield - forecasting and mitigating space weather effects on high-voltage power transmission systems, Natural Hazards, 2009, in press.

Pulkkinen, A., L. Rastätter, M. Kuznetsova, M. Hesse, A. Ridley, J. Raeder, H.J. Singer, and A. Chulaki, Systematic evaluation of ground and geostationary magnetic field predictions generated by global magnetohydrodynamic models, *Space Weather*, 2009, in press.

Ramesh, R., Kathiravan, C., Kartha, S.S., and **Gopalswamy**, N., Radioheliograph observations of metric type II bursts and the kinematics of coronal mass ejections, *ApJ*, 2009, submitted.

Reale, F., **Klimchuk**, J.A., Parenti, S., Testa, P., XRT detection of hot plasma in active regions and nanoflare heating, Proceedings of the Hinode II Meeting (*ASP Conf. Ser.* Vol. XX), ed. B. Lites, 2009, in press.

Reginald, N.L., O.C. **St.Cyr**, J.M. **Davila**, D.M. **Rabin**, M. Guhathakurta, and D.M. Hassler, Electron temperature maps of low solar corona: ISCORE results from the total solar eclipse of 29 March 2006 in Libya, *Solar Phys.*, 2009, submitted.

Robertson, I.P., S. Sembay, T.J. **Stubbs**, K. **Kuntz**, M. **Collier**, T. Cravens, **S. Snowden**, H. **Hills**, F. **Porter**, P. Travnicek, J. Carter, and A. Read, Solar wind charge exchange observed through the lunar atmosphere, *Geophys. Res. Lett.*, 2008GL035170, 2009, in press.

Sarantos, M, R.M. **Killen**, W.E. McClintock, E.T. Bradley, M. **Benna**, and J A. **Slavin**, Limits to Mercury's magnesium exosphere following MESSENGER's second flyby, *Geoph. Res. Lett.*, 2009, submitted.

Schmelz, J.T., V.L. Kashyap, S.H. Saar, B.R. **Dennis**, P.C. Grigis, L. Lin, E.E. De Luca, G.D. **Holman**, L. Golub, and M.A. Weber, Some like it hot: Coronal heating observations from Hinode X-ray telescope and RHESSI, *ApJ*, 704:863(869), October 2009, submitted.

Schmidt, J.M., and **Ofman**, L., Global simulation of an EIT wave, *The Astrophysical Journal*, 2009, submitted.

Schuck, P., The photospheric energy and helicity budgets of the flux-injection hypothesis, *Astrophysical Journal*, 2009, submitted.

Selwa, M., Murawski, K., Solanki, S.K., **Ofman**, L., Excitation of vertical kink waves in a solar coronal arcade loop by a periodic driver, *Astronomy and Astrophysics*, 2009, submitted.

Shanmugaraju, A, Moon, Y.-J., Cho, K.S., Bong, S.C., **Gopalswamy**, N., Akiyama, S., Yashiro, S., Umapathy, S., and Vrsnak, B., Quasiperiodic oscillations in LASCO CME speeds, *ApJ*, 2009, submitted.

Sibeck, D.G. and R.Q. Lin, Concerning the motion of flux transfer events generated by component reconnection across the dayside magnetopause, *J. Geophys. Res.*, 2009, submitted.

Sittler Jr., E.C., R.E. **Hartle**, A.S. **Lipatov**, J.F. **Cooper**, C. Bertucci, A.J. Coates, K. Szego, M. **Shappirio**, D.G. **Simpson**, and J.E. Wahlund, Saturn's magnetospheric interaction with Titan as defined by Cassini encounters T9 and T18: New results, *Plan. Sp. Sci.*, 2009, in press.

Sittler Jr, E. C., R.E. **Hartle**, J.F. **Cooper**, A. **Lipatov**, R.E. Johnson, C. Bertucci, A.J. Coates, C. Arridge, K. Szego, M. **Shappirio**, D.G. **Simpson**, R. Tokar and D.T. Young, Saturn's magnetosphere and properties of upstream flow at titan: preliminary results, *Plan. Sp. Sci.*, 2009, submitted.

Spence, H.E., A. Case, M.J. Golightly, T. Heine, J.B. Blake, P. Caranza, W.R. Crain, J. George, M. Lalic, A. Lin, M.D. Looper, J.E. Mazur, D. Salvaggio, J.C. Kasper, T.J. **Stubbs**, M. Doucette, P. Ford, R. Foster, R. Goeke, D. Gordon, B. Klatt, J. O'Connor, M. Smith, T. Onsager, C. Zeitlin, L. Townsend, and Y. Charara, CRaTER: The Cosmic Ray Telescope for the Effects of Radiation experiment on the Lunar Reconnaissance Orbiter Mission, *Space Sci. Rev.*, SPAC560, 2009, in press.

Strong, K.T., and J.L.R. **Saba**, The solar cycle from a whole-Sun perspective, poster given at the STEREO-3/SOHO-22 Workshop: Three Eyes on the Sun – Multi-spacecraft studies of the corona and impacts on the heliosphere, The De Vere Royal Bath Hotel, Bournemouth, Dorset, England, 2009 Apr 27–May 01, submitted.

Su, Y., G.D. **Holman**, B.R. **Dennis**, A.K. **Tolbert**, and R.A. **Schwartz**, A test of thick-target nonuniform ionization as an explanation for breaks in solar flare hard X-ray spectra, *Ap. J*, 2009, in press.

Summerlin, E.J., Dilusive acceleration of particles at collisionless magnetohydrodynamic shocks, Houston, Texas: Rice University, Ph. D. thesis, 2009, submitted.

Sych, R.A., Nakariakov, V.M., Anfinogentov, S., **Ofman**, L., Web-based data processing system for automated detection of oscillations with applications to solar atmosphere, *Solar Physics*, 2009, submitted.

Truhlik V., L. Triskova, and D. **Bilitza**, Variations of daytime and nighttime electron temperature and heat flux in the topside ionosphere for low and high solar activity, *J. Atmos. Sol.-Terr. Phys*, 2009, submitted.

Turk-Katircioglu, F., Z. Kaymaz, and D.G. **Sibeck**, Magnetosheath cavities: Case studies using Cluster observations, *Ann. Geophys.*, 2009, in press.

Walker, R.J., J. **Merka**, T.A. King, T. Narock, S.P. Joy, L.F. Bargatze, P. Chi, and J. Weygand (2009), The Virtual Magnetospheric Observatory. IGY+50 Proceedings/ *CODATA Data Science Journal*, 2008, submitted.

Wang, T.J., **Ofman**, L., **and Davila**, J.M., Hinode/EIS observations of propagating slow magnetoacoustic waves in a coronal loop, Proceedings of the second Hinode science meeting, M. Cheung, B. Lites, T. Magara, J. Mariska, and K. Reeves (eds.), *ASP Conference Series*, 2009, in press.

Wang, T., J.W. **Brosius**, R.J. **Thomas**, D.M. **Rabin**, and J. M. **Davila**, Absolute radiometric calibration of the EUNIS-06 SW channel using its lab-calibrated LW channel and theoretically predicted insensitive line ratios, *Ap. J.*, 2009, submitted.

Warmuth. A., G.D. **Holman**, B.R. **Dennis**, G. Mann, H. Aurass, and R.O. **Milligan**, Rapid changes of electron acceleration characteristics at the end of the impulsive phase of an x-class solar flare, *ApJ*, 699:917(922), July 2009, submitted.

Webb, P. A., M. M. Kuznetsova, M. Hesse, and L. Rastaetter, Ionosphere-Thermosphere models at the community coordinated modeling center, *Radio Sci.*, 2009, in press.

Williams, P.E., and M. Cuntz., A method for the treatment of supergranule advection by giant cells, *Astron. Astrophys.*, 2009, in press.

Xie, H., O.C. **St. Cyr**, N. **Gopalswamy**, S. **Yashiro**, J. Krall, M. **Kramar** and J. **Davila**, On the origin, 3D structure and dynamic evolution of CMEs near solar minimum, *Solar Phys.*, 2009, in press.

Yashiro, S. and N. **Gopalswamy**, Statistical relationship between solar flares and coronal mass ejections, in Universal Heliophysical Processes, *Proceedings of IAU Symposium 257*, p. 233-243, 2009, in press.

Yashiro, S., G. **Michalek**, and N. **Gopalswamy**, A comparison of coronal mass ejections identified by manual and automatic methods, *Annales Geophysicae*, Volume 26, Issue 10, pp.3103-3112, 2008, in press.

Zenitani, S., M. **Hesse**, and **A. Klimas**, Relativistic Two-fluid simulations of guide field magnetic reconnection, *Astrophys. J.*, 2009, in press.

Zenitani, S., M. **Hesse**, and A. **Klimas**, Two-Fluid MHD simulations of relativistic magnetic reconnection, *Astrophys. J.*, 2009, in press.

Zhang Y-L., L.J. Paxton, and D. **Bilitza**, Near real-time assimilation of auroral peak E-region density and equatorward boundary in IRI, *Adv. Space Res*, 2009, submitted.

Presentations

Airapetian, V., **Ofman**, L., **Sitter Jr.**, E.C., **Kramar**, M., Probing Thermodynamic and Kinematic Properties of a Coronal Streamer Event Formed During the Solar Minimum, SPD meeting, Boulder, CO, USA, June 14-18, 2009 (poster).

Airapetian, V. and **Klimchuk**, J.A. "Models of Impulsively Heated Solar Active Regions," presented at Coronal Loops 4 Workshop, Florence, Italy, June 2009 (poster).

Abbo, Lucia, Leon **Ofman**, Silvio Giordano, Roberto Lionello, and Zoran Mikic, Streamers study at solar minimum: combination of UV observations and numerical modeling, poster presented at Solar Wind 12, in Saint-Malo, France, June 21-26, 2009.

Amm, O., J. Weygand, V. Angelopoulos, B. Beheshti, E. Steinmetz, M. Engebretson, A. Viljanen, **A. Pulkkinen**, H. Gleisner, H. Frey, D. Mende, Equivalent ionospheric currents from the GIMA, Greenland, MACCS, and THEMIS ground magnetometer arrays, presented at Fall AGU meeting, San Francisco, California, December 15-19, 2008. (poster)

Barnes, Robin, Daniel Morrison, Michele Weiss, E Lis Immer, Matthew Potter, Robert Holder, Dennis Patrone, Chris Colclough, Robert **McGuire**, Robert **Candey**, Dieter **Bilitza**, Bernard **Harris**, Janet Kozyra, Peter Fox, Ron Heelis, James Russell, "Enabling Science Research with Coordinated Data From SuperDARN and VITMO", presented at Fall AGU meeting, San Francisco, California, December 15-19, 2008.

Benson, R.F. and S.F. **Fung**, ISIS topside-sounder plasma-wave investigations as guides to desired Virtual Wave Observatory (VWO) data search capabilities, paper SA51B-07 presented at the Fall AGU meeting, San Francisco, California, December 15-19, 2008.

Bilitza, Dieter, Robin Barnes, Robert **Candey**, Bernard **Harris**, Robert Holder, E Lis Immer, Robert **McGuire**, Daniel Morrison, Dennis Patrone, Matthew Potter, Michele Weiss, "Leveraging Capabilities in the Community: CDAWeb Data and Services within VITMO", presented at Fall AGU meeting, San Francisco, California, December 15-19, 2008.

Breech, BA, Goldstein, M.L., Roberts, A., and Usmanov, A., First Steps Towards a Simplified Model for Sub-Alfvénic Flows in the Corona, presented at Solar Wind 12, in Saint-Malo, France, June 21-26, 2009 (poster).

Breech, B., Cranmer, S.R., Matthaeus, W.H., Kasper, J.C., and Oughton, S., Studying the Heating of the Solar Wind Through Electron and Proton Effects, presented at Solar Wind 12, in Saint-Malo, France, June 21-26, 2009 (poster).

Boardsen, S. A., J. A. Slavin, B. J. Anderson, M. H. Acuna, H. Korth and S. C. Solomon, (2008), Narrow-Band Ultra-Low-Frequency Wave Observations During the October 6, 2008, Flyby of Mercury, presented at Fall AGU meeting, San Francisco, California, December 15-19, 2008.

Boardsen, S.A., J.A. **Slavin**, B.J. Anderson, M.H. Acuna, H. Korth, and S.C. Solomon (2009), Comparison of Ultra-Low-Frequency Waves at Mercury under Northward and Southward IMF, presented at Fall AGU meeting, San Francisco, California, December 15-19, 2008.

Brosius, J.W., Observation of the Conversion From Explosive to Gentle Chromospheric Evaporation During a Solar Flare.

Brown, L.E., M.E. Hill, R.B. Decker, J.F. **Cooper**, S.M. Krimigis, and J.D. Vandegriff, New access and analysis tools for Voyager LECP data, presented at Fall AGU meeting, San Francisco, California, December 15-19, 2008 (poster).

Brown, L.E., M.E. Hill, R.B. Decker, J.F. **Cooper**, S.M. Krimigis, and J.D. Vandegriff, Heliophysics Research Using Virtual Observatories, presented at Fall AGU meeting, San Francisco, California, December 15-19, 2008 (poster).

Burlaga, L.F., Heliospheric observations (SHINE), A Voyage through the Heliosphere, presented Parker Lecture at Fall 2008 AGU Meeting, Dec. 15-19, 2008.

Burlaga, L.F., Compressible "Turbulence" in the Heliosheath, presented at SHINE 2009 Conference, Wolfville, Nova Scotia, August 2009.

Burlaga, L.F., Radial and solar cycle variations of magnetic fields in the heliosheath, presented at SOHO 23 Meeting, Northeast Harbor, Maine, Sept. 2009.

Buzulukova, N., M. **Fok**, M. **Kuznetsova**, A. **Pulkkinen**, L. **Rastaetter**, P. Brandt, G. Toth, Inner magnetosphere--global MHD coupled code: Initial results, presented at Fall AGU meeting, San Francisco, California, December 15-19, 2008 (poster).

Buzulukova, N., M. **Fok,** D. McComas, P. Brandt, J. Goldstein, P. Valek, J. Alquiza, First stereoscopic views of the ring current from TWINS energetic neutral atoms imagers, presented at AGU Joined Assembly, Toronto, Canada, May 2009.

Buzulukova, N., Ring current : global imaging and global modeling

Buzulukova, N., M. **Fok**, Global imaging and global modeling of ring current, presented at TRIO-CINEMA meeting, Kyung Hee University, Korea, 19-23 October 2009.

Candey , Robert M., Reine A. **Chimiak**, Bernard T. **Harris**, Todd King, Robert E. **McGuire**, Thomas W. **Narock** (2008), "Heliophysics Event List Manager (HELM)", presented at Fall AGU meeting, San Francisco, California, December 15-19, 2008.

Chamberlin, P. C., T. N. Woods, F. G. Eparvier, D. Judge, and L. Didkovsky, Observing changes in the solar EUV spectral irradiance, presented at SPD Meeting, Boulder, CO, USA, May 2009.

Chamberlin, P. C., T. N. Woods, and M. Haberreiter, Solar cycle minimum measurements of the solar extreme ultraviolet (EUV) spectral irradiance on April 14, 2008, presented at IAU, Rio de Janeiro, Brazil, Aug 2009.

Chamberlin, P. C., T. N. Woods, F. G. Eparvier, A. R. Jones, Next generation X-Ray Sensors (XRS) for the NOAA GOES-R series satellites, presented at SPIE Optics and Photonics, San Diego, CA, Aug 2009.

Chamberlin, P. C., Solar Flares, presented at LASP REU Summer School, Boulder, CO, June 2009.

Chamberlin, P. C., G. Lu, Z. Sternovsky, P. Withers, and T. N. Woods, Using the Flare Irradiance Spectral Model (FISM) to study the response of the Earth, Mars and Moon to solar flares, presented at EGU General Assembly, Vienna, Austria, April 2009.

Chamberlin, P. C., T. N. Woods, F. G. Eparvier, R. A. Hock, A. R. Jones, D. L. Woodraska, Absolute calibration of CU/LASP's X-Ray and EUV spectrometers, presented at Solar EUV-IR Workshop, Freiburg, Germany, April 2009.

Chamberlin, P. C., T. N. Woods, F. G. Eparvier, R. A. Hock, A. R. Jones, D. L. Woodraska, Extreme ultraviolet measurements from LASP/CU, presented at Solar EUV-IR Workshop, Freiburg, Germany, April 2009.

Chamberlin, P. C., T. N. Woods, J. Harder, R. A. Hock, M. Snow, The extreme ultraviolet contributions to the Solar Irradiance Reference Spectra (SIRS), presented at Fall AGU, San Francisco, California, December 15-19, 2008.

Chamberlin, P. C., The Flare Irradiance Spectral Model (FISM) and its contributions to space weather research and instrument design, presented at Naval Research Laboratory Seminar, Washington, DC, Oct., 2008.

Christian, E.R., Observing the Outskirts of the Heliosphere: The Interstellar Boundary Explorer (IBEX) Mission, presented at 31st ICR Conference, Łódz', Poland, July 7-15, 2009 (poster).

Christian, E.R., G.A. **de Nolfo**, W.R. Binns, M.H. Israel, J.W. **Mitchell**, T. **Hams**, J.T. **Link**, M. **Sasaki**, A.W. Labrador, R.A. Mewaldt, E.C. Stone, C.J. Waddington, M.E. Wiedenbeck, Identifying galactic cosmic ray origins with Super-TIGER, presented at 31st ICR Conference, Łódz', Poland, July 7-15, 2009.

Collado-Vega, Y., R. Kessel, R. **Boller**, V. **Kalb**, Statistical study of magnetopause boundary region vortices observed during modeled solar wind conditions, presented at 9th International School for Space Simulations (ISSS-9), Saint-Quentin-en-Yvelines, France, July 3-10, 2009.

Collado-Vega, Y., R. Kessel, R. **Boller**, V. **Kalb,** Statistical study of magnetopause boundary region vortices observed during modeled solar wind conditions, presented at Fall AGU, San Francisco, California, December 15-19, 2008.

Collier, M. R., T. J. **Stubbs**, H. K. **Hills**, J. Halekas, W. M. **Farrell**, G. Delory, J.**Espley**, and **P. Webb,** Lunar potential determination using Apollo-era data and modern measurements and models, P51D-03, presented at Fall AGU meeting, San Francisco, California, December 15-19, 2008.

Cooper, J. F., Innermost Van Allen Radiation Belt for High Energy Protons at Saturn, presented at DPS/AAS 2008 Conference, Ithaca, NY, Oct. 10-15, 2008 (oral).

Cooper, J.F., K. Kauristie, A.T. Weatherwax, G.W. Sheehan, R.W. Smith, I. Sandahl, N. Østgaard, S. Chernouss, B. J. **Thompson**, L. Peticolas, M. H. **Moore**, D. A Senske, L. K. Tamppari, and E. M. **Lewis**, IHY-IPY Outreach on Exploration of Polar and Icy Worlds in the Solar System, DPS/AAS 2008 Conference, Ithaca, NY, Oct. 10-15, 2008 (poster).

Cooper, J.F., J.D. Richardson, M.E. Hill, and S. J. **Sturner**, Impact of universal plasma and energetic particle processes on icy bodies of the Kuiper belt and the Oort cloud**,** presented at AGU Chapman Conference on Universal Heliophysical Processes (IHY), Savannah, Georgia, Nov. 10-14, 2008 (poster).

Cooper, J.F., K. Kauristie, A.T. Weatherwax, G.W. Sheehan, R.W. Smith, T.D. **Cline**, E.M. **Lewis**, and G. Haines-Stiles, IPY Science and Outreach in Polar Partnership, presented at Polar Gateways Arctic Circle Sunrise 2008 Conference at the Top of the World, , Barrow, Alaska, January 23-29, 2008.

Cooper, J.F., K. Kauristie, A.T. Weatherwax, G.W. Sheehan, R.W. Smith, T. D. **Cline**, E.M. **Lewis**, and G. Haines-Stiles, IPY Science and Outreach in Polar Partnership, presented at Fall AGU meeting, San Francisco, California, December 15-19, 2008 (talk).

Cooper, J.F., N. **Lal**, R.E. **McGuire**, A. Szabo, T.W. **Narock**, T. P. Armstrong, J. W. Manweiler, J. D. Patterson, M. E. Hill, J. D. Vandegriff, R. B. McKibben, C. Lopate, and C. Tranquille, Virtual Energetic Particle Observatory (VEPO), presented at Fall AGU meeting, San Francisco, California, December 15-19, 2008 (poster).

Cooper, J.F., N. **Lal**, R.E. **McGuire**, A. Szabo, T.W. **Narock**, T. P. Armstrong, J. W. Manweiler, J. D. Patterson, M. E. Hill, J. D. Vandegriff, R. B. McKibben, C. Lopate, and C. Tranquille, Heliophysics Research Using Virtual Observatories, presented at Fall AGU meeting, San Francisco, California, December 15-19, 2008 (poster).

Cooper, J.F., K. Kauristie, A.T. Weatherwax, G.W. Sheehan, R.W. Smith, I. Sandahl, N. Østgaard, S. Chernouss, B. J. **Thompson**, L. Peticolas, M. H. **Moore**, D. A Senske, L. K. Tamppari, and E. M. **Lewis**, IHY-IPY Outreach on Exploration of Polar and Icy Worlds in the Solar System, presented at Polar Gateways Arctic Circle Sunrise Conference 2008, Barrow, Alaska, January 23-29, 2008 (poster).

Cooper, P.D., J.F. **Cooper**, E.C. **Sittler**, M.H. **Burger**, S.J. **Sturner**, and A.M. Rymer, Saturn magnetospheric impact on surface molecular chemistry and astrobiological potential of Enceladus, presented at Fall AGU meeting, San Francisco, California, December 15-19, 2008.

Cooper, J. F., Limiting Charged Particle Flux Spectrum at the Heliopause and Beyond, presented at EGU General Assembly, Vienna, Austria, April 19-24, 2009.

Cooper, J.F., Comparative Magnetospheric Contributions to Polar Atmospheric Outflows at Enceladus, and Ganymede, presented at EGU General Assembly, Vienna, Austria, April 19-24, 2009.

Cooper, J.F., Neutral injection sources and transport in Saturn's inner radiation belt, presented at Magnetospheres of the Outer Planets Conference 2009, Cologne, Germany, July 27-31, 2009 (poster).

Cooper, J.F., E.C. **Sittler**, R. E. **Hartle**, and R. E. Johnson, Plasma ion and exospheric gas composition objectives for EJSM, presented at Europa Jupiter System Mission Instrument Workshop, Johns Hopkins University Applied Physics Laboratory, July 15-17, 2009 (poster).

Cooper, J.F., R.E. **Hartle**, and E.C. **Sittler**, Origins for Surface and Exospheric Composition of Ganymede, presented at European Planetary Science Congress, Sept. 13-18, 2009, Potsdam, Germany, 2009.

Coyner, A.J.; **Davila**, J.M.; **Brosius**, J.W.; **Ofman**, L., Analysis of active region and quiet Sun spectra from SERTS-99 observations, presented at SPD meeting, Boulder, CO, USA, June 14-18, 2009 (poster).

Coyner, A.J.; **Davila**, J.M.; **Ofman**, L. Constraints On Coronal Non-thermal Velocities From SERTS 1991-1997 Observations, presented at SPD meeting, Boulder, CO, USA, June 14-18, 2009 (poster).

Delory, G.T., J.S. Halekas, W.M. **Farrell**, T.J. **Stubbs**, The electrical environment of the Moon, AE23A-04, presented at Fall AGU meeting, San Francisco, California, December 15-19, 2008.

Espley, J. W. Freeman, R. R. **Vondrak**, and J. Kasper, Lunar surface potential increases during terrestrial bow shock traversals, presented at the 2nd Annual Lunar Science Forum, NASA Lunar Science Institute, NASA Ames Research Center, Moffett Field, CA, July 21–23, 2009.

Faden, J., R.S. Weigel, E.E. West, and J. **Merka** (2008). Autoplot: a Browser for Science Data on the Web., presented at Fall AGU meeting, San Francisco, California, December 15-19, 2008.

Farrell, W.M., G.T. Delory, R.M. **Killen,** R.P. Lin, J.S. Halekas, S. Bale, D. Krauss-Varban, R.R. **Vondrak,** M.R. **Collier,** J. **Keller,** T. **Jackson,** R. **Hartle,** M. **Hesse,** M. **Sarantos,** R. Elphic, A. Colaprete, T.J. **Stubbs,** D.M. Hurley, J. Marshall, R. Hodges, D.A. **Glenar,** W. Paterson, H.E. Spence, L. **Bleacher,** H. **Weir,** M. Horanyi, M. **Dube,** M. Hyatt, J. Kasper, and Y. Saito, Dynamic Response of the Environment At the Moon (DREAM): A NLSI team exploring the solar-lunar connection, presented at the 2nd Annual Lunar Science Forum, NASA Lunar Science Institute, NASA Ames Research Center, Moffett Field, CA, July 21–23, 2009.

Farrell, W.M., T.J. **Stubbs,** J.S. Halekas, G.T. Delory, M.R. **Collier,** and R.R. **Vondrak,** The anticipated electrical environment within permanently shadowed lunar craters, presented at the 2nd Annual Lunar Science Forum, NASA Lunar Science Institute, NASA Ames Research Center, Moffett Field, CA, July 21–23, 2009.

Feofilov , G., A.A. **Kutepov,** B.T. **Marshall,** W.D. **Pesnell,** R.A. **Goldberg,** L.L. **Gordley,** and J.M. **Russell,** Temperature and water vapor measured by SABER/TIMED and implications for mesospheric ice clouds, presented at Fall AGU meeting, San Francisco, California, December 15-19, 2008.

Feofilov, A.G., S.V. Petelina, **A.A. Kutepov,** W.D. **Pesnell,** and R.A. **Goldberg,** Water vapor, temperature, and ice particles in polar mesosphere as measured by SABER/TIMED and OSIRIS/ODIN instruments, presented at the 11th Scientific Assembly of IAGA 2009, Sopron, Hungary, August 2009.

Feofilov, G., S. V. Petelina, **A. A. Kutepov,** W. D. **Pesnell,** and R. A. **Goldberg,** Water vapor, temperature, and ice particles in the polar mesosphere as measured by SABER/TIMED and OSIRIS/Odin instruments, presented at the 11th Scientific Assembly of IAGA, Sopron, Hungary, August 2009 (convenor of symposium)

Feofilov, G., A.A. **Kutepov,** W.D. **Pesnell,** and R.A. **Goldberg,** Non-lte diagnostics of broadband infrared emissions from the mesosphere and lower thermosphere, presented at Atmospheric Studies by Optical Remote Sensing, Kiev, Ukraine, August 2009.

Feofilov, A.G., A.A. Kutepov, B.T. Marshall, W.D. **Pesnell,** R.A. Goldberg, L.L. Gordley, and J.M. Russell, Temperature and water vapor measured by SABER/TIMED and implications for mesospheric ice clouds, presented at Fall AGU meeting, San Francisco, California, December 15-19, 2008.

Fung, S.F., The Virtual Wave Observatory (VWO), presented at Fall AGU meeting, San Francisco, CA, December 15-19, 2008.

Fung, S. F., Magnetospheric Field-aligned electron density measurements by IMAGE RPI, presented at a Workshop on Advances in Space Plasmas, in honor of the 70th birthday of Dr. Dennis Papadopoulos, University of Maryland College PARK, January 15, 2009 (invited).

Fung, S.F., Tan, L.C., X. Shao, A.S. Sharma, Acceleration of magnetospheric relativistic electrons by ultra-low frequency waves during storm recovery: A comparison between two cases observed by CLUSTER, presented at the 17th CLUSTER Workshop, Uppsala University, Uppsala, Sweden, May 12-15, 2009.

Fok, M.C., Ring current modeling: Approaches, status and outstanding challenges, presented at GEM Workshop, Snowmass, Colorado, June 2009.

Fok, M.C., A. **Glocer**, and Q. **Zheng**, Recent developments in the radiation belt environment model, presented at International Living With A Star Workshop, Ubatuba-SP, Brazil, October 2009.

Fok, M.C., A. **Glocer**, Q. **Zheng** Recent Developments in the radiation belt environment model, presented at International Living With A Star Workshop, Ubatuba-SP, Brazil, October 2009.

Fok, M., N. **Buzulukova**, D. McComas, P. Brandt, J. Goldstein, P. Valek, J. Alquiza, The role of plasma sheet conditions in ring current formation and energetic neutral atom emissions: TWINS results and CRCM comparison, presented at AGU Joined Assembly, Toronto, Canada, May 2009.

Galkin, I.A., S **Fung**, T.A.King, B.W. Reinisch, Registering Active and Passive IMAGE RPI Datasets with the Virtual Wave Observatory, presented at Fall AGU meeting, San Francisco, California, December 15-19, 2008.

Gedalin, M., Balikhin, M., and L. **Ofman**, Collisionless relaxation of downstream ion distribution at shocks: Theory, simulations, and observations, EGU General Assemby, Vienna, Austria, April 19-24, 2009 (talk).

Gilbert, H., **Kilper**, G., Alexander, D., **Kucera**, T., "Using Prominence Mass Inferences in Different Coronal Lines to Obtain the He/H Abundance," presented at SPD meeting, Boulder, CO, USA, June 14-18, 2009.

Gilbert, H., G. **Kilper**, T. **Kucera**, D. Alexander. Using Prominence Mass Inferences in Different Coronal Lines to Obtain the He/H Abundance, presented at SPD meeting, Boulder, CO, USA, June 14-18, 2009.

Gilbert, H., G. **Kilper**, T. **Kucera**, D. Alexander. Using prominence mass inferences in different coronal lines to obtain the He/H abundance, presented at Fall AGU meeting, San Francisco, California, December 15-19, 2008.

Glenar, D.A., T.J. **Stubbs**, A. Colaprete, D.T. Richard, and G.T. Delory, Optical scattering processes observed at the Moon: Predictions for the LADEE Ultraviolet/Visible

Spectrometer, presented at the 2nd Annual Lunar Science Forum, NASA Lunar Science Institute, NASA Ames Research Center, Moffett Field, CA, July 21-23, 2009.

Glenar, D.A., T.J. **Stubbs**, and R.R. **Vondrak**, A reanalysis of Apollo light scattering observations: Implications for the spatial distribution of lunar exospheric dust, presented at the 2nd Annual Lunar Science Forum, NASA Lunar Science Institute, NASA Ames Research Center, Moffett Field, CA, July 21-23, 2009.

Goldberg, R.A., The unusual northern polar summer of 2002, University of Colorado, August 3, 2009.

Goldberg, R.A., A.G. **Feofilov**, A.A. **Kutepov**, W.D. **Pesnell**, R. Lateck, and J.M. Russell III, Temperature trends in the polar mesosphere between 2002-2007 using TIMNED/SABER data, presented at Fall AGU meeting, San Francisco, California, December 15-19, 2008.

Goldberg, R. A., A. G. **Feofilov**, A. A. **Kutepov**, W. D. **Pesnell**, F. S. **Schmidlin,** and J. M. Russell III, The unusual northern polar summer of 2002, presented at the 19th ESA Symposium on European Rocket and Balloon Programmes and Related Research, Bad Reichenhall, Germany, June 2009.

Glocer, A., G. Toth, Y. Ma, T. Gombosi, D. Welling, J. Zhang, L. Kistler, M-C **Fok**, Modeling iono-spheric outflows and their effect on the magnetosphere, presented at GEM Workshop, Snowmass, Colorado, June 2009.

Glocer, A., M. **Fok**, G. Toth, Modeling the radiation belts during a geomagnetic storm, presented at Spring AGU Meeting, Toronto, Canada, Mar, 2009

Glocer, A., M. **Sarantos**, J. **Slavin**, M. **Fok**, Three dimensional MHD modeling of Mercury's magnetosphere during ICMEs, presented at 2009 European Planetary Science Congress, Potsdam, Germany, September, 2009.

Gurman, J.B., Bogart, R., Spencer, J., Hill, F., Suarez Sola, I., Reardon, K., **Hourcle,** J., **Hughitt,** K., Martens, P., and Davey, A., The Virtual Solar Observatory: Where Do We Go from Here?, presented at AAS/SPD meeting, Boulder, CO, USA, June 14-18, 2009.

Habbal, S.R., A. N. **Daw**, H. Morgan, J. Johnson, M. Druckmuller, H. Druckmullerova, I. Scholl, M. B. Arndt, and A. Pevtsov, The role of heavy ions as coronal diagnostics: recent results from total solar eclipse observations, presented at Fall AGU meeting, San Francisco, California, December 15-19, 2008.

Habbal, S.R., A. **Daw**, H. Morgan, J. Johnson, M. Druckmuller, and V. Rusin, The curious case of the Fe XI 789.2 nm line, presented at AAS/SPD meeting, Boulder, CO, USA, June 14-18, 2009.

Halekas, J.S., G.T. Delory, T.J. **Stubbs**, W.M. **Farrell**, and R.P. Lin, Developing a predictive capability for lunar surface charging during solar energetic particle events, #1357, presented at the Lunar and Planetary Science Conference XL, Lunar and Planetary Institute, Houston, TX, March 23–27, 2009.

Halekas, J.S., G.T. Delory, R.P. Lin, T.J. **Stubbs**, and W.M. **Farrell**, The effects of solar energetic particle events on the lunar plasma environment, P31B-1389, presented at Fall AGU meeting, San Francisco, California, December 15-19, 2008.

Hills, H.K., M.R. **Collier**, W.M. **Farrell**, and T.J. **Stubbs**, Review of ALSEP SIDE results and data products, P31B-1391, presented at Fall AGU meeting, San Francisco, California, December 15-19, 2008.

Holman, G.D., Return Current Losses Revisited, BAA.S., 41, 853, 2009.

Hughitt, V.K., J. **Ireland**, M.J. Lynch, P. **Schmeidel**, G. **Dimitoglou**, D. **Mueller**, B. **Fleck,** Helioviewer: A Web 2.0 Tool for visualizing heterogeneous heliophysics data, presented at Fall AGU meeting, San Francisco, California, December 15-19, 2008 (poster).

Imber, S.M., **Slavin**, J.A., Glassmeier, K.-H., Angelopoulos, V., THEMIS observations of travelling compression regions driven by earthward-moving flux ropes, presented at the 11th Scientific Assembly of IAGA 2009, Sopron, Hungary, August 2009.

Imber, S.M., **Slavin**, J.A., Glassmeier, K.H., Angelopoulos, V., THEMIS observations of travelling compression regions driven by earthward-moving flux ropes, presented at GEM Workshop, Snowmass, Colorado, June 2009.

Imber, S.M., **Slavin,** J. A., Glassmeier, K.H., Angelopoulos, V., THEMIS multi-point observations of travelling compression regions: First tail season results, presented at THEMIS Team Meeting, March 2009.

Ireland, J., M. S. Marsh, T.A. **Kucera**, C.A. **Young**, Automated detection of oscillating areas in the solar atmosphere, presented at Fourth Solar Image Processing Workshop, Baltimore, MD, 26-30 October 2008 (poster).

Ireland, J., M.S. Marsh, T.A. **Kucera**, C.A. **Young**, Automated detection of oscillations in extreme ultraviolet imaging data: effect of background trend removal, presented at Fall AGU meeting, San Francisco, California, December 15-19, 2008 (poster).

Ireland, J., Helioviewer: Discovery for everyone everywhere, presented at AIA Science Team Meeting, LMSAL Palo Alto, 20-22 April, 2009 (talk).

Ireland, J., V. K. **Hughitt,** D. **Müeller,** G. Dimitoglou, P. Schmiedel, B. **Fleck,** The Helioviewer Project: Discovery for everyone everywhere, presented at AAS/SPD meeting, Boulder, CO, USA, June 14-18, 2009.

Ireland, J., **Marsh,** M. S., **Kucera,** T. A., **Young,** A., "Automated Detection of Oscillating Areas in the Solar Atmosphere," presented at AAS/SPD meeting, Boulder, CO, USA, June 14-18, 2009.

Ireland, J., M. S. Marsh, T. A. **Kucera,** C. A. **Young,** Automated detection of oscillating areas in the solar atmosphere, presented at AAS/SPD meeting, Boulder, CO, USA, June 14-18, 2009.

Israelevich, P; **Ofman,** L., Hybrid Simulation of Parallel Electric Field Excitation of by a standing Alfvén wave, presented at EGU General Assembly, Vienna, Austria, April 19-24, 2009 (poster).

Joy, S.P., R. J. Walker, T. King, J. **Merka,** L.F. Bargatze, J. Weygand, P. Chi, J. Mafi, T. **Narock,** and R.L. McPherron (2008). The science centered approach of the Virtual Magnetospheric Observatory, presented at Fall AGU meeting, San Francisco, California, December 15-19, 2008.

Karpen, J. T., S. K. **Antiochos,** C. R. DeVore, & M. G. Linton, "A numerical investigation of unsheared flux cancellation," presented at Evershed Memorial Symposium on "Magnetic Coupling between the Interior and the Atmosphere of the Sun" in Bangalore, India, December 2-5, 2008

Karpen, J. T., C. R. DeVore, S. K. **Antiochos,** & M. G. Linton, "2D and 3D numerical simulations of flux cancellation," presented at AAS/SPD meeting, Boulder, CO, USA, June 14-18, 2009.

Kepko, L., "Periodic reconnection in the midtail plasmasheet," presented at Non-linear magnetosphere workshop, Vina Del Mar, Chile, January, 2009.

Kepko, L., "Flow, aurora and Pi2 associations observed by THEMIS," presented at Dartmouth College, February, 2009.

Kepko, L., "Flow, aurora and Pi2 associations observed by THEMIS," presented at the 11th Scientific Assembly of IAGA 2009, Sopron, Hungary, August 2009.

Kepko, L., "Redline observations of substorm onset," presented at GEM Workshop, Snowmass, Colorado, June 2009.

Khazanov, G.V., and W. **Lyatsky,** Forecasting and Monitoring of Key Geospace Parameters, presented at AGU Joined Assembly, Toronto, Canada, May 2009.

Khazanov, G.V., W. Lyatsky, and J. **Kozyra,** Solar cycle dependence of relativistic electrons at geostationary orbit, presented at Fall AGU meeting, San Francisco, California, December 15-19, 2008.

King, T., R. Walker, J. **Merka**, and T. **Narock** (2008). The Virtual Observatory Experience - Meeting User and Data Provider Needs, presented at Fall AGU meeting, San Francisco, California, December 15-19, 2008.

King, T.A., J.R. **Thieman**, D.A. **Roberts**, and J. **Merka** (2009). Recent advances of the SPASE data model, presented at EGU General Assembly, Vienna, Austria, April 19-24, 2009 (oral).

King, T.A., J. **Merka,** T. **Narock**, R.Walker, and L. Bargatze (2009). Registry framework: Front-to-back, presented at Earth and Space Science Informatics Workshop. University of Maryland, Baltimore County, August 3–5, 2009.

King, J.H., N.E. Papitashvili, OMNIWeb-Plus and the evolution of NSSDC and SPDF value-added Space Physics Data Products and Services, presented at Fall AGU meeting, San Francisco, California, December 15-19, 2008.

Kilper, G., H. **Gilbert**, D. Alexander, Mass composition in pre-eruption quiet Sun filaments, presented at AAS/SPD meeting, Boulder, CO, USA, June 14-18, 2009.

Kilper, G. K., Mass Composition and Dynamics in Quiet Sun Prominences, presented at Doctoral Thesis, Rice University, 2009.

Kirk, Michael S., and W.D. **Pesnell**. Automated detection of polar coronal holes in the EUV, presented at AAS/SPD meeting, Boulder, CO, USA, June 14-18, 2009.

Klimchuk, J.A., "Coronal Loop models and those annoying observations," presented at Second Hinode Science meeting, Boulder, Colorado, October 2008 (invited keynote talk).

Klimchuk, J.A.,"Heating and Dynamics of the Corona"), presented at Solar-C Science Definition Meeting, Tokyo, Japan, November 2008 (invited).

Klimchuk, J.A.,"Coronal Heating: The Loops Guidepost", presented at Evershed Memorial Symposium on "Magnetic Coupling between the Interior and the Atmosphere of the Sun" in Bangalore, India, December 2-5, 2008 (invited).

Klimchuk, J.A.,"The Angry Sun: Explosions in the Corona", presented at Physics Research Lab., Ahmedabad, India, December 2008 (invited colloquium).

Klimchuk, J.A.,"Heating of the Solar Corona and its Loops", presented at Physics Dept., Univ. of New Hampshire, April 2009 (invited colloquium).

Klimchuk, J.A., "Impulsive Heating of Coronal Loops," presented at International Space Science Institute (Team Parenti), Bern, Switzerland, January 2009.

Klimchuk, J.A.,"Nanoflare Heating of the Solar Corona," presented at GSFC HSD Division seminar, January 2009.

Klimchuk, J.A., "Observations of Nanoflare Produced Hot (~ 10 MK) Plasma," presented at AAS/SPD meeting, Boulder, CO, USA, June 14-18, 2009.

Klimchuk, J.A., "Loop Lifetimes and Durations and Their Relationship to Nanoflare Storms," presented at Coronal Loops 4 Workshop, Florence, Italy, June 2009.

Klimchuk, J.A., "The Existence and Origin of Turbulence in Solar Active Regions," presented at Coronal Loops 4 Workshop, Florence, Italy, June 2009.

Klimchuk, J.A., "Nanoflare Heating of the Corona," presented at IAU, Rio de Janeiro, Brazil, Aug 2009.

Koval, A., A. Szabo, Interplanetary shock shapes: comparison of local and global parameters, presented at Fall AGU meeting, San Francisco, California, December 15-19, 2008 (poster).

Kramar, M., J. **Davila**, H. **Xie**, S. **Antiochos**, On The 3D Structure of the Pre- and After CME coronal streamer belt, presented at AAS/SPD meeting, Boulder, CO, USA, June 14-18, 2009.

Kutepov, U. Berger, **A. Feofilov**, A.Medvedev, Additional radiative cooling of the mesopause region due to small-scale temperature fluctuations associated with gravity waves. presented at the 11th Scientific Assembly of IAGA 2009, Sopron, Hungary, August 2009.

Leamon, R. J., Solar Wind Sources and Coronal Holes: Predicting and Visualising the Source of Earth-Directed Solar Wind

Leamon, R. J. and S. W. McIntosh, The Center-to-Limb Variation of TRACE Travel-Times, AGU Fall Meeting 2008, abstract #SH41A-1610

Leamon, R. J., The Magnetic Region of Influence, Seminar, Stanford University, May 2009 (also given at Lockheed Martin Solar and Astrophysics Laboratory, Palo Alto)

Leamon, R. J. and S. W. McIntosh, How the Solar Wind Ties to its Photospheric Origins, AAS SPD , abstract #31.01

Lipatov, A., and J.F. **Cooper**, Europa's plasma environment: 3D hybrid kinetic simulation, presented at 9th International School for Space Simulations (ISSS-9), Saint-Quentin-en-Yvelines, France, July 3-10, 2009.

Lipatov, A.S., J.F. **Cooper**, Europa's plasma environment: 3D hybrid kinetic simulation, presented at European Planetary Science Congress, Potsdam, Germany, Sept. 13-18, 2009.

Liu, W., **Wang**, T., **Dennis**, B.R., and **Holman**, G. D., Episodic X-ray Emission Accompanying the Activation of an Eruptive Prominence: Evidence of episodic magnetic reconnection, BAA.S., 41, 848, 2009.

Liu, W., Petrosian, V., **Dennis**, B.R., and **Holman**, G. D., Unshearing motions, asymmetries, and correlations of conjugate hard X-ray footpoints in the 2003 October 29 X10 flare: an Imaging Spectroscopic Study, BAA.S., 41, 849, 2009.

Lyatsky, W., and G.V. **Khazanov**, Alfvén and Ion Wings over Auroral Arcs, presented at the THEMIS SWG team meeting, Sep 14-16, 2009, Annapolis, Maryland.

Lyatsky, W., G.V. **Khazanov**, et al., Polar magnetic indices: A new key to correlate magnetic variations in the Earth's hemispheres, presented at the 11th Scientific Assembly of IAGA 2009, Sopron, Hungary, August 2009.

Lyatsky, W., G.V. **Khazanov**, and **Lytskaya**, S., Seasonal variation of geomagnetic activity and interhemispheric currents, presented at Fall AGU meeting, San Francisco, California, December 15-19, 2008.

Lyatsky, W., and G.V. **Khazanov**, Effects of solar wind density on processes in the magnetosphere, presented at Huntsville 2008 Workshop:"The Physical Processes for Energy and Plasma Transport across Magnetic Boundaries," Huntsville, 26-31 October, 2008

Lytskaya, S., **Lyatsky**, W., and G.V. **Khazanov**, Relationship between Substorm Activity and Magnetic Disturbances in Two Polar Caps, „The Physical Processes for Energy and Plasma Transport across Magnetic Boundaries", Huntsville, 26-31 October, 2008.

Maddox, M., M. **Hesse**, M. **Kuznetsova**, L. **Rastaetter**, P. **MacNeice**, P. **Jain**, J. **Garneau**, D. **Berrios**, A. **Pulkkinen**, D. **Rowland**, The Integrated Space Weather Analysis System, presented at Fall AGU meeting, San Francisco, California, December 15-19, 2008 (poster).

Martin, Steven, O de la Beaujardiere, D Hunton, G Wilson, P Roddy, R Coley, Ron Heelis, G Earle, P Straus, P Bernhardt, Ken **Bromund**, Robert **Candey**, Robert **Pfaff**, Doug **Rowland**, R Holzworth, Ramona Kessel, "Community access to the C/NOFS Satellite Data -- Facilitating new opportunities for space weather research", presented at Fall AGU meeting, San Francisco, California, December 15-19, 2008.

McGuire, R., D **Bilitza**, R **Candey**, R **Chimiak**, J F **Cooper**, S **Fung**, B **Harris**, R **Johnson**, J **King**, T **Kovalick**, H **Leckner**, M **Liu**, N **Papitashvili**, D **Roberts**, Services,

Perspectives and directions of the Space Physics Data Facility, presented at Fall AGU meeting, San Francisco, California, December 15-19, 2008.

McGuire, Robert, Dieter **Bilitza**, Robert **Candey**, Reine **Chimiak**, John **Cooper**, Shing **Fung**, Bernard **Harris**, Rita **Johnson**, Joseph **King**, Tamara **Kovalick**, Howard **Leckner**, Natalia **Papitashvili**, Aaron **Roberts** (2008), "The User Community and a Multi-Mission Data Project: Services, Experiences and Directions of the Space Physics Data Facility", presented at Fall AGU meeting, San Francisco, California, December 15-19, 2008.

McGuire, Robert, Dieter **Bilitza**, Robert **Candey**, Reine **Chimiak**, John **Cooper**, Shing **Fung**, Bernard **Harris**, Rita **Johnson**, Joseph **King**, Tamara **Kovalick**, Howard **Leckner**, Michael **Liu**, Natalia **Papitashvili**, Aaron **Roberts**, "Services, Perspectives and Directions of the Space Physics Data Facility, presented at Fall AGU meeting, San Francisco, California, December 15-19, 2008 (poster).

McGuire, R., D. **Bilitza**, R. **Candey**, R. **Chimiak**, J.F. **Cooper**, S. **Fung**, B. **Harris**, R. **Johnson**, J. **King**, T. **Kovalick**, H. **Leckner**, M. **Liu**, N. **Papitashvili**, D. **Roberts**, Heliophysics Research Using Virtual Observatories, presented at Fall AGU meeting, San Francisco, California, December 15-19, 2008.

McIntosh. S.W., **Leamon**, R. J. and B. de Pontieu, The Spectroscopic Footprint of the Fast Solar Wind, AGU Fall Meeting 2008, abstract #SH41A-1612.

Merka, J., **A. Szabo**, T. W. **Narock**, R. J. Walker, T. King, J. A. **Slavin**, S. **Imber**, H. Karimabadi, and J. Faden (2009). An example of using the Virtual Heliospheric and Magnetospheric Observatories for a substorm study, presented at EGU General Assembly, Vienna, Austria, April 19-24, 2009.

Merka, J., **A. Szabo**, T. W. **Narock**, R. J. Walker, T. King, J. A. **Slavin**, S. **Imber**, H. Karimabadi, and J. Faden (2008). Using the Virtual Heliospheric and Magnetospheric Observatories for geospace studies, presented at Fall AGU meeting, San Francisco, California, December 15-19, 2008.

Milam, **B.**, **R.** Pfaff, **and** G. Khazanov, UPC Orion-The next era in providing access to space for heliospherics missions, presented at Directions in Ionosphere-Thermosphere-Mesosphere research, February 10-12, 2009, Redondo Beach, CA.

Moore, T.E., The auroral linkage, presented at Huntsville 2008 Workshop:"The Physical Processes for Energy and Plasma Transport across Magnetic Boundaries," Huntsville, 26-31 October, 2008

Moore, T. E., Plasma sheet circulation pathways, the nonlinear magnetosphere, presented at Viña del Mar, Chile, 19-23 Jan 2009.

Moore, T.E., Plasma heating by convective pick-up, presented at Spring AGU Meeting, Toronto, Canada, Mar, 2009

Moore, T. E., Role of internal plasma sources in planetary magnetospheres, presented at Asia Oceania Geophysical Society Annual Meeting, Singapore, 10-14 Aug., 2009.

Morrison, Daniel, E Lis Immer, Rose Daley, Dennis Patrone, Matthew Potter, Robert Holder, Robin Barnes, Chris Colclough, Stu Nyland, J Sam Yee, Elsayed Talaat, James Russell, Ron Heelis, Janet Kozyra, Dieter **Bilitza**, Robert **McGuire**, Robert **Candey**, Peter Fox (2008), "Enabling Visual Search and Discovery with the Virtual ITM Observatory", presented at Fall AGU meeting, San Francisco, California, December 15-19, 2008.

Morrison, Daniel, E Lis Immer, Rose Daley, Dennis Patrone, Matthew Potter, Robert Holder, Robin Barnes, Chris Colclough, Stu Nyland, J Sam Yee, Elsayed Talaat, James Russell, Ron Heelis, Janet Kozyra, Dieter **Bilitza**, Robert **McGuire**, Robert **Candey**, Peter Fox (2008), "Enabling New Discovery with the Virtual ITM Observatory", presented at Fall AGU meeting, San Francisco, California, December 15-19, 2008.

Morrison, Daniel, Michele Weiss, E Lis Immer, Dennis Patrone, Matthew Potter, Robert Holder, Robin Barnes, Chris Colclough, Stu Nyland, J Sam Yee, Elsayed Talaat, Dieter **Bilitza**, Robert **McGuire**, Robert **Candey**, Bernard **Harris**, James Russell, Ron Heelis, Janet Kozyra, Peter Fox, "Performing Science Research with the Virtual ITM Observatory", presented at Fall AGU meeting, San Francisco, California, December 15-19, 2008.

Mount, E. E., S. Ronald, N. Pope, A. N. **Daw**, and A. G. Calamai, Optical fluorescence of long lived states in NO^+, presented at 75th Annual Meeting of the Southeastern Section of the American Physical Society APS, Raleigh, NC, October 30–November 1 2008.
Mueller, D., **Dimitoglou**, G., Hughitt, V. K., **Ireland**, J., Wamsler, B., Fleck, B, A Novel Approach to discovery and access to solar data in the petabyte age, presented at AAS/SPD meeting, Boulder, CO, USA, June 14-18, 2009.

Narock, T. W., V. Yoon, J. **Merka**, and A. **Szabo**, Semantic e-Science in Space Physics - A case study, presented at Spring AGU Meeting, Toronto, Canada, Mar, 2009

Narock, T., J. King, J. Merka, and A. Szabo (2008). A view of a unfied heliophysics data environment, presented at Fall AGU meeting, San Francisco, California, December 15-19, 2008.

Narock, T., V. Yoon, **J. Merka, and A. Szabo** (2009). Semantic e-science in space physics: A case study. presented at Fall AGU meeting, San Francisco, California, December 15-19, 2008.

Nigro, G., **Klimchuk**, J. A. et al., "Onset and nonlinear development of current sheet instabilities," presented at Coronal Loops 4 Workshop, Florence, Italy, June 2009.

Nitta, N., Aschwanden, M., Freeland, S., Lemen, J., Wuelser, J., **Zarro**, D.M., "The CME-Flare Relation Revisited With STEREO Observations", presented at AAS/SPD meeting, Boulder, CO, USA, June 14-18, 2009.

Ofman, L., Three-dimensional MHD models of waves in active regions, presented at Royal Observatory of Belgium, Brussels, Belgium, March 31, 2009.

Ofman, L., Hybrid models of Waves in Solar Wind Plasma, presented at Department of Geophysics and Planetary Sciences, Tel Aviv University, Israel, May 18, 2009.

Ofman, L., Coronal seismology, or why we need automated detection of waves in the solar corona, presented at Fourth Solar Image Processing Workshop, Baltimore, MD, 26-30 October 2008 (talk).

Ofman, L., Acceleration and heating of solar wind ions by turbulent wave spectrum, presented at Fall AGU meeting, San Francisco, California, December 15-19, 2008 (poster).

Ofman, L., Acceleration and heating of the solar wind (poster), presented at Space Climate Symposium 3 and Space Climate School, in Saariselkae, Finnish Lapland, March 15-22, 2009.

Ofman, L., Waves in coronal loops: what can we learn from 3D MHD models, presented at Workshop on MHD waves and seismology of the solar atmosphere, in Leuven, Belgium, April 6-8, 2009 (talk).

Ofman, L., Fast solar wind - connecting the corona and the heliosphere, presented at STEREO-3/SOHO-22 Workshop: Three Eyes on the Sun - Multi-spacecraft studies of the corona and impacts on the heliosphere, The De Vere Royal Bath Hotel, Bournemouth, Dorset, England, April 27 - May 1, 2009 (talk).

Ofman, L., New models of coronal streamers, presented at the 11th Scientific Assembly of IAGA 2009, Sopron, Hungary, August 2009 (talk).

Ofman, L., **Kramar**, M., Modeling the slow solar wind during solar minimum, presented at SOHO 23 Meeting, Northeast Harbor, Maine, Sept. 2009.

Pap, J., Sun-climate relation, presented at the World Federation of Scientists in Erice, Sicily, August 2009 (invited talk).

Pap, J., participated at the LWS Focus Team meeting in Boulder in early 2009 and gave forty minute presentation J Pap research, presented at Fall AGU meeting, San Francisco, California, December 15-19, 2008 (poster).

Parenti, S., **Klimchuk**, J.A., et al., "Observing the Hot Coronal Plasma with the EUI/HRI On Board Solar Orbiter," presented at Solar Orbiter Workshop, Sorrento, Italy, May 2009.

Patsourakos, S. and **Klimchuk**, J.A., "EIS Observations of Hot Lines in Active Regions: Constraints on Coronal Heating," presented at Second Hinode Science meeting, Boulder, Colorado, October 2008

Patsourakos, S. and **Klimchuk**, J.A., "Spectroscopic observations of hot lines constraining coronal heating in solar active regions," presented at AAS/SPD meeting, Boulder, CO, USA, June 14-18, 2009.

Patsourakos, S. and **Klimchuk**, J.A., "Spectroscopic observations of hot lines constraining coronal heating in solar active regions," presented at Coronal Loops 4 Workshop, Florence, Italy, June 2009.

Pesnell, W.D., Predicting solar cycle 24 with geomagnetic precursors, presented at AAS/SPD meeting, Boulder, CO, USA, June 14-18, 2009.

Pope, N., S. Ronald, E. Mount, **A. Daw**, A. Calamai, Rate Coefficients of Singly Ionized Triatomic Hydrogen Using An RF Ion Trap, presented at American Physical Society, 40th Annual Meeting of the APS Division of Atomic, Molecular and Optical Physics, May 19-23, 2009.

Pulkkinen, A., L. **Rastätter**, M. **Kuznetsova** and A. **Chulaki,** GEM Challenge 2008: Ground magnetic field perturbations, presented at mini GEM, San Francisco, California, December 14, 2009 (invited talk).

Pulkkinen, A., N. **Buzulukova,** L. **Rastaetter,** M. **Kuznetsova**, A. Viljanen, and R. Pirjola, First-principles-based modeling of geomagnetically induced currents at mid- and low-latitudes, presented at Fall AGU meeting, San Francisco, California, December 15-19, 2008 (talk).

Pulkkinen, A., A. **Taktakishvili**, D. Odstrcil and W. Jacobs, Novel approach to geomagnetically induced current forecasts based on remote solar observations, presented at Space Weather Workshop, Boulder, Colorado, April 28 - May 1, 2009 (poster).

Pulkkinen, A., L. **Rastätter**, M. **Kuznetsova** and A. **Chulaki**, GEM Challenge: Ground magnetic field perturbations (update), presented at GEM Workshop, Snowmass, Colorado, June 2009 (invited talk).

Pulkkinen, A., L. **Rastätter**, M. **Kuznetsova,** M. **Hesse and** A. **Chulaki**, Geospace models: What works and what does not?, presented at Heliophysics Science Division Director's Seminar, NASA/GSFS, August 21, 2009. (talk)

Pulkkinen, A., A. **Taktakishvili**, and T. Oates, Automatic determination of the conic coronal mass ejection model parameters, presented at George Mason University, Space Sciences Seminar, September 30, 2009. (invited talk)

GSFC Heliophysics Science Division 2009 Science Highlights

Raestaetter, L., M. **Hesse**, M. **Kuznetsova**, A. **Pulkkinen**, T.I. Gombosi, Modeled cross-polar cap potential response after sudden IMF changes, presented at Fall AGU meeting, San Francisco, California, December 15-19, 2008 (poster).

Rezac, L, **Feofilov**, **A.**, **Kutepov, A., Pesnell**, W.D., **Goldberg**, R.A., Russell, J.M., Self-consistent Diagnostics of SABER MLT Limb Radiances in the 15 and 4.3 um Channels, presented at Fall AGU meeting, San Francisco, California, December 15-19, 2008

Richard, D. T., D. A. **Glenar**, T. J. **Stubbs**, S. S. Davis, and A. Colaprete, "But still, like dust, I'll rise": Scattering signatures of complex lunar particulates, presented at the 2nd Annual Lunar Science Forum, NASA Lunar Science Institute, NASA Ames Research Center, Moffett Field, CA, July 21–23, 2009.

Ronald, S., E. Mount, N. Pope, A. N. **Daw**, and A. G. Calamai, Rate Coefficients for H^{3+} Production Measured in an RF Ion Trap, presented at 75th Annual Meeting of the Southeastern Section of the American Physical Society APS, Raleigh, NC, October 30–November 1 2008.

Saba, J. L. R., and K. T. **Strong**, A bursty description of the solar cycle, presented at Space Weather Workshop, Boulder, Colorado, April 28 - May 1, 2009 (poster).

Saba, J. L. R., and K. T. **Strong**, Intercomparison of recent solar activity cycles, presented at Space Weather Workshop, Boulder, Colorado, April 28 - May 1, 2009 (poster).

Saba, J. L. R., and K. T. **Strong**, Comparing the evolution of solar cycles 21, 22, and 23 via a sensitive magnetic ratio, presented at SOHO-23: Understanding a Peculiar Solar Minimum, Northeast Harbor, Maine, Sept. 2009 (poster and contributed talk).

Sarantos, M., R. M. Killen, A. S. Sharma, and A. E. Potter, The dynamic lunar exosphere: clues from sodium, presented at Fall AGU meeting, San Francisco, California, December 15-19, 2008.

Sarantos, M, J. A. **Slavin**, M. **Benna**, T. H. Zurbuchen, S. M. Krimigis, D. Schriver, P. Trávníček, and S. C. Solomon, Comparison of models for Mercury's pickup ions with MESSENGER measurements for northward and southward IMF, presented at EGU General Assembly, Vienna, Austria, April 19-24, 2009.

Sarantos, M., and J. A. **Slavin**, 2009. On the possible formation of Alfvén wings at Mercury during encounters with coronal mass ejections, presented at EGU General Assembly, Vienna, Austria, April 19-24, 2009.

Sarantos, M., R. M. **Killen**, M. **Benna**, R.E. **Hartle**, and A. S. Sharma, The Lunar Exosphere: expectations for LADEE measurements, presented at EGU General Assembly, Vienna, Austria, April 19-24, 2009.

GSFC Heliophysics Science Division 2009 Science Highlights

Sarantos, M., R. M. **Killen**, R.E. **Hartle**, M. **Benna**, and A. S. Sharma, The Lunar Exosphere: observations, models, and expectations for LADEE measurements, presented at NASA 2009 Lunar Science Forum, NASA Ames Research Center,CA, July 2009.

Schmidlin, F. S., **Goldberg**, R. A., Remotely Sensed Temperature Variability, presented at the 11th Scientific Assembly of IAGA 2009, Sopron, Hungary, August 2009.

Schroeder, P. C., **A. Szabo**, A. Davis, G. Ho, J. Kasper, **J. Merka**, **T. Narock**, J. Raines, **K. Rash**, **D. A. Roberts**, and J. Vandegriff (2009), Providing data in a virtual world, presented at EGU General Assembly, Vienna, Austria, April 19-24, 2009.

Schwartz, R.A.; **Zarro**, D.; Csillaghy, A.; **Dennis**, B.; **Tolbert**, A. K.; **Etesi**, L. VSO For Dummies, presented at AAS/SPD meeting, Boulder, CO, USA, June 14-18, 2009.

Selwa, M. A., **Ofman**, L., The role of AR topology on excitation, trapping and damping of individual loop oscillations (poster), AGU Fall Meeting, in San Francisco, California, USA, December 15-19, 2008.

Selwa, M.A., **Ofman**, L., 3D numerical simulations of coronal loops oscillations (poster), BUKS2009: Workshop on MHD waves and seismology of the solar atmosphere, in Leuven, Belgium, April 6-8, 2009.

Selwa, M. A., **Ofman**, L., Solanki, S.K., **Tongjiang Wang**, 3D numerical simulations of coronal loops oscillations, presented at STEREO-3/SOHO-22 Workshop: Three Eyes on the Sun - Multi-spacecraft studies of the corona and impacts on the heliosphere, The De Vere Royal Bath Hotel, Bournemouth, Dorset, England, April 27 - May 1, 2009 (poster).

Shao, X, S. F. **Fung**, L.C. Tan, A. S. Sharma, Relativistic electron acceleration by compressional-mode Ultra-Low Frequency waves, presented at the Conference on Modern Challenges in Nonlinear Plasma Physics, Sani Resort, Greece, June 15-19, 2009.

Sittler Jr, E.C., **Ofman** L., **Airapetian**, V., **Kramar**, M., Development of solar wind model driven by empirical heat flux and pressure terms, presented at Solar Wind 12, in Saint-Malo, France, June 21-26, 2009 (poster).

Sittler, E.C.; **Ofman**, L.; **Selwa**, M. A.; **Kramar**, M., Development of solar wind model driven by empirical heat flux and pressure terms, presented at Fall AGU meeting, San Francisco, California, December 15-19, 2008 (talk).

Sittler Jr, E.C., R.E. **Hartle**, J.F. **Cooper**, R.E. Johnson, H. T. Smith, M. D. **Shappiro**, and D. J. **Simpson**, Methane group ions in Saturn's outer magnetosphere, presented at DPS/AAS 2008 Conference, Ithaca, NY, Oct. 10-15, 2008 (Titan, poster).

Sittler Jr., E.C. J.F. **Cooper**, R. E. **Hartle**, P. **Mahaffy**, N. **Paschalidis**, M. **Coplan**, T. A. **Cassidy**, and R. E. Johnson, Plasma and ion neutral composition measurements for Europa and Ganymede, presented at presented at Europa Jupiter System Mission Instrument Workshop, Johns Hopkins University Applied Physics Laboratory, July 15-17, 2009 (poster).

Sittler, E.C., R.E. **Hartle**, J.F. **Cooper**, A. **Lipatov**, C. Bertucci, A.J. Coates, C. Arridge, K. Szego, M. **Shappirio**, D.G. **Simpson**, R. Tokar, and D.T. Young, Saturn's magnetosphere and properties of upstream flow at Titan: Preliminary results, presented at Magnetospheres of the Outer Planets Conference 2009, Cologne, Germany, July 27-31, 2009

Sittler, E., J. **Cooper**, R. **Hartle**, P. **Mahaffy**, T. Cassidy, and R. Johnson, Europa's and Ganymede's surface composition by measuring pickup ions: Model calculations, presented at European Planetary Science Congress, Potsdam, Germany, Sept. 13-18, 2009.

St. Cyr, O.C., D. Young, W.D. **Pesnell**, A. Lecinski, and J. Eddy, Recent studies of the behavior of the Sun's white-light corona over time, presented at Fall AGU meeting, San Francisco, California, December 15-19, 2008.

Strong, K. T., and J. L. R. **Saba**, The solar cycle from a whole-Sun perspective, presented at STEREO-3/SOHO-22 Workshop: Three Eyes on the Sun - Multi-spacecraft studies of the corona and impacts on the heliosphere, The De Vere Royal Bath Hotel, Bournemouth, Dorset, England, April 27 - May 1, 2009 (poster).

Strong, K. T., and J. L. R. **Saba**, A hypothesis on solar cycle to cycle interactions, presented at AAS/SPD meeting, Boulder, CO, USA, June 14-18, 2009 (poster).

Strong K., and J. **Saba**, "When Did Cycle 23 Go Wrong?", presented at the European Week of Astronomy & Space Science, University of Herfordshire, Hatfield, UK. 19-24 April 2009 (Invited Talk).

Strong K., "The New Sun." presented at the Waterlooville Technical College, Hampshire, UK. 16 April 2009 (public lecture).

Strong K., "The Sun's Role in Climate Change, presented at the Sir Patrick Moore Planetarium, Chichester, UK. 17 April 2009 (public lecture).

Strong K., and J. **Saba**, "The Sun from a Whole Sun Perspective", presented at STEREO-3/SOHO-22 Workshop: Three Eyes on the Sun - Multi-spacecraft studies of the corona and impacts on the heliosphere, The De Vere Royal Bath Hotel, Bournemouth, Dorset, England, April 27 - May 1, 2009 (poster).

Strong, K. T., and J. L. R. **Saba**, When and where did Cycle 23 go wrong?, presented at SOHO-23: Understanding a Peculiar Solar Minimum, Northeast Harbor, Maine, Sept. 2009 (poster).

Strong K., and J. **Saba**, "Where and When did Cycle 23 go Wrong?", presented at the SHINE 2009 session on "Recent Unprecedented Weak Solar Wind : A Solar Glitch or a

Harbinger for the Next Maunder Minimum?" in Wolfville, Nova Scotia, 3-7 August 2009 (poster).

Stubbs, T. J., The interaction of the space plasma environment with the surface of the Moon, presented at the Planetary Astronomy Lunch Series (PALS), University of Maryland, College Park, November 13, 2008 (invited seminar).

Stubbs, T. J., D. A. **Glenar**, M. R. **Collier**, W. M. **Farrell**, J. S. Halekas, G. T. Delory, and R. R. **Vondrak**, On the possible role of dust in the lunar ionosphere, presented at the 2nd Annual Lunar Science Forum, NASA Lunar Science Institute, NASA Ames Research Center, Moffett Field, CA, July 21–23, 2009.

Stubbs, T. J., Planetary Science Update: NLSI teams and future lunar missions, presented at the Lunar Airborne Dust Toxicity Assessment Group (LADTAG) meeting, Houston, TX, April 20–21, 2009.

Stubbs, T. J., D. A. **Glenar**, D. T. Richard, and A. Colaprete, Predictions for the optical scattering at the Moon, as observed by the LADEE UV/VIS spectrometer, #2348, presented at the Lunar and Planetary Science Conference XL, Lunar and Planetary Institute, Houston, TX, March 23–27, 2009.

Stubbs, T.J., W.M. **Farrell**, J.S. Halekas, G.T. Delory, M. R. **Collier**, D.H. **Berrios**, and R.R. **Vondrak**, Lunar surface charging in the magnetotail, P31B-1390, presented at Fall AGU meeting, San Francisco, California, December 15-19, 2008.

Szabo, A., A. Koval, Relationship of Interplanetary Shock Micro and Macro Characteristics: A Wind Study, presented at Fall AGU meeting, San Francisco, California, December 15-19, 2008 (poster).

Su, Y., **Holman**, G.D., **Dennis**, B.R., **Tolbert**, A.K., and **Schwartz**, R.A., A Test of Thick-Target Nonuniform Ionization as an Explanation for Breaks in Solar Flare Hard X-ray Spectra, BAA.S., 41, 854, 2009.

Tan, LC, X. Shao, A.S. Sharma, S. F. **Fung**, Acceleration of Magnetospheric Relativistic Electrons by Ultra-Low Frequency Waves: A Comparison Study, presented at AGU Joined Assembly, Toronto, Canada, May 2009.

Truhlik, V., J. M. Grebowsky, L. Triskova, D. **Bilitza**, and R. **Benson**, Resurrection of a unique plasmapause ion composition data base, presented at EGU General Assembly, Vienna, Austria, April 19-24, 2009 (paper #3752).

Toth, G., A. **Glocer**, T. Gombosi, Modeling multi-ion hydrodynamics, presented at SIAM Conference on Computational Science and Engineering, Miami,FL, March 2-6, 2009.

Walker R.,J. **Merka,** T. **King**, T. Narock, S. **Joy,** L.F. **Bargatze,** P. **Chi,** J. **Weygand,** The Virtual Magnetospheric Observatory, International Polar Year, International Symposium: Fifty Years after IGY, presented in Tsukuba City, Japan, 10–13 November, 2008.

Walker, R.J., T. King, J.M. Weygand, **J. Merka**, L.F. Bargatze, P. Chi, J. Ma, T.W. **Narock**, R. L. McPherron, and S. Joy (2009). The science centered approach of the Virtual Magnetospheric Observatory, presented at EGU General Assembly, Vienna, Austria, April 19-24, 2009.

Wang, T.J., **Ofman**, L., **Davila**, J., Propagating slow magnetoacoustic waves in coronal loops observed by Hinode/EIS, presented at AAS/SPD meeting, Boulder, CO, USA, June 14-18, 2009 (talk).

Wang, T.J., L. **Ofman**, and J.M. **Davila**, Propagating slow magnetoacoustic waves in a coronal loop, presented at Second Hinode Science meeting, Boulder, Colorado, October 2008 (poster).

Williams, P.E., Semi-empirical studies of solar supergranulation and related phenomena, presented at Heliophysics Science Division Director's Seminar, NASA/GSFC, October 3, 2008.

Williams, P.E., and W.D. **Pesnell**, Analysis of photospheric convection flows over a solar cycle, presented at AAS/SPD meeting, Boulder, CO, USA, June 14-18, 2009.

Williams, P.E., and W.D. **Pesnell**, Photospheric manifestations of supergranules during the last two solar minima, presented at SOHO-23: Understanding a Peculiar Solar Minimum, Northeast Harbor, Maine, Sept. 2009.

Williams, Peter E., W. D. **Pesnell**, Analysis of Photospheric Convection Flows Over Solar Cycle", presented at AAS/SPD meeting, Boulder, CO, USA, June 14-18, 2009.,

Zank, G. P., Gang Li, and Olga Verkhoglyadova, "Particle Acceleration at Interplanetary Shocks" presented at Heliophysics Science Division Director's Seminar, NASA/GSFC, April 24, 2009.

Zhang, H., D. G. Sibeck, Q.-G. Zong, J. P. McFadden, K.-H. Glassmeier, J. W. Bonnell, A. Roux, "THEMIS Observations of A Series of Hot Flow Anomalies", presented at the 2nd International Space Weather Conference, Nanjing,China, October 17-21, 2009 (invited talk).

Zhang, H., Q.-G. Zong, D. G. Sibeck, T. A. Fritz, J. P. McFadden, K. H. Glassmeier, D. Larson, "Dynamic Motion of Bow Shock and Magnetopause and the Magnetospheric Response-THEMIS Observations", presented at Asia Oceania Geosciences Society 6nd annual meeting, Singapore, 11–15 August 2009 (invited talk).

Zhang, H., D. G. Sibeck, Q.-G. Zong, J. P. McFadden, S. B. Mende, K. H. Glassmeier, and K. Yumoto, "Global Magnetospheric Response to an Interplanetary Shock: THEMIS Spacecraft and Ground Magnetometer Observations", presented at the THEMIS SWG team meeting, Sep 14-16, 2009, Annapolis, Maryland (contributed talk).

Zhang, H., D. G. Sibeck, Q.-G. Zong, J. P. McFadden, K.-H. Glassmeier, J. W. Bonnell, A. Roux, "THEMIS Observations of A Series of Hot Flow Anomalies", presented at GEM Workshop, Snowmass, Colorado, June 2009 (contributed talk).

Zhang, H., D. G. Sibeck, Q.-G. Zong, J. P. McFadden, K.-H. Glassmeier, J. W. Bonnell, A. Roux, "THEMIS Observations of A Series of Hot Flow Anomalies", presented at THEMIS SWG meeting, Boulder, CO, March 23-25, 2009 (contributed talk).

Zheng, Qiuhua, Mei-Ching **Fok**, "Investigation of energy and pitch angle cross diffusion effects on the outer radiation belts using a kinetic Radiation Belts Environmental model," presented in Toronto, Canada, May 2009.

APPENDIX 3: OPERATIONAL HSD MISSIONS

Interstellar Boundary Explorer (IBEX)

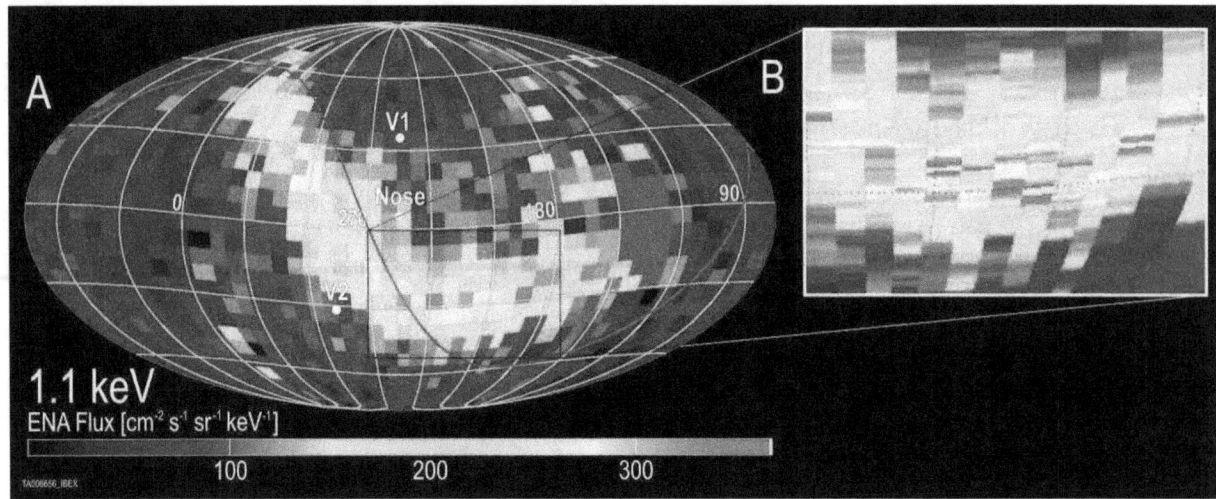

Background: The solar system moves through a part of the galaxy referred to as the local interstellar medium (LISM). It is built up from material released from the stars of the Milky Way galaxy through stellar winds, novas, and supernovas. The interstellar medium has considerable structure, and the region where the solar wind interacts with the LISM is very complex. IBEX uses two ENA detectors to map out this interaction region at several energies. One of the first IBEX sky maps, accumulated over six months of scanning, is shown above. The bright ribbon of ENA emission is not predicted by any models and is still not understood. IBEX images reveal global properties of the interstellar boundaries that separate the heliosphere from the local interstellar medium. Because IBEX provides global maps of the interstellar interaction, IBEX observations are highly complementary to, and synergistic with, the detailed single-direction measurements provided by the Voyager satellites.

Scientific Goals: IBEX's sole science objective is to discover the global interaction between the solar wind and the interstellar medium. IBEX achieves this objective by taking a set of global ENA images that answer four fundamental science questions:
- What is the global strength and structure of the termination shock?
- How are energetic protons accelerated at the termination shock?
- What are the global properties of the solar wind flow beyond the termination shock and in the heliotail?
- How does the interstellar flow interact with the heliosphere beyond the heliopause?

GSFC Role: An HSD scientist is a Co-I on IBEX, and the Deputy Mission Scientist is also in HSD.

Status: IBEX is in its prime mission and released its first all-sky maps in five papers in the journal *Science* on 2009 October 15.

Communications/Navigation Outage Forecasting System (C/NOFS)

The C/NOFS Spacecraft

Background: The C/NOFS mission includes a satellite designed to investigate and forecast scintillations in Earth's ionosphere. It was launched on a Pegasus-XL rocket on 2008 April 17 into LEO configuration with an inclination of 13°, a perigee of 400 km, and an apogee of 850 km. The satellite, which is operated by the USAF Space Test Program, will allow the US military to predict the effects of ionospheric activity on signals from communication and navigation satellites, outages of which could potentially cause problems in battlefield situations.

C/NOFS makes comprehensive measurements of vector DC and wave electric fields, magnetic fields, plasma density and temperature, ion drifts, neutral winds, and lightning detector counts, and includes GPS scintillation and radio beacon experiments. Combined with strong modeling and ground-based observing components, this mission promises to demonstrably advance scientific understanding of Equatorial Spread F (ESF) irregularities and their conditions for growth.

CINDI Instrument on C/NOFS: The Coupled Ion Neutral Dynamic Investigation (CINDI) is a Mission of Opportunity investigation on C/NOFS sponsored by NASA and designed and built by the University of Texas at Dallas. CINDI involves two instruments that measure the concentration and kinetic energy of the ions and neutral particles in space as the satellite passes through them. This information will be used in building models to understand the various structures in the ionosphere, such as plasma depletions and associated turbulence in the nightside, low-latitude ionosphere. These structures can interfere with radio signals between Earth and spacecraft in orbit, thus causing errors in tracking and loss of communication.

GSFC Role on C/NOFS: HSD built the C/NOFS Vector Electric Field Instrument (VEFI), which consists primarily of an electric field detector that utilizes three orthogonal 20-m tip-to-tip double-probe antennas. VEFI measures DC electric fields, which cause the bulk plasma motion that drives the ionospheric plasma to be unstable. Additionally, it measures the quasi-DC electric fields within the plasma density depletions to reveal the motions of the depletions relative to the background ionosphere. VEFI also measures the vector AC electric field, which characterizes the ionospheric disturbances associated with spread-F irregularities. An FGM, an optical lightning detector, and a fixed-bias Langmuir probe are also included in the VEFI instrument package. HSD also manages the MO&DA funding for CINDI, and provides Project Scientist support.

GSFC Heliophysics Science Division 2009 Science Highlights

Significant Milestones in FY09: The instruments have been commissioned and are returning excellent data. Among the many initial results revealed by C/NOFS instruments is the finding that the nightside, low-latitude ionosphere is highly structured even during solar minimum and that the majority of spread-F depletions observed thus far occur post-midnight.

Example of DC electric fields gathered with GSFC's electric field detector.

Aeronomy of Ice in the Mesosphere (AIM)

Noctilucent clouds

Background: The AIM spacecraft was launched from a Pegasus rocket on 2007 April 25 into a 600-km orbit. It has observed three Northern Hemisphere seasons and two Southern Hemisphere seasons already.

Scientific Goals: AIM is the first satellite mission dedicated to the study of polar mesospheric clouds (PMCs), also known as noctilucent clouds, and it makes measurements that can provide information on how these clouds form and vary. AIM addresses the following questions:

- Are there temporal variations in PMCs that can be explained by changes in solar irradiance and particle input?
- What changes in mesospheric properties are responsible for north/south differences in PMC features?
- What controls interannual variability in PMC season duration and latitudinal extent?
- What is the mechanism of teleconnection between winter temperatures and summer hemisphere PMCs?
- What is the global occurrence rate of gravity waves outside the PMC domain?

Despite a significant increase in PMCs research in recent years, relatively little is known about the basic physics of these clouds at "the edge of space" and why they are changing. They have increased in brightness over time, are being seen more often, and appear to be occurring at lower latitudes than ever before. It has been suggested (and debated) that these changes are linked to global climate change.

GSFC Role: Dr Jackman is the Project Scientist, Dr Sigworth is the deputy Project Scientist, and Dr Cuevas is the Mission Director of the AIM mission.

The Cloud Imaging and Particle Size (CIPS) experiment shows that clouds are highly variable from orbit to orbit and day to day. "Ice voids" were observed that look like tropospheric features.

Significant Project Milestones in FY09:

- Temperature change has been found to be the dominant factor in controlling season onset, variability during the season, and season end for the PMCs.
- Meteoric smoke particles were found to be the most likely source of nucleation sites for ice particles in the PMCs.
- PMCs are greatly affected by atmospheric waves leading to localized heating and "ice voids" as shown in the figure above.
- AIM entered its extended mission phase in June 2009.

Time History of Events and Macroscale Interactions During Substorms

Background: With an array of spacecraft covering the near-Earth solar wind, magnetosheath, and inner and outer magnetosphere, as well as a network of ground observatories tracking geomagnetic perturbations and auroral signatures night after night, THEMIS currently serves as the cornerstone of the HGO, enabling researchers to understand the magnetospheric response to ever-varying solar wind conditions. THEMIS also serves as a bridge between the ISTP and MMS/RBSP eras, enabling studies of this response as a function of solar cycle.

Scientific Goals: The primary objective of the mission is to determine the cause of geomagnetic substorms. The mission employed five identical spacecraft and an array of ground-based all-sky imagers and magnetometers to pinpoint when and where substorm onset occurs in the magnetotail. Secondary and tertiary objectives of the mission include understanding the processes that accelerate thermal plasmas to form the ring current and radiation belts, and obtaining a detailed understanding of the solar wind-magnetosphere interaction via simultaneous measurements of the solar wind, foreshock, magnetosheath, and magnetopause.

GSFC Role: GSFC provides the Project Scientist and participates in scientific analysis.

Significant Project Milestones in FY09: The prime phase of the mission concluded in September 2009. Press conferences on "Double-cusp reconnection" at the 2008 Fall AGU, "Million-amp space twisters" at the 2009 EGS meeting, and "Pinpointing the substorm epicenter with Pc1 waves" at the 2009 Spring AGU meeting were widely covered, as was the publication of "The origin of radiation belt hiss" in *Science*. *GRL* highlighted field line resonance results and put dayside chorus results on its cover.

Mission Status: All instruments and spacecraft are fully operational. The same is true for the ground observatories. The apogees of the outermost two spacecraft are currently being raised to lunar distance. Following lunar flybys in December 2009 to February 2010, these two spacecraft will first enter Lissajous orbits paralleling the Moon next fall. They will then enter permanent lunar orbits. The mission involving the two THEMIS spacecraft at the Moon will be named "ARTEMIS". The remaining three innermost spacecraft will continue the core THEMIS mission; albeit with slightly different orbits that will enable determination of both meridional and azimuthal gradients. Existing datasets are widely disseminated; this will continue for forthcoming datasets. The THEMIS team holds a weekly science teleconference to which all scientists are invited.

GSFC Heliophysics Science Division 2009 Science Highlights

Solar Terrestrial Relations Observatory (STEREO)

Background: The twin STEREO spacecraft – A and B – were launched on 2006 October 26 from Kennedy Space Center aboard a Delta 7925 launch vehicle. Each spacecraft used close flybys of the Moon to escape into orbits about the Sun near 1 AU; one spacecraft (A) now leads Earth, while the other (B) trails. As viewed from the Sun, the two spacecraft separate at about 44° to 45° per year. Each STEREO spacecraft is equipped with an almost identical set of optical, radio, and in situ particle and field instruments provided by US and European investigators.

An unearthly eclipse of the Sun by the Moon as viewed STEREO B.

Scientific Goals: The purposes of the STEREO mission are to:

- Understand the causes and mechanisms of CME initiation
- Characterize the CME propagation through the inner heliosphere to Earth
- Discover the mechanisms and sites of energetic-particle acceleration in the low corona and the interplanetary medium
- Develop a 3D time-dependent model of the magnetic field topology, temperature, density, and velocity structure of the ambient solar wind.

GSFC Role: GSFC, with Swales Aerospace, provided the inner coronagraph, COR1, for the STEREO Sun–Earth Connection Coronal and Heliospheric Investigation (SECCHI). The HSD manages the mission, through the Project Scientist's office and the STEREO Science Center. The Science Center is the focal point for STEREO science coordination as well as for E/PO; it processes the space-weather-beacon data, and archives STEREO telemetry, mission support data, higher-level instrument data, and analysis software. HSD science team members provide software for SECCHI and engage in science analysis.

Significant Project Milestones in FY09:
- Tomographic reconstruction of 3D electron densities in the inner corona (1.5 – 4.0 solar radii) from COR1 (Kramar *et al.* 2009, *Solar Phys.*, 10.1007/s11207-009-9401-2)
- Observations of EUV wave reflection from a coronal hole (Gopalawamy *et al.* 2009, *ApJ*, 691, L123)
- Improved COR1 calibration by W. Thompson

Twenty-eight STEREO-related papers were published in a topical issue of the journal *Solar Physics*, "STEREO Science Results at Solar Minimum," in May 2009.

Hinode

Background: Hinode is a mission of the Japan Aerospace Exploration Agency (JAXA) with US, UK, ESA, and Norwegian collaboration. It was launched on an M-V rocket from Uchinoura Space Center, Japan, on 2006 September 22. The satellite was maneuvered to the quasi-circular Sun-synchronous orbit over the day/night terminator, which allows near-continuous observation of the Sun. Hinode was planned as a three-year mission to explore the magnetic fields of the Sun. It consists of a coordinated set of optical, extreme ultraviolet (EUV), and X-ray instruments (the Solar Optical Telescope, the EUV Imaging Spectrometer, and the X-ray Telescope) to investigate the interaction between the Sun's magnetic field and its corona. Each of the instruments has the highest angular resolution ever achieved in a solar instrument for its spectral band.

Scientific Goals: Hinode investigates the interaction between the Sun's magnetic field and the corona. The result will be an improved understanding of the mechanisms that power the solar atmosphere and drive solar eruptions. This information will tell how the Sun generates magnetic disturbances and high-energy particle storms that propagate from the Sun to Earth and beyond; in this sense, Hinode will help scientists predict space weather. It is using its three instruments together to unravel basic information to understand:

Hinode has obtained high-resolution images of the poles of the Sun, which show resolved spicules approximately 300 km across.

- How energy generated by magnetic-field changes in the lower solar atmosphere (photosphere) is transmitted to the upper solar atmosphere (corona)
- How that energy influences the dynamics and structure of the upper atmosphere
- How energy transfer and atmospheric dynamics affect on interplanetary space

GSFC Role: Three members of the Hinode operations team are based at GSFC and several HSD scientists are involved in the analysis of Hinode data. Sten Odenwald and Ravi Grant perform Hinode E/PO activities at GSFC. The SDAC/VSO at GSFC serves Hinode data to the solar community. The Hinode project is managed by MSFC.

Mission Status: Hinode's X-band transmitter signal began to experience irregularities in December 2007. Because of increased irregularities in February, the Hinode team is now performing downlink with the backup S-band antenna. Additional JAXA, NASA, and Norwegian ground stations now provide more downlink opportunities. New compression algorithms are being used to optimize the data downlink via available telemetry.

GSFC Heliophysics Science Division 2009 Science Highlights

Solar Radiation and Climate Experiment (SORCE)

Background: To continue to monitor the Sun's radiative energy output and to better understand radiative forcing of the Earth's climate, NASA launched the SORCE satellite on 2003 January 25. The satellite flies at an altitude of 640 km in a 40° inclination orbit. SORCE carries four instruments that have greatly improved the accuracy and precision of solar total and spectral irradiance measurements. All instruments acquire data during each of the satellite's 15 daily orbits, producing data products on timescales as short as five minutes, but more commonly one to four times per day. The PI (Dr Woods) and mission control are located at the University of Colorado Laboratory for Atmospheric and Space Physics.

Scientific Goals: To answer the questions:
- What is the absolute value of the TSI?
- How does the Sun's spectral irradiance vary, and what are the impacts on terrestrial climate?
- What aspects of solar variability influence the Earth's atmosphere and how?

GSFC Role:
- Project Scientist, Dr Cahalan (Code 613.2)
- Deputy Project Scientist, Dr Rabin (HSD)
- Data archiving (GSFC Earth Science Data and Information Center)

Significant Project Milestones in FY09:
- Discovered that the solar spectral irradiance in the visible does not vary in phase with the TSI over the solar cycle. This requires new studies in solar heating of the Earth's atmosphere and surface.
- Established reference spectra for the 2008 solar cycle minimum using simultaneous observations throughout the X-ray, UV, visible, and IR regions as well as TSI.

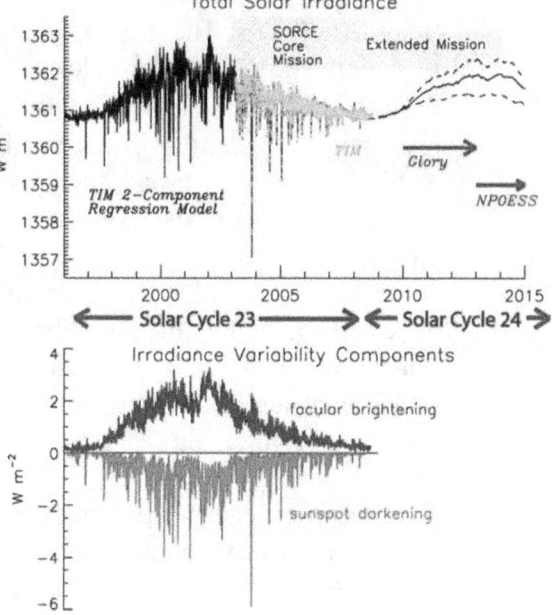

SORCE has tracked the decline of solar irradiance in solar cycle 23 and solar cycle minimum in 2008, as shown in the upper panel by the TIM measurements of TSI, indicated by the green symbols. From these observations, the sunspot and facular sources of TSI variations have been identified and shown in the lower panel. Estimates of TSI before the SORCE launch in 2003 are computed from sunspot and facular data and are shown as the solid black line in the upper panel. Provided that SORCE continues for four more years, SORCE will track the rise and maximum of solar cycle 24, permitting significant overlap with NASA's Glory mission with its TIM instrument.

Ramaty High Energy Solar Spectroscopic Imager (RHESSI)

Background: RHESSI has recorded almost 50,000 X-ray flares since its launch on 2002 February 5, and continues to operate successfully. The single RHESSI instrument is an imaging spectrometer observing the Sun in X-rays to γ-rays with time resolutions of a few seconds.

Over 11,000 flares with detectable emission have been identified above 12 keV, ~950 above 25 keV, and 30 above 300 keV, with 18 showing γ-ray line emission. In addition over 25,000 microflares have been detected above 6 keV.

Scientific Goals: The primary scientific goal of RHESSI is to understand the energy release and particle acceleration during solar flares. This is achieved through X-ray and γ-ray imaging spectroscopy with high angular and energy resolution over the broad energy range from 3 keV to 17 MeV. The focus of the extended mission in FY09 is to integrate new RHESSI flare observations on the rise towards solar maximum with the observations of STEREO, Hinode, SDO, and Fermi, enabling new studies of energy release and particle acceleration processes in flares and CMEs that are more comprehensive than have been previously possible. These include studies of the following topics:

- Processes leading up to the flare/CME trigger point
- Initiation of the energy release itself, possibly best revealed by the nonthermal effects even in the weakest microflares
- Location of the electrons and ions in more of the large γ-ray flares
- Location and properties of the coronal hard X-ray sources seen in many flares
- Detailed temporal and spatial comparisons between flares and associated CMEs

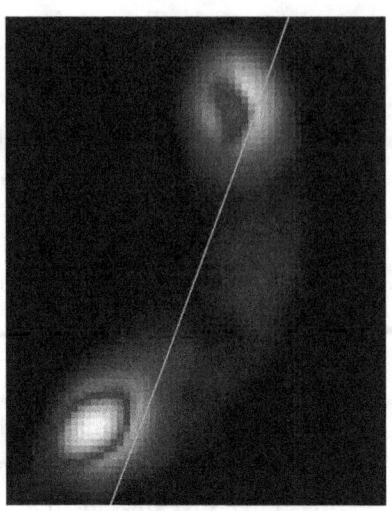

RHESSI X-ray image of a flare loop at the limb of the Sun (yellow curve) in four energy bands showing dispersion in source locations resulting from density and temperature gradients in the loop. The energy bands are 3-6 keV (violet), 6-12 keV (blue), 12-25 keV (green), and 25-50 keV (orange/white). The loop foot points (orange/white) are on the visible disk of the Sun. The flare occurred on 2002 Nov. 28 at 04:36 UT.

GSFC Role: HSD supplied the flight-qualified cryocooler to maintain the germanium detectors at their operating temperature below 100 K, as well as the flight tungsten grids vital to the Fourier-transform imaging technique. HSD has participated in the mission operations phase through monitoring instrument performance, developing data analysis software, and publishing results in the scientific literature. GSFC is also the main site for data distribution to the international scientific community and for the archiving the RHESSI data.

Significant Project Milestones in FY09:
Fifty international participants attended the 9th RHESSI science workshop that was held in Genoa, Italy, on 2009 September 2–5.

GSFC Heliophysics Science Division 2009 Science Highlights

Thermosphere, Ionosphere, Mesosphere, Energetics, and Dynamics (TIMED)

Background: The mesosphere, lower thermosphere, ionosphere (MLTI) region is a gateway between Earth's environment and space, where the Sun's energy is first deposited into Earth's environment. TIMED is focusing on a portion of this atmospheric region located approximately 60–180 km above the surface. The TIMED spacecraft was launched on 2001 December 7 from Vandenberg Air Force Base, California, aboard a Delta II launch vehicle into a 625-km circular orbit with a 74.1° inclination.

TIMED observes effect of solar eclipse of 2006 March 29 on the thermosphere.

Scientific Goals: TIMED goals are to characterize the physics, dynamics, energetics, thermal structure, and composition of Earth's MLTI region. The extended mission (2013) objectives are:

- To characterize and understand the solar cycle-induced variability of the MLTI region
- To address the processes related to human-induced variability of the mesosphere-lower thermosphere
- Quantify the solar EUV irradiance, the primary energy input to the MLTI region

GSFC Role: HSD administered and monitored 17 grants and contracts in 2008, while GSFC has provided oversight of satellite activities. HSD scientists are making fundamental contributions to the interpretation of SABER data including several presentations and publications.

Significant Project Milestones in FY09

- A major breakthrough this year was the development of an algorithm for the first extraction of mesospheric water vapor from the SABER data.
- TIMED went through the senior review.
- SEE released its Version 1.0 data, which includes an updated long-term degradation correction as well as a new algorithm for measurements from 0.1 to 27 nm.

Cluster

Background: Launched in 2000, Cluster is a revolutionary mission investigating the magnetosphere and near-Earth solar wind. The mission's uniqueness stems from its use of four spacecraft to distinguish the spatial and temporal properties of geospace boundaries. The Cluster spacecraft formation allows 3D snapshots of plasma structures and measurements of gradients in key plasma parameters. The capability to vary inter-spacecraft separations enables magnetospheric investigations on different spatial scales, making it an active research tool. Cluster's multiscale era continued this past year wherein, near apogee, three spacecraft formed an equilateral triangle 10,000 km on a side. While the plane of the triangle could be specified for each target, the fourth spacecraft would be placed at a perpendicular distance of 20 km to 10,000 km, depending on the scientific focus. In 2010, the science focus will be on the auroral acceleration region. ESA has approved another mission extension through 2012 with future scientific focus to be finalized at the Science Operations Working Group meetings.

Scientific Goals: The goals of the extended mission include using the four Cluster spacecraft to:
- Characterize solar wind and plasma sheet turbulence
- Study the structure of the bow shock
- Determine exterior cusp structure
- Determine triggering and evolution of the auroral acceleration region
- Determine the extent of magnetotail bursty bulk flows and the width of reconnection regions
- Investigate E-fields in tail dynamics
- Study the plasma environment of radiation belts in the inner plasmasphere
- Determine chorus properties during magnetic storms and their effectiveness in accelerating MeV electrons, important for RBSP mission planning.
- Study solar wind-magnetosphere interaction and substorm processes with THEMIS

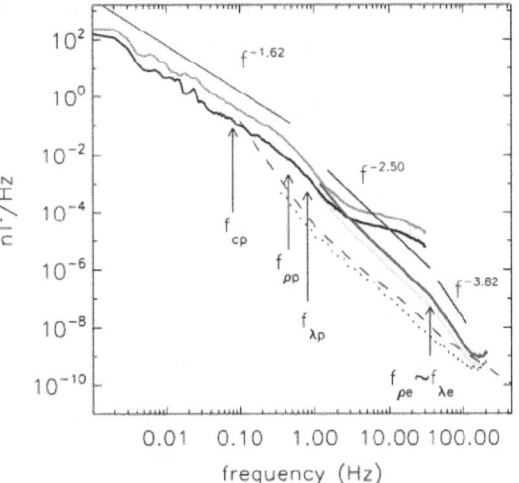

Parallel (black) and perpendicular (red) magnetic spectra from Cluster FGM (f < 33 Hz) and STAFF data (1.5 < f < 225 Hz, green and blue, correspondingly), showing first evidence of magnetic turbulence cascade for two decades above f ~ 0.4 Hz. The second breakpoint at f ~ 35 Hz and a steeper spectrum of $f^{-3.9}$ indicate cascade and dissipation of kinetic Alfvén waves at electron scales [Sahraoui et al., 2009].

GSFC Role: Since launch, the Project Office led by the US Cluster Project Scientist at GSFC manages and coordinates the work of US instrument teams and investigators. HSD scientists are involved in the data analysis.

Significant Project Milestones in FY09: Cluster achieved the first direct measurement of solar wind magnetofluid turbulence down to spatial scales that characterize electron kinetic physics. The observations suggest that, for the intervals analyzed, the fluctuations are part of a cascade characterized as strong kinetic Alfvénic turbulence.

Two Wide-Angle Imaging Neutral-Atom Spectrometers (TWINS)

Background: TWINS stereoscopically images the magnetosphere and the charge exchange ENAs over a broad energy range (~1–100 keV) by using two identical instruments on a pair of widely spaced high-altitude, high-inclination spacecraft. TWINS was launched into two nadir-pointing Molniya orbits at 63.4° inclination with perigees of 7.2 R_E and apogees of 1000 km respectively.

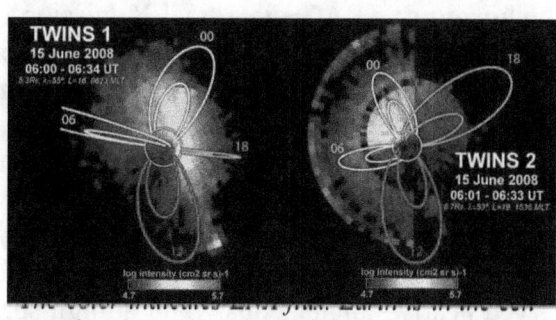

ter of each image, surrounded by dipole magnetic field lines at 4 and 8 R_E equatorial crossing-points.

Its measurement strategy is to use a neutral-atom imager and the Lyman-α detector, both mounted on a rotating actuator platform to allow 360° azimuthal view. The TWINS instrumentation is essentially the same as the Medium-Energy Neutral Atom (MENA) instrument on the IMAGE mission. This instrumentation consists of a neutral-atom imager covering the ~1–100 keV energy range with 4° × 4° angular resolution, 1-min time resolution, and a simple Lyman-α imager to monitor the geocorona.

Scientific Goals:
- Establish the global connectivities and causal relationships between processes in different regions of Earth's magnetosphere
- Determine the structure and evolution of the storm-time magnetosphere
- Understand the energization and transport of magnetospheric plasma populations
- Characterize the storm-time sources and sinks of energetic magnetospheric plasma

GSFC Role: The project scientist from HSD participates in weekly science teleconferences, offers advice on achieving TWINS science goals, and assists in interpreting ENA images. HSD scientists participate in the analysis and interpretation of the TWINS data.

Significant Project Milestones in FY09:
- 2008 November: ENA emissions from low altitude were analyzed and found to have similar energy spectral properties to precipitating ions observed by DMSP satellites.

Transition Region And Coronal Explorer (TRACE)

Background: TRACE is a SMEX mission designed to study the connections between fine-scale magnetic fields and the associated plasma structures on the Sun in a quantitative way by observing the photosphere, the transition region, and the corona. It was launched by a Pegasus rocket on 1998 April 1 into a Sun-synchronous orbit to get near-continuous observations of the Sun. With TRACE, these temperature domains can be observed nearly simultaneously and at high spatial resolution. TRACE operations will be terminated following intercalibration with the Atmospheric Imaging Assembly (AIA) instrument on SDO which is manifested for launch in February 2010.

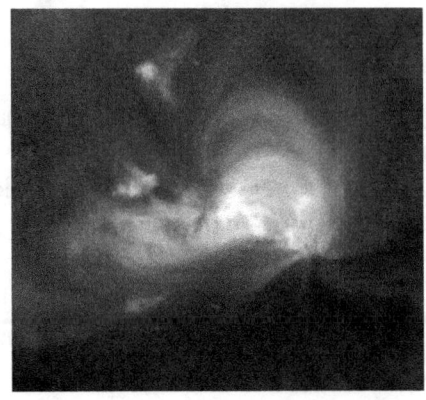

A new-cycle active region (NOAA AR11005) observed by TRACE on 2008 October 14, in Fe XII 195 Å.

Scientific Goals: To simultaneously capture high spatial and temporal resolution images of the transition region. The TRACE data will provide quantitative observational constraints on the models and thus stimulate real advances in understanding the transition region. The data also allow scientists to follow the evolution of magnetic field structures from the solar surface to the corona, investigate the mechanisms of the heating of the outer solar atmosphere, and investigate the triggers and onset of solar flares and CMEs.

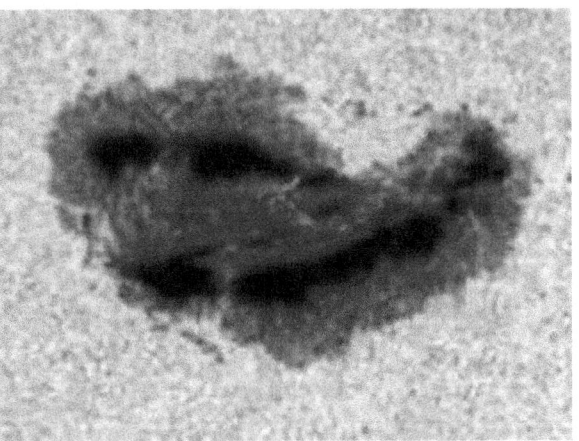

TRACE identified one possible source of the magnetic stress that causes flares: sunspots that rotate, storing energy in the magnetic field.

GSFC Role: GSFC provides mission management (mission scientist and resource analyst), mission operations, scientific operations, and mirrored archive/data access facilities for the TRACE SMEX mission. Mission operations are carried out in the SMEX MOC in Building 3, nearby which is the TRACE EOF. TRACE data are served by the PI team and via the VSO.

Significant Project Milestones in FY09: TRACE science operations were funded only through mid-FY09 on the assumption that SDO launch and TRACE-AIA intercalibration could occur before the end of the fiscal year. SDO launch delays, however, have led to novel planning and budgeting by the SSMO, PI team, and Mission Scientist to accommodate the current SDO launch schedule. This has resulted in TRACE science operations being continued but scaled back during lengthening eclipse periods.

Advanced Composition Explorer (ACE)

Background: ACE was launched in August 1997 carrying six high-spectral-resolution instruments designed to measure the elemental, isotopic, and ionic charge-state composition of energetic nuclei from solar wind to cosmic ray energies, and three instruments to provide the interplanetary context for these studies. Since January 1998, ACE has been in orbit about the L1 point, 1.5 million km sunward of Earth. Data from ACE are used to study the acceleration and transport of solar, interplanetary, and galactic particles with unprecedented precision.

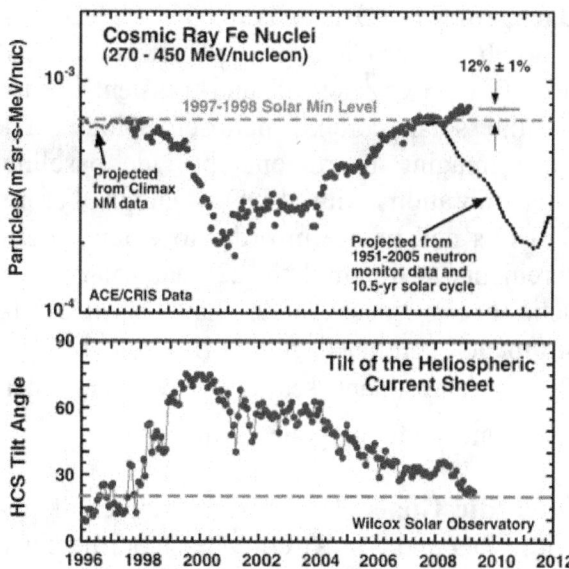

Scientific Goals: The prime objective of ACE is to measure and compare the composition of several samples of matter, including the solar corona, the solar wind, and other interplanetary particle populations, the local interstellar medium, and galactic matter. The scientific questions for the extended mission are:

- How do the compositions of the Sun, solar wind, solar particles, interstellar medium, and cosmic rays differ, and why?
- How does the solar wind originate and evolve through the solar system?
- What is the structure of CMEs and other transients, and how do they evolve?
- How are seed particles fractionated and selected for acceleration to high energies?
- How are particles accelerated at the Sun, in the heliosphere, and in the galaxy?
- How are energetic particles transported in the heliosphere and the galaxy?
- What causes the solar wind, energetic particles, and cosmic rays to vary over the solar cycle?
- How does the solar wind control the dynamic heliosphere?
- How does the heliosphere interact with the interstellar medium?
- How do solar wind, energetic particles, and cosmic rays contribute to space weather over the solar cycle?
- What solar and interplanetary signatures can be used to predict space weather?

GSFC Role: Mission operations support for ACE is provided by the SSMO Project Office, Code 444. The ACE Deputy Project Scientist is in HSD. An HSD scientist is an instrument scientist for the Solar Isotope Spectrometer (SIS) and for the Cosmic-Ray Isotope Spectrometer (CRIS).

Mission Status: The objectives of the ACE Mission for FY09 through FY13 were summarized by the ACE Science Working Team in a proposal for a NASA Headquarters Senior Review in the Spring of 2008.

Solar and Heliospheric Observatory (SOHO)

Background: SOHO has a white-light coronagraph that provides a Sun–Earth line view of both the evolution of, and transient events in, the solar corona; helioseismology and EUV imaging instruments provide baseline intercalibration with SDO analogs before SOHO's end of life in order to extend measurements to a complete, 22-year solar magnetic cycle; spectrometers and particle instruments continued monitoring of the H I Lyman-α resonant scattering corona, solar wind, and solar energetic particles.

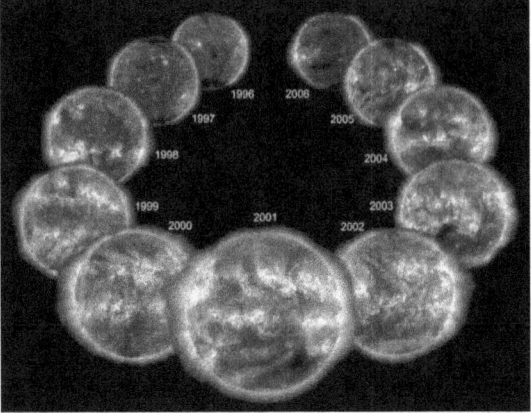

Eleven years of EUV observations of the Sun

Scientific Goals: There are a number of goals associated with SOHO: in conjunction with SDO and STEREO, understand the causes and mechanisms of CME initiation, and the propagation of CMEs through the heliosphere; continue to monitor the TSI; monitor the H I Lyman-α corona in order to improve scientific understanding of solar wind acceleration and the distributions of seed particles accelerated as solar energetic particles; continue the measurement of interstellar winds; continue the search for global solar g-modes; and provide operational predictions of solar energetic particles during manned space missions.

GSFC Role: GSFC provides project management, mission operations, scientific operations, an analysis facility, and archive/data access facilities for the SOHO mission. The SDAC houses, among other data sets, all SOHO data other than the Michelson Doppler Imager helioseismology archive and serves it to the worldwide scientific community via the Internet, through both the SOHO archive interface and the VSO. SOHO has been in operation long enough (14 years) that there is no funding for science in the project budget.

Significant Project Milestones in FY09
- Continued helioseismology and coronal observations through the longest "dry spell" for solar activity in a century
- Measured the lowest TSI since spaceborne observations began in the late 1970's
- Transitioned to automated spacecraft operations for all routine contacts (everything except momentum management, stationkeeping, and spacecraft rolls)
- Migrated EOF Core System (ECS) operations to a sustainable platform
- Initiated planning for EOF and Experimenters' Analysis Facility (EAF)/SDAC move to Bldg. 21.

Mission Status: We continue to await the ESA Scientific Programme Committee decision (now scheduled for early October) on the availability of funding for the ESA share of SOHO scientific and technical management.

Wind

Background: Wind is a comprehensive solar wind laboratory for long-term in situ solar wind measurements. Wind is a spin-stabilized spacecraft launched in 1994 November 1 and placed in a halo orbit around L1, more than 200 R_E upstream of Earth, to observe the unperturbed solar wind that is about to impact the magnetosphere of Earth. Wind, together with ACE and SOHO, provide the 1-AU baseline for inner and outer heliospheric missions.

Wind provides a third point of solar wind observations enhancing the science return of the STEREO mission as well as continuing to monitor the solar wind input for geospace studies.

Scientific Goals: The primary science objectives of the Wind mission are as follows:

- Provide complete plasma, energetic-particle, and magnetic-field measurements for magnetospheric and ionospheric studies
- Investigate basic plasma processes occurring in the near-Earth solar wind
- Provide baseline, 1-AU, ecliptic-plane observations for inner and outer heliospheric missions

GSFC Role: Three of the still-functioning seven instruments were developed at GSFC, namely the Magnetic Field Investigation (MFI; A. Szabo, PI), the electron analyzer of the Solar Wind Experiment (SWE; K. Ogilvie, PI), and the high-energy particle instrument (Energetic Particles, Acceleration, Composition, and Transport—EPACT; T. Von Rosenvinge, PI). Moreover, a significant portion of the radio and plasma waves instrument was provided by GSFC (M. Kaiser, PI). The mission is also managed from GSFC with A. Szabo serving as the Project Scientist and M. Collier serving as the Deputy Project Scientist.

Significant Project Milestones in FY09:

- Compelling evidence that the solar wind is heated by an Alfvén-cyclotron dissipation mechanism based on the 1-AU study of solar wind hydrogen and helium temperatures and their differential flows.
- Evidence for backscattering of energetic particles from interplanetary shocks

Mission Status: Wind will remain at L1 indefinitely and has sufficient consumables left for several more decades of operation.

GSFC Heliophysics Science Division 2009 Science Highlights

Geotail

Background: Geotail was launched in 1992 as a United States-Japan mission. It crosses all boundaries through which solar wind energy, momentum, and particles must pass to enter the magnetosphere. Knowledge of the physical processes operating at these boundaries is vital to understanding the flow of mass and energy from the Sun to Earth's atmosphere. The long-lived Geotail spacecraft continues to provide critical and unique geospace measurements essential to fulfilling the key objectives of the HGO at minimal cost.

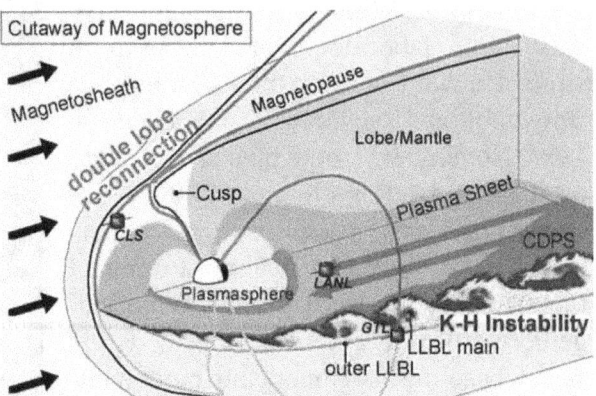

Advances in magnetospheric physics often result from multi-spacecraft coordination as shown here. Geotail made measurements of the Kelvin-Helmholtz instability in the low-latitude dusk boundary layer while Cluster simultaneously detected high-latitude reconnection near the noon meridian.

Scientific Goals: During the current extended mission, the Geotail science focuses are:

- Providing extensive coverage of the magnetospheric boundary layer to delineate mechanisms controlling the entry and transport of plasma into the magnetosphere that is then energized to produce magnetic storms
- Providing supplementary measurements to THEMIS to reveal the spatial and temporal scales of substorm phenomena in the magnetotail
- Providing near-Earth plasma and magnetic field measurements as Geotail spends about 35% of its time in the solar wind
- Providing an important complementary data source for validation of global simulations
- Providing observations that define the location and physics of tail magnetic reconnection and particle acceleration
- Determining energetic-particle environments up to, and including, penetrating γ-rays

GSFC Role: GSFC provides ground-data-system support for Geotail. Deep Space Network (DSN) telemetry data are transferred to the GSFC data system, which performs the initial reduction and merging with trajectory data provided by the Japanese. Data for the experiments are stored and can be accessed by both Japanese and American experimenters. Key parameters are produced directly from the DSN playback data at GSFC and are available from the Coordinated Data Analysis Web (CDAWeb).

Voyager

Background: The Voyager spacecraft continue their epic journey of discovery, traveling through a vast unknown region of the heliosphere on their way to the interstellar medium. Both Voyagers are now traversing the heliosheath, with the first crossings of the heliopause and the first in situ observations of the interstellar medium still to come. The twin Voyager 1 and 2 spacecraft continue exploring where no spacecarft from Earth has flown before. Now in the 32nd year after their 1977 launches, Voyager 1 and 2 are 16 and 13 billion km from the Sun, respectively, and they are approaching the boundary region—the heliopause—where the Sun's dominance of the environment ends and interstellar space begins. Voyager 1, more than three times the distance of Pluto, is farther from Earth than any other human-made object and speeding outward at more than 65,000 km/hr. Both spacecraft are still sending scientific information about their surroundings through the DSN.

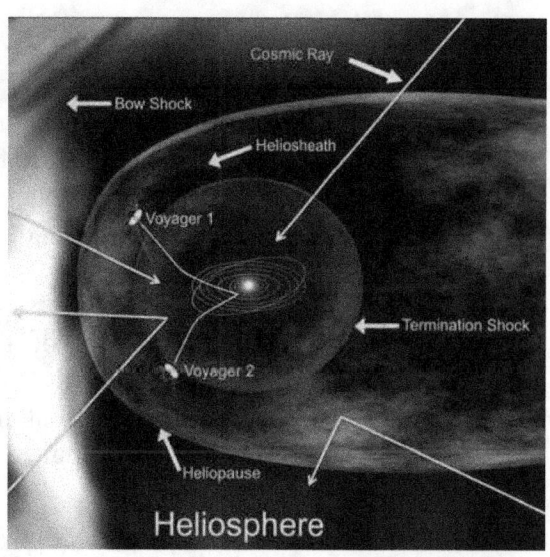

An artist's concept of the heliosphere, a magnetic bubble that partially protects the solar system from cosmic rays.

Scientific Goals: The goals for the Voyager spacecraft are to explore the interaction of the heliosphere with the local interstellar medium and to study the heliosheath. Major mysteries remain unresolved, such as the size and shape of the heliosphere and the source and acceleration mechanism for the anomalous cosmic rays. The Voyager Interstellar Mission (VIM), in combination with IBEX, should be able to answer some of these questions. The nature of the turbulence and the dynamics of major solar wind structures downstream of the termination shock will also be examined by the VIM.

GSFC Role: GSFC's principal contribution is through the Magnetometer (MAG) instrument.

Significant Project Milestones in FY09: The two Voyager spacecraft have been heading out of the solar system for over 30 years. Voyager 2 crossed the termination shock in 2007 at about 11 billion kilometers from the Sun, and Voyager crossed the termination shock in 2004. The turbulence in the heliosheath is unlike that observed in the supersonic solar wind or anywhere on Earth. It is highly compressive and contains a great variety or coherent structures as well as chaotic variations. Correcting for the decreasing polar magnetic fields near the Sun associated with the unique solar cycle 23, it was found that the radial gradient of the magnetic field strength in the inner heliosheath is between 0.017 and 0.055 micro-Gauss/AU. This observation is consistent with a recent 3D MHD model of the magnetized flow in the heliosheath.

APPENDIX 4: FUTURE MISSIONS

Solar Dynamics Observatory (SDO)

Background: SDO is the first LWS mission. It will use telescopes to study the Sun's magnetic field, the interior of the Sun, and changes in solar activity. Some of the telescopes will take pictures of the Sun, while others will view the Sun as a star.

Scientific Goals: The primary goal of the SDO mission is to understand—driving towards a predictive capability—the solar variations that influence life on Earth and humanity's technological systems by determining:
- How the Sun's magnetic field is generated and structured
- How this stored magnetic energy is converted and released into the heliosphere and geospace in the form of solar wind, energetic particles, and variations in the solar irradiance

The SDO spacecraft

GSFC Role: GSFC built the spacecraft, designed and built the dedicated ground system, and Dean Pesnell (HSD) is the Project Scientist. Several HSD people are part of the science investigation and help with E/PO. After launch, SDO will be managed by GSFC.

Mission Status: In FY09, SDO made ready for its launch. In August SDO was shipped to the AstroTech facility near the Kennedy Space Center, where testing of the observatory continues. Due to delays in the Atlas V launch vehicle manifest, SDO remains on the ground waiting for a launch date, now expected in early 2010. During the hiatus the science teams have worked on building the data system required to handle the enormous data flow SDO will generate.

SDO is built, tested, and ready to launch.

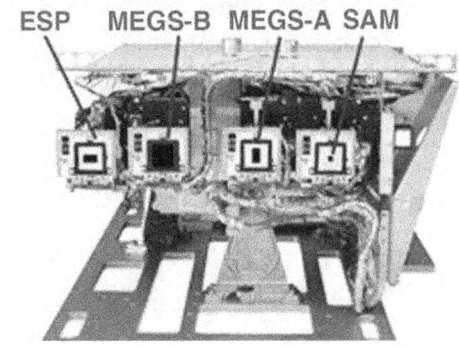

The EVE experiment will measure solar EUV irradiance with unprecedented spectral resolution, temporal cadence, and precision.

Radiation Belt Storm Probes (RBSP)

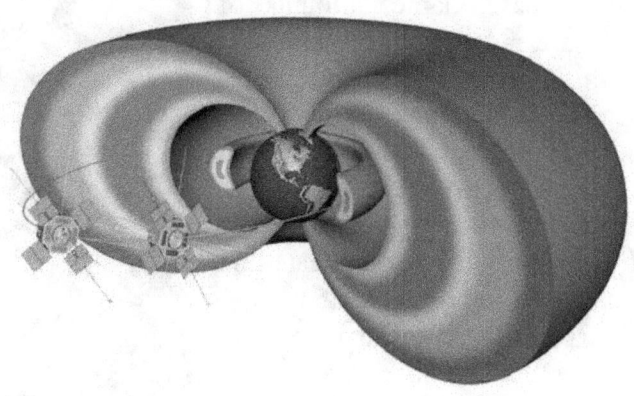

Background: Charged particles in the Van Allen radiation belts of the inner magnetosphere have energies sufficient to pose a hazard to both astronauts and spacecraft. A host of processes have been proposed to account for the variations in strength of the radiation belts as a function of time and location. The distances separating the RBSP spacecraft will vary over the course of this two-spacecraft mission, enabling researchers to distinguish between spatial and temporal effects, determine cause and effect, and identify the spatial extent of phenomena relevant to radiation belt physics. Both spacecraft will carry beacons to provide real-time observations of the radiation belts to space weather forecasters.

Scientific Goals: The objectives of the RBSP mission are to determine:
- Which physical processes produce radiation belt enhancement events
- What are the dominant mechanisms for relativistic electron loss
- How ring current and other geomagnetic processes affect radiation belt behavior

GSFC Role: GSFC retains overall technical authority for the mission, but applies "light-touch" management over this mission, which has been assigned to JHU/APL. All Project Science responsibilities have been assigned to JHU/APL.

Significant Project Milestones in FY09: The mission entered Phase C/D on 2009 January 1 following a successful KDP-C review.

Mission Status: Spacecraft and instruments are currently undergoing critical design reviews, preparatory for the mission CDR scheduled for December 2009. The planned launch date is in May 2012.

Magnetospheric MultiScale (MMS)

Background: The details of how nature releases large amounts of energy are not well understood, although the conversion of magnetic energy from reconnection of oppositely directed fields into heated plasma and energetic particles appears to play a central role. It has not been possible to replicate the typical conditions in the laboratory, or to diagnose the physical processes that control the release of energy, particularly at the small spatial scales where electrons become unmagnetized.

Magnetic reconnection at the magnetospheric boundary

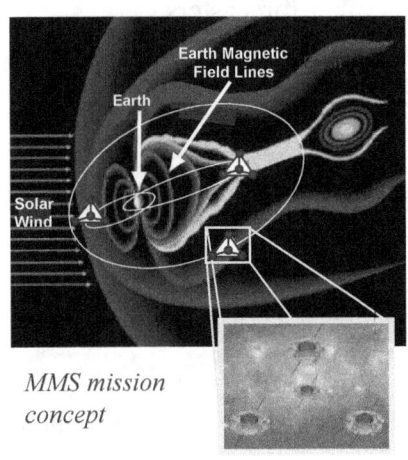

MMS mission concept

Scientific Goals: Observations of reconnection require the 3D determination of the magnetic field and plasma in the vicinity of the site of particle energization. The overall goal of the Magnetospheric Multiscale (MMS) mission is to understand the microphysics of reconnection by determining the kinetic processes responsible for its initiation and evolution. MMS will determine the role played by electron inertial effects and turbulent dissipation in driving magnetic reconnection and will also determine the rate of magnetic reconnection and the parameters that control it.

GSFC Role: MMS is a major GSFC product and involves many resources in HSD. Overall, the project is managed at GSFC and the four spacecraft will be built at GSFC. The Project Scientist, Dr Goldstein, and the two Deputy Project Scientists, Dr Adrian and Dr Le, are members of HSD, as is the Lead Investigator for the Fast Plasma Instrument (FPI), Dr Moore. The Dual Electron Sensors will be built and tested at GSFC. The Japanese-built Dual Ion Sensors are the responsibility of HSD. HSD scientists are also members of the Fields, Theory and Modeling, and Interdisciplinary Science Teams.

Mission Status: MMS is now in Phase C and is proceeding with the final design and initial hardware builds. A major renovation and expansion of the HSD laboratory facilities is nearing completion. A CDR will be held later this year.

Solar Orbiter

Background: The Sun's atmosphere and the heliosphere are unique regions of space, where fundamental physical processes common to solar, astrophysical, and laboratory plasmas can be studied in detail and under conditions impossible to reproduce on Earth. The results from SOHO and Ulysses have enormously advanced scientific understanding of the solar corona, the associated solar wind, and the 3D heliosphere. However, the point has been reached where in situ measurements much closer to the Sun, combined with high-resolution imaging and spectroscopy at high latitudes, promise to bring about major breakthroughs.

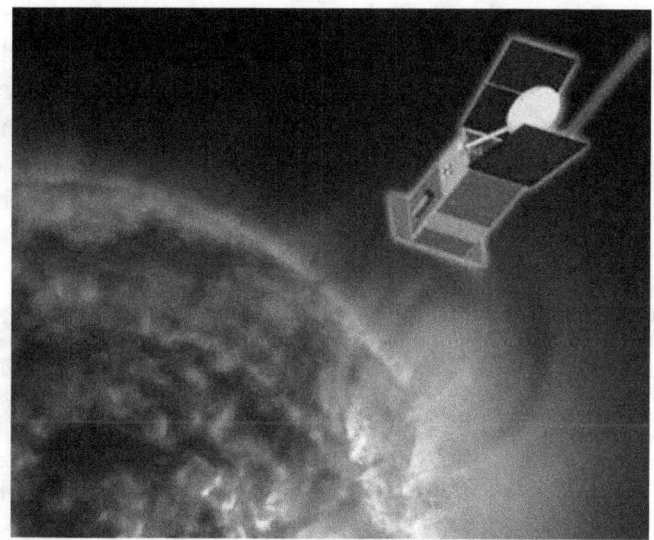

Scientific Goals: The broad science objective is to gain a better understanding of solar activity and variability. In practice, it means extending scientific knowledge of the solar interior to higher latitudes and greater depths, beyond what SDO can achieve. Studying the near-surface layers at high latitudes is a prime objective. Probing the deep interior with sufficient accuracy may, however, require longer observation durations than will be possible. Demonstrating the concept of stereoscopic helioseismology is also an important objective. SDO and/or ground-based facilities, together with Solar Orbiter, would make a most powerful combination.

GSFC Role: GSFC is hosting the US Solar Orbiter Project Office and managing four PI teams and the launch vehicle. Hardware for two of those US-contributed instruments (SPICE and SWA-HIS) are being provided by GSFC Co-I's.

Mission Status: The Solar Orbiter mission is in Phase A (formulation). It is participating in ESA's Cosmic Visions M-class competition, and a down-select among the competitors is to be announced in February 2010.

GSFC Heliophysics Science Division 2009 Science Highlights

Solar Probe Plus (SP+)

Background: SP+ is humanity's first visit to the Sun to explore the complex and time-varying interplay of the Sun and Earth, which affects human activity. SP+ will determine where and what physical processes heat the corona and accelerate the solar wind to its supersonic velocity. A combined remote-sensing and in situ sampling from within the solar corona itself will provide a "ground truth" never before available from astronomical measurements made from spacecraft in Earth's orbit or Lagrangian points. SP+ is currently under study as part of NASA's SMD.

The baseline mission provides for 24 perihelion passes inside 0.16 AU (35 R_S), with 19 passes occurring within 20 R_S of the Sun. The first near-Sun pass occurs 3 months after launch, at a heliocentric distance of 35 R_S. Over the next several years, successive Venus gravity assist (VGA) maneuvers gradually lower the perihelia to ~9.5 R_S—by far the closest any spacecraft has ever come to the Sun. The spacecraft completes its nominal mission with three passes, separated by 88 days, at this distance.

Artist concept of the SP+ spacecraft from the SP+ STDT report.

Scientific Goals: Although the SP+ science objectives remain the same as those established for Solar Probe in 2005, the new mission design differs dramatically from the 2005 design (as well as from all previous Solar Probe mission designs since the 1970s). The 2005, and earlier, missions involved one or two flybys of the Sun at a perihelion distance of 4 R_S by a spacecraft placed into a solar polar orbit by means of a Jupiter gravity assist. In contrast, SP+ remains nearly in the ecliptic plane and makes many near-Sun passes at increasingly lower perihelia.

GSFC Role: GSFC will provide a mission scientist and will participate in instrument proposals.

Mission Status: A NASA Announcement of Opportunity soliciting instrument proposals is expected in late 2009.

APPENDIX 5: ACRONYM LIST

1D	One Dimensional
2D	Two Dimensional
3D	Three Dimensional
AAS	American Astronomical Society
AB	Batchelor of Arts
ACE	Advanced Composition Explorer
ACES	Auroral Current and Electrodynamics Structure (sounding rocket)
ACRIM	Active Cavity Radiometer Irradiance Monitor
AE	Atmosphere Explorer
AFRL	Air Force Research Laboratory
AGU	American Geophysical Union
AIA	Atmospheric Imaging Assembly (on SDO)
AIM	Aeronomy of Ice in the Mesosphere (satellite)
AIP	Astrophysical Institute of Potsdam
AISRP	Applied Information Systems Research Program
ALI-ARMS	Accelerated Lambda Iterations for Atmospheric Radiation and Molecular Structure (model)
AMS	American Meteorological Society
APL	Applied Physics Laboratory (of JHU)
ARMS	Adaptive Refined MHD Solver
ASI	All-Sky Imagers
ASIC	Application-Specific Integrated Circuit
ASTID	Astrobiology Instrument Development (NASA program)
ATC	Advanced Technology Center (of Lockheed Martin)
AU	Astronomical Unit, the Earth-Sun distance, $\sim 1.5 \times 10^6$ km
BA	Batchelor of Arts
BATSE	Burst and Transient Source Experiment
BATS-R-US	Block-Adaptive-Tree-Solarwind-Roe-Upwind-Scheme
BP	Bright Point (Coronal)
BS	Batchelor of Science
BSc	Batchelor of Science
CAPS	Cassini Plasma Spectrometer
CAS	Chinese Academy of Sciences
CAWSES	Climate and Weather of the Sun-Earth System (international SCOSTEP program)
CBC	Canadian Broadcasting Corporation
CCD	Charge-Coupled Device
CCMC	Community Coordinated Modeling Center (at GSFC)
CDAP	Cassini Data Analysis Program
CDAW	Coordinated Data Analysis Workshop
CDAWeb	Coordinated Data Analysis Web (at NASA/GSFC SPDF)

CDF	Common Data Format
CDR	Critical Design Review
CDS	Coronal Diagnostic Spectrometer (on SOHO)
CESR	Centre d'Etudes Spatiale des Rayonments
CETP	Centre d'Etudes des Environnements Terrestre et Planétaire (France)
CFC	Combined Federal Campaign
CGRO	Compton Gamma Ray Observatory
CINDI	Coupled Ion-Neutral Dynamic Investigations (NASA Mission of Opportunity, part of C/NOFS payload)
CIPS	Cloud Imaging and Particle Size Experiment (on AIM)
CIT	California Institute of Technology
CLARREO	Climate Absolute Radiance and REfractivity Observatory
CME	Coronal Mass Ejection
CMU	Carnegie Mellon University
C/NOFS	Communications/Navigation Outage Forecasting System (USAF)
CNRS	Centre National de la Recherche Scientifique (France)
COHOWeb	Coordinated Heliospheric Observations Web (NSSDC interface)
Co-I	Co-Investigator
COI	Cone of Influence
COR1	Inner coronagraph on STEREO SECCHI
COSPAR	Committee on Space Research
CRCM	Comprehensive Ring Current Model
CRIS	Cosmic Ray Isotope Spectrometer (on ACE)
CRESST	Center for Research and Exploration in Space Science & Technology
CRRES	Combined Release and Radiation Effects Satellite
CRS	Cosmic Ray Subsystem (on Voyager)
CTIP	Coupled Thermosphere-Ionosphere-Plasmasphere (model)
CUA	Catholic University of America
DC	Direct Current
DDCS	Deputy Director's Council of Science
DEM	Differential Emission Measure
DES	Dual Electron Spectrometer (component of MMS FAST)
DSCOVR	Deep Space Climate ObserVatory
DL	Double Layer
DoD	Department of Defense (US)
DMSP	Defense Meteorological Satellites Program
DREAM	Dynamic Response of the Environment At the Moon
DSN	Deep Space Network
Dst	Disturbance storm time [index]
EAF	Experimenters' Analysis Facility
ECCS	European Cooperation for Space Standardization
ECHOES	Electron Concentration vs. Height from an Orbiting

	Electromagnetic Sounder
EIS	Extreme-ultraviolet (EUV) Imaging Spectrometer (on Hinode)
EIT	EUV Imaging Telescope (on SOHO)
ENA	Energetic Neutral Atoms
ENLIL	Not an acronym; a time-dependent 3D MHD model of the heliosphere, named after the Sumerian god of the wind
EOF	Experimenters' Operations Facility
EPACT	Energetic Particles Acceleration, Composition, and Transport (on Wind)
EPI	Energetic Particle Instrumentation (to fly on SP+)
EPI-HI	EPI High-energy Instrument
E/PO	Education and Public Outreach
E/POESS	Education and Public Outreach for Earth and Space Science
EPRI	Electrical Power Research Institute
ESA	European Space Agency
ESA	Empirical Shock Arrival
ESF	Equatorial Spread F
EUNIS	Extreme-Ultraviolet Normal-Incidence Spectrograph
EUV	Extreme Ultraviolet
EUVE	EUV Explorer
EUVS	EUV Sensor (in GOES-R EXIS package)
EVE	Extreme ultraviolet Variability Experiment (on SDO)
EXIS	EUV and X-ray Irradiance Sensors (on GOES-R; includes EUVS and updated XRS)
FAC	Field Aligned Current
FAST	Fast Auroral Snapshot Explorer
FASTSAT	Fast, Affordable, Science and Technology Satellite
FGM	Fluxgate Magnetometer (e.g., on ST5 and Cluster)
FIP	First Ionization Potential
FISM	Flare Irradiance Spectral Model
FITS	Flexible Image Transport System
FOK RB	Radiation Belt model by M-C Fok
FOK RC	Ring Current model by M-C Fok
FPI	Fast Plasma Instrument (on MMS)
FRBR	Functional Requirements for Bibliographic Data
FTE	Full-Time Equivalent
FTE	Flux Transfer Event
FTS	Fourier Transform Spectrometer (on ACE)
FUV	Far-UltraViolet
FYS	First Year Seminar
GBM	GLAST Burst Monitor
GEC	Geospace Electrodynamics Connections (mission concept)
GEM	Geospace Environment Modeling
GEST	Goddard Earth Science and Technology

GSFC Heliophysics Science Division 2009 Science Highlights

GI	Guest Investigator
GIC	Geomagnetically Induced Current
GLAST	Gamma Ray Large Area Space Telescope (former name of Fermi)
GME	Goddard Medium Energy Experiment (on IMP-8)
G/MOWG	Geospace Management Operations Working Group
GMU	George Mason University
GOES	Geostationary Operational Environmental Satellite
GPS	Global Positioning System
GRIPS	Gamma-Ray Imager/Polarimeter for Solar flares (balloon mission)
GRL	Geophysical Research Letters
GRS	Gamma Ray Spectrometer (SMM)
GSFC	Goddard Space Flight Center
GSRP	Graduate Student Researchers Program
G-TECHS	Goddard Thermal Electron Capped Hemisphere Spectrometer
GUMICS	Grand Unified Ionosphere-Magnetosphere Coupling Simuation
GUVI	Global Ultraviolet Imager (on TIMED)
HAO	High Altitude Observatory (of NCAR)
HDMC	Heliophysics Data and Modeling Consortium
HEAO	High-Energy Astrophysical Observatory
HELEX	Heliophysical Explorers (NASA-ESA mission)
HELM	Heliophysics Event List Manager (VxO project)
HENA	High Energy Neutral Atom imager (on IMAGE)
HET	High Energy Telescope (of STEREO IMPACT)
HGI	Heliophysics Guest Investigator
HGO	Heliophysics Great Observatory
HI	Heliospheric Imager (on STEREO)
HMI	Helioseismic and Magnetic Imager (on SDO)
HPEG	High-Precisions Electric Gate
HSD	Heliophysics Science Division
HST	Hubble Space Telescope
HTS	High-Temperature Superconductor
HYDRA	Not an acronym; hot plasma analyzer (on Polar)
HXRBS	Hard X-Ray Burst Spectrometer
IAU	International Astronomical Union
IBEX	Interstellar Boundary Explorer
ICESTAR	Interhemispheric Conjugacy Effects in Solar Terrestrial and Aeronomy Research
IDL	Interactive Data Language (for data and image analysis)
IGY	International Geophysical Year
IHY	International Heliophysical Year
IKI	Russian Space Research Institute
ILWS	International Living with a Star (heliophysics program)
IMACS	Imaging [Multi-Aperture] Spectrograph of Coronal electrons
IMAGE	Imager for Magnetopause-to-Aurora Global Exploration (satellite)

IMF	Interplanetary magnetic field
IMP	Interplanetary Monitoring Platform
IMPACT	In situ Measurements of Particles and CME Transients (instrument on STEREO)
IMS	Ion Mass Spectrometer
INMS	Ion-Neutral Mass Spectrometer
IP	Interplanetary
IPY	International Polar Year
IR	Infrared
IRAD	Independent Research and Development
IRI	International Reference Ionosphere
IRIS	Interface Region Imaging Spectrograph (future solar SMEX)
ISAS	Japan's Institute for Space and Aeronautical Science
ISCORE	Imaging Spectrograph of Coronal Electrons
ISEE	International Sun Earth Explorer
ISIS	International Satellites for Ionospheric Studies
ISO	International Standards Organization
ISTP	International Solar Terrestrial Physics
ISWI	International Space Weather Initiative
IT	Information Technology
IT	Ionosphere-Thermosphere (region)
ITSP	Ionosphere-Thermosphere Storm Probes
IUE	Internal Ultraviolet Explorer
JAXA	Japan Aerospace Exploration Agency
JGR	Journal of Geophysical Research
JHU	Johns Hopkins University
JPL	Jet Propulsion Laboratory (of CIT)
JSC	Johnson Space Center
JSOC	Joint Science and Operations Center
KDP	Key Decision Point (in NASA approval process)
KDP-C	KDP for moving project to Phase C
L1	First Lagrangian point (Sun-Earth gravitational balance point)
LADEE	Lunar Atmosphere and Dust Environment Explorer
LADTAG	Lunar Airborne Dust Toxicity Advisory Group
LASCO	Large Angle and Spectrometric Coronagraph (on SOHO)
LENA	Low-Energy Neutral Atom (imager on IMAGE)
LEO	Low Earth Orbit
LFM	Lyon-Fedder-Mobarry (MHD code)
LISM	Local Interstellar Medium
LLBL	Low-Latitude Boundary Layer
LMATC	Lockheed Martin Advanced Technology Center
LMSAL	Lockheed Martint Solar & Astrophysics Laboratory
LOC	Local organizing committee
LOS	Line-of-Sight

LPP	Laboratoire de Physique des Plasmas (of CNRS)
LPW	Langmuir Probe and Waves (on MAVEN)
LSU	Louisiana State University
LTE	Local Thermodynamic Equilibrium
LW	Long Wavelength
LWS	Living With a Star (NASA HSD program)
MACS	Goddard's Multi-Aperture Coronal Spectrograph
MaCWAVE	Mountain and Convective Waves Ascending Vertically (program)
MAG	Magnetometer (on Voyager)
MagCon	Magnetospheric Constellation
MAS	MHD About a Sphere
MAVEN	Mars Atmosphere and Volatile EvolutioN (Mars Scout 2013)
MC	Magnetic Cloud
MCAT	Magnetic Cloud Analysis Tool
MDI	Michelson Doppler Imager (on SOHO)
MDP	
MDR	Mission Definition Review
MEGS	Multiple EUV Grating Spectrograph (part of SDO EVE)
MENA	Medium-Energy Neutral Atom (instrument on IMAGE)
MESSENGER	Mercury Surface, Space Environment, Geochemistry and Ranging
MFI	Magnetic Field Investigation (on Wind)
MGS	Mars Global Surveyor
MHD	MagnetoHydroDynamic(s)
MINI-ME	Miniature Imager Neutral Ionospheric atoms and Magnetospheric (on the Space Test Program spacecraft)
MIT	Massachusetts Institute of Technology
MK	Mega Kelvin (10^6 K)
MLSO	Mauna Loa Solar Observatory
MLT	Mesosphere and Lower Thermosphere
MLTI	Mesosphere and Lower Thermosphere/Ionosphere
MMC	Magnetosphere Mission Concept
MMS	Magnetospheric MultiScale
MO&DA	Mission Operations and Data Analysis
MOC	Mission Operations Center
MOR	Mission Operations Room
MOSES	Multi-Order Solar EUV Spectrograph (sounding rocket)
MOWG	Management Operations Working Group
MRoI	Magnetic Range of Influence
MS	Master of Science
MSc	Master of Science
MSFC	Marshall Space Flight Center
MSTIDS	Medium-Scale Traveling Ionospheric DisturbanceS
MSQS	Magnetospheric State Query System
NAC	NASA Advisory Council

GSFC Heliophysics Science Division 2009 Science Highlights

NAS	National Accademy of Sciences
NASA	National Aeronautics and Space Admimnistration
NCAR	National Center for Atmospheric Research (in Boulder, CO)
NEAR	Near Earth Asteroid Rendezvous
NESC	NASA Engineering and Safety Center
NGDC	NOAA Geophysical Data Center
NHK	Nippon Hōsō Kyōkai (Japanese Broadcasting Corporation)
NIR	Near InfraRed
NOAA	National Oceanographic and Atmospheric Administration
NPP	NASA Postdoctoral Program
NRAO	National Radio Astronomy Observatory
NRC	National Rescarch Council
NRL	National Research Laboratory
NSF	National Science Foundation
NSO	National Solar Observatory
NSSDC	National Space Science Data Center (at GSFC)
OMNI	Not an acronym; OMNI data are spacecraft interspersed, near-Earth solar-wind data
ONR	Office of Naval Research
OpenGGCM	Open Geospace General Circulation Model
OPR	Outer Planets Research
PDR	Preliminary Design Review
PEACE	Plasma Electron and Current Experiment (on Cluster)
PFSS	Potential Field Source Surface
PhD	Doctor of Philosophy
PI	Principal Investigator
PIC	Particle-in-Cell (in simulation code)
PICARD	Not an acronym; a French solar irradiance, helioseismology, and metrology mission named after astronomer Jean Picard.
PISA	Plasma Impedance Spectrum Analyzer (Space Test Program s/c)
PLP	Planara Langmuir Probe (on C/NOFS)
PMC	Polar Mesospheric Clouds
PSU	Pennsylvania State University
PWG	Polar-Wind-Geotail
R_E	Radius of the Earth
R_S	Radius of the Sun
R&D	Research and Development
RAS	Royal Accademy of Science
RAS	Russian Accademy of Sciences
RAISE	Rapid Acquisition Imaging Spectrograph Experiment
RBE	Radiation Belt Environment
RBSP	Radiation Belt Storm Probes
RB	Radiation Belt

GSFC Heliophysics Science Division 2009 Science Highlights

RC	Ring Current
RHESSI	Ramaty High Energy Solar Spectroscopic Imager
ROSES	[NASA] Research Opportunities in Space and Earth Sciences
RPI	Radio Plasma Imager (on IMAGE)
RPWS	Radio and Plasma Wave Science (instrument on Cassini)
RQ	Radio Quiet
SABER	Sounding of the Atmosphere using Broadband Emission Radiometry (instrument on TIMED)
SBIR	Small Business Innovative Research
SCIFER	Sounding of the Cleft Ion Fountain Energization Region
SCOSTEP	Scientific Committee on Solar-Terrestrial Physics
SDAC	Solar Data Analysis Center (at GSFC)
SDAT	Science Data Analysis Tool
SDO	Solar Dynamics Observatory
SECCHI	Sun–Earth Connection Coronal and Heliospheric Imager (on STEREO)
SECEF	Sun–Earth Connection Education Forum
SED	Sun-Earth Day
SEE	Solar EUV Experiment (on TIMED)
SERB	Space Experiment Review Boards
SERTS	Solar Extreme-ultraviolet Research Telescope and Spectrograph
SESDA	Space and Earth Science Data Analysis
SESI	Science and Engineering Student Intern
SHINE	Solar, Heliospheric, and Interplanetary Environment
SI	Stellar Imager
SIGGRAPH	Special Interest Group on GRAPHics and Interactive Techniques
SIM	Spectral Irradiance Monitor (on SORCE)
SIS	Science Information Systems
SIS	Solar Isotope Spectrometer (on ACE)
SMD	Science Mission Directorate
SMEX	Small Explorer
SMM	Solar Maximum Mission
SOC	Science Operations Center
SOC	Scientific Organizing Comittee
SOHO	Solar and Heliospheric Observatory
SOLSTICE	Solar Stellar Irradiance Comparison Measurement (on SORCE)
SORCE	Solar Radiation and Climate Experiment (satellite)
SOT	Solar Optical Telescope (on Hinode)
SP+	Solar Probe Plus
SSDO	Space Science Data Operations
SPASE	Space Physics Archive Search and Extract
SPD	Solar Physics Division of the AAS
SPDF	Space Physics Data Facility
SPICE	Spectral Imaging of the Solar Environment (for Solar Orbiter)
SPSO	Science Proposal Support Office

GSFC Heliophysics Science Division 2009 Science Highlights

SSAC	Space Science Advisory Committee
SSC	STEREO Science Center
SSDO	Space Science Data Operations
SSMO	Space Science Mission Operations
SSN	Sunspot Number
SSREK	Solar System Radio Explorer Kiosk
STAFF	Spatial Temporal Analysis of Field Fluctuations (on Cluster)
ST5	Space Technology 5
STDT	Science and Technology Definition Team
STEREO	Solar Terrestrial Relations Observatory
STP	Solar Terrestrial Probe
SUMI	Solar Ultraviolet Normal Magnetograph Investigation
SUNBEAMS	Students Enthusiastic About Science and Maths
SVS	Scientific Visualizations Studio (at GSFC)
SW	Short Wavelength
SW	Solar Wind
SWAC	Space Weather Action Center
SWA-HIS	Solar Wind Analyzer–Heavy Ion Spectrometer
SWAN	Space Weather Awareness at NASA
SWAVES	Not an acronym; instrument on STEREO (see WAVES)
SWE	Solar Wind Experiment (on Wind)
SWICS	Solar Wind Ionic Composition Spectrometer (on Ulysses)
SWMF	Space Weather Modeling Framework
SWMF/GM+IM+R	SWMF/Global Magnetosphere + Inner Magnetosphere + Ring Current Radiation Belt
SWMF-IH	SWMF – Inner Heliosphere
SWMF-SC	SWMF – Solar Corona
SWRI	SouthWest Research Institute
SXT	Soft X-Ray Telescope (on Yohkoh)
TECHS	Thermal Electron Capped Hemisphere Spectrometer (on SCIFER)
TES	Thermal Emission Spectrometer (on MGS)
TGF	Terrestrial Gamma-ray Flashes
THEMIS	Time History of Events and Macroscale Interactions during Substorms (fleet of 5 spacecraft)
TIDE	Thermal Ion Dynamics Experiment (on Polar)
TIGER	Trans-Ion Galactic Element Recorder (balloon-borne instrument)
TIM	Total Irradiance Monitor (on SORCE, and to fly on Glory)
TIMS	Technical Information and Management Services
TIMED	Thermosphere Ionosphere Mesosphere Energetics and Dynamics
TOPIST	Topside Ionospheric Scaler with True height (algorithm)
TR&T	Targeted Research and Technology (in LWS program)
TRACE	Transition Region and Coronal Explorer
TRICE	Twin Rockets to Investigate Cusp Electrodynamics
TSC	Technology Steering Committee
TSSM	Titan Saturn System Mission

GSFC Heliophysics Science Division 2009 Science Highlights

TTI	Thermospheric Temperature Imager (on Space Test Program spacecraft)
TWINS	Two Wide-angle Imaging Neutral-Atom Spectrometers
UAH	University of Alabama Huntsville
UCB	University of California, Berkeley
UCLA	University of California, Los Angeles
UK	United Kingdom
ULF	Ultra-Low Frequency
UMCP	University of Maryland, College Park
UMd	University of Maryland
UMBC	University of Maryland, Baltimore County
UN	United Nations
UNBSS	United Nations Basic Space Science
UNH	University of New Hampshire
US	United States
USAF	United States Air Force
USNA	United States Naval Academy
USRA	Universities Space Research Association
USU-GAIM	Utah State University Global Assimilation and Ionospheric Model
UTD	University of Texas at Dallas
UV	UltraViolet
UVSC	UV Spectro-Coronagraph (on SOHO)
VDF	Velocity Distribution Function
VEFI	Vector Electric Field Instrument
VEPO	Virtual Energetic Particles Observatory
VERIS	VEry high angular Resolution Imaging Spectrometer (sounding rocket)
VGA	Venus Gravity Assist
VHO	Virtual Heliospheric Observatory
VIM	Voyager Interstellar Mission
VIRGO	Variability of Solar Irradiance and Gravity Oscillations (on SOHO)
VIS	Visible Imaging System (on Polar)
VISIONS	VISualizing Ion Outflow via Neutral atom imaging during a Substorm
VITMO	Virtual Ionospheric/Thermospheric/Mesospheric Observatory
VLA	[NRAO] Very Large Array (of radio telescopes)
VLF	Very Low Frequency
VMO/G	Virtual Magnetic Observatory at Goddard
VMR	Volume Mixing Ratio
VRML	Virtual Reality Modeling Language
VSO	Virtual Solar Observatory
VUV	Vacuum UltraViolet (0.1–190 nm)
VWO	Virtual Wave Observatory
VxO	Virtual discipline Observatory

GSFC Heliophysics Science Division 2009 Science Highlights

WAVES	Not an acronym; a radio and plasma-wave instrument (on Wind)
WDC	World Data Center
Wind	Not an acronym; NASA spacecraft in the Global Geospace Science Program
WINDMI	Wind-Driven Magnetosphere-Ionosphere (model)
WSA	Wang-Sheeley-Arge (model)
XMM	X-ray Multi-Mirror Mission
XRS	X-Ray Sensor (on GOES)
XRT	X-Ray Telescope (on Hinode)

GSFC Heliophysics Science Division 2009 Science Highlights

BACK COVER KEY

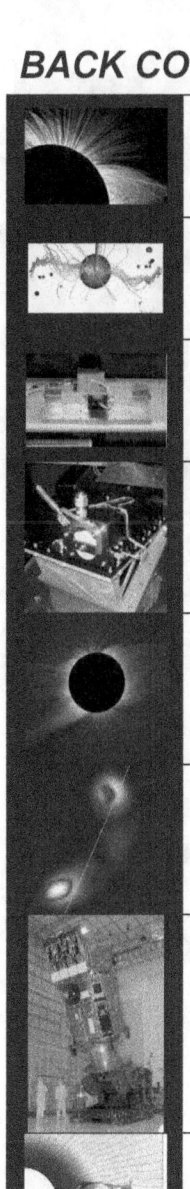

	The acceleration of the solar wind remains a mystery, with many different mechanisms competing to explain the outflow from the Sun. Recent results show that waves and turbulence are not responsible.
	Flux-rope formation has been observed in the coronal emission during solar flares but computer simulations now imply the same occurs in the Earth's magnetosphere.
	The PISA instrument prior to delivery to MSFC for integration onto the FASTSat spacecraft.
	The Thermospheric Temperature Imager to be launched on the FASTSat spacecraft in spring 2010.
	An overlay of Fe XI (red), Fe XIII (blue), and Fe XIV (green) images of the solar corona taken during the 2008 total solar eclipse. Such data have been used to derive the first maps of the 2D distribution of coronal electron temperature and ion charge state.
	RHESSI images the high-energy footpoints of a solar flare loop on the western limb of the Sun. This demonstrates the acceleration of electrons and ions to high energies during the flare process.
	SDO being integrated and tested at GSFC. This massive solar observatory will be launched into a geosynchronous orbit to provide near-continuous coverage of solar coronal activity, magnetic field variability, and EUV emissions. The mission will produce over 2 Tbytes of data every day.
	The results from a 2.5D MHD model for the initiation and propagation of a CME through the solar wind.
	Children having fun while learning about the Sun and other stars at one of our many E/PO events.
	Recent MHD simulations have shown that nonrelativistic reconnection mechanisms can carry over into relativistic plasmas.
	In 2012 NASA plans to launch the Radiation Belt Storm Probes mission to help us better understand the population of the Earth's radiation belts and the dynamics of the particles involved.

Back Cover Caption:
Heliophysics image highlights from 2009. For details of these images, see the key on Page 239.